T0304490

Modelling ^{1}H NMR Spectra of Organic Compounds

Modelling ^{1}H NMR Spectra of Organic Compounds

Theory, Applications and NMR Prediction Software

RAYMOND J. ABRAHAM

Chemistry Department, The University of Liverpool, UK

MEHDI MOBLI

IBM Institute for Molecular Biosciences, The University of Queensland, St Lucia, Australia

A John Wiley and Sons, Ltd, Publication

This edition first published 2008
© 2008 John Wiley & Sons Ltd

Registered office

John Wiley & Sons Ltd, The Atrium, Southern Gate, Chichester, West Sussex,
PO19 8SQ, United Kingdom

For details of our global editorial offices, for customer services and for information about how to apply for permission
to reuse the copyright material in this book please see our website at www.wiley.com.

Library of Congress Cataloging-in-Publication Data

Abraham, R. J. (Raymond John), 1933–
    Modelling ^{1}H NMR spectra of organic compounds: theory and applications /
    Raymond Abraham, Mehdi Mobli.
        p.   cm.
    Includes bibliographical references and index.
    ISBN 978-0-470-72301-2 (cloth)
    1. Proton magnetic resonance spectroscopy.   2. Organic compounds—spectra.
    3. Organic compounds—Structure.
    I. Mobli, Mehdi.   II. Title.
    QD96.P7A27 2008
    543′.66—dc22

                                                                2008026793

British Library Cataloguing in Publication Data

A catalogue record for this book is available from the British Library

ISBN 978-0-470-72301-2   (H/B)

Set in 10/12pt Times by Integra Software Services Pvt. Ltd, Pondicherry, India
Printed and bound in Great Britain by CPI Antony Rowe, Chippenham, Wiltshire.

Contents

Preface

This book serves as an introduction to the structure dependence of ^{1}H chemical shifts and couplings and to the methods used to predict and model ^{1}H NMR spectra. The prediction of ^{1}H NMR chemical shifts to the accuracy required by the structural chemist is still a challenge for the theoretician. We present here a collection of ^{1}H chemical shifts in molecules of known conformation which cover most of the structure types and functional groups in organic chemistry. This data set is used to refine a simple mechanistic theory of ^{1}H chemical shifts and this also provides a useful breakdown of the interactions responsible for ^{1}H chemical shifts. Chapter 1 introduces the basic concepts of NMR and of the interactions responsible for ^{1}H chemical shifts in organic compounds. Chapter 2 gives the interpretation of ^{1}H coupling patterns, the analysis of ^{1}H spectra and a review of HH and HF couplings with up-to-date theory. Chapter 3 details the methods used to predict ^{1}H chemical shifts, and Chapters 4–7 give the observed and calculated ^{1}H chemical shifts of a variety of organic compounds. Chapter 8 considers the effects of structure calculations, rate processes and solvent on ^{1}H spectra and Chapter 9 details the use and application of the computer programs in the accompanying CD. These programs allow the user (a) to draw a molecule on the screen and minimize the conformational energy, (b) to calculate from this file the ^{1}H couplings and chemical shifts and (c) to predict and display the resulting ^{1}H NMR spectrum. It is also an introduction to the CHARGE routine which calculates the ^{1}H chemical shifts and couplings of an organic compound from the molecular structure. To obtain the ^{1}H NMR spectrum from the chemical shifts and couplings the secular determinant has to be solved and the frequencies and intensities of the spectral transitions calculated. This is achieved by the CALAC routine, a development of the LAOCOON program (see Chapter 2). The CHARGE routine is included in two computer programs, HNMRSPEC and NMRPredict, both of which calculate the ^{1}H NMR spectrum for a given compound from the couplings and chemical shifts given by the CHARGE routine. In HNMRSPEC this is performed by the CALAC routine. The CHARGE routine requires an accurate 3-dimensional molecular structure as input and this is provided by the PCMODEL program. Using the PCMODEL and HNMRSPEC programs plus the plotting program HPLOT the user can display the ^{1}H NMR spectrum of any given organic compound. The NMRPredict package includes both ^{1}H and ^{13}C spectral predictions. A neural network approach is used to predict ^{13}C chemical shifts. It includes two ^{1}H spectral prediction packages. One is a data based incremental prediction. The other incorporates both PCMODEL and the CHARGE routine. There is also a plotting routine to give the ^{1}H spectrum from the chemical shifts and couplings. The PCMODEL program is used here to generate the conformational profile of the

compound studied. The CHARGE routine then predicts the ^{1}H chemical shifts and couplings of the conformations of the compound and the weighted average of these values is
used in the NMRPredict software to calculate and display the resulting ^{1}H NMR spectrum.
The PCMODEL, HNMRSPEC, CALAC, HPLOT and NMRPredict programs are provided
on the CD with this book, with detailed examples of molecular input and output for all of
them. There are three appendices, all databases of observed and calculated ^{1}H chemical
shifts. A1 consists of 394 molecules of various types, A2 of 113 substituted benzenes and
A3 of 65 substituted pyridines. The appendices are all on the CD-ROM.

A number of people have helped in the research described in this book. Drs Tony Thomas
and Lee Griffiths have assisted in this research for many years, both as research students
and later as research collaborators. It is a pleasure to acknowledge their contributions and
also those of other members of RJA's research group in Liverpool over the last 30 years,
Drs Julie Fisher, Brian Hudson, Alan Rowan, Paul Smith, Ian Haworth, Guy Grant, Mark
Warne, Mark Edgar, Nick Ainger, Marcos Canton, Matthew Reid, Manuel Perez, Rodothea
Koniotou and Jonathan Byrne. We acknowledge also the technical help of Dr Paul Leonard
and Mrs Sandra Otti and RJA acknowledges the award of a Leverhulme Trust Emeritus
Fellowship to assist in the writing of this book. We also acknowledge the encouragement,
patience and fortitude of Barbara and Louise, without which this book would not have been
written.

Who should read this book?

The book is aimed at graduate and industrial scientists and would be of use to any student
or researcher who is using ^{1}H NMR spectroscopy as an aid in determining the structure
and conformation of organic and bio-organic molecules.

Raymond J. Abraham and Mehdi Mobli

1

Introduction to ^{1}H NMR Chemical Shifts

1.1   Historical Background

^{1}H NMR spectroscopy began in 1945 when two groups of physicists, Bloch, Hansen and Packard[1] at Stanford and Purcell, Torrey and Pound[2] at Harvard, first detected the radiofrequency signal from atomic nuclei when placed in a magnetic field. They shared the 1952 Nobel Prize in Physics for this work. This was ^{1}H NMR as they used the hydrogen nuclei in water and paraffin wax to obtain their signals. As the technology developed other nuclei were found to exhibit NMR signals, but the resonance frequency of these signals depended on the chemical environment of the nuclei. This was first observed by Knight[3] in metals and metal salts and later by Dickinson[4] for the ^{19}F nuclei in fluorocompounds. Also Proctor and Yu[5] observed two signals in the ^{14}N spectrum of ammonium nitrate. They attributed this unexpected result to some 'nasty chemical effect'. Thus the phenomenon of nuclear chemical shifts was discovered. Further advances in magnet resolution allowed the historic experiment of Arnold, Dharmatti and Packard,[6] when they resolved the three types of hydrogen atoms in ethanol (Figure 1.1), the first example of ^{1}H chemical shifts and this illustrated the immense potential of ^{1}H NMR in structural organic chemistry.

Since this original discovery ^{1}H NMR spectroscopy is now widely used in all scientific disciplines from physics to medicine and is now even part of the high school syllabus. It is also the most common and powerful analytical tool of the research scientist. The detection of the hydrogen atom ^{1}H resonances in a molecule was possible since this isotope has a spin of $^1/_2$, is magnetically active, has a high natural abundance and is present in most organic compounds. The other nucleus of general interest for the organic chemist, the carbon ^{12}C isotope, has zero spin and therefore no magnetic moment. The ^{13}C nucleus has spin $^1/_2$ but has a natural abundance of only ca. 1 %. For this reason it took another two decades for the first ^{13}C NMR spectrum with acceptable quality to be produced.

Modelling ^{1}H NMR Spectra of Organic Compounds: Theory, Applications and NMR Prediction Software
Raymond Abraham and Mehdi Mobli   © 2008 John Wiley & Sons, Ltd

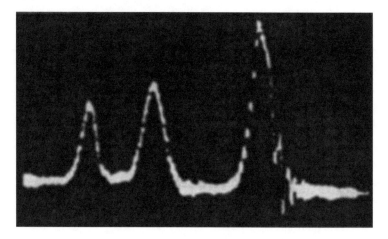

Figure 1.1 *The first NMR spectrum of ethanol (from Arnold, Dharmatti and Packard,[6] reproduced by permission of the American Institute of Physics).*

In the subsequent years immense advances in instrumentation, such as cryomagnets and extensive computational facilities, led to the development of Fourier-Transform (FT) NMR spectrometers as well as multidimensional techniques. At this point NMR methods had become so advanced that they were being used for study of biomolecules, complex chemical matrices and even for imaging of living organisms. The significance of these developments in NMR have been recognized with several Nobel Prizes being awarded to researchers in this field. In 1991 Richard Ernst received the prize for his work on multidimensional techniques, in 2002 Kurt Wüthrich for his work on determining the structure of biomolecules using NMR and most recently in 2003 Paul Lauterbur and Peter Mansfield for their work on imaging (MRI) of living organisms.

In this book we consider one important aspect of ^{1}H NMR, the solution spectra of organic compounds. Recent progress in this area has been directed towards obtaining NMR data more rapidly and enhancing the sensitivity of the equipment. Improved automated methods combined with automatic sample changers have enabled both one dimensional and multidimensional experiments to be performed with minimal interaction with the instruments. This has led to the development of high throughput systems including liquid chromatography (LC)–NMR systems coupled with solid phase extraction (SPE) methods.[7,8] Sensitivity has been improved significantly through the use of cryoprobes and also by using probes and NMR tubes with smaller dimensions (currently probes for NMR tubes with a 1 mm diameter are available).[7]

Utilizing these methods it is possible to run up to a 1000 samples per day. The major bottleneck of the process therefore lies not in the acquisition of data but rather the interpretation and assignment of the spectra produced. The development of tools for automatic assignment of spectral data is therefore highly desirable and this is particularly the case for ^{1}H NMR, the most common NMR spectra. Unfortunately ^{1}H chemical shifts have proved to be the most difficult to predict as well. Protons tend to be in the periphery of the molecule and can therefore easily be influenced by non-bonded interactions such as neighbouring groups (intramolecular) or neighbouring solute and solvent molecules (intermolecular). The narrow range of proton resonances, typically 10–15 ppm, reinforces the need for accurate predictions of their chemical shifts and this is the main objective of this work.

When we consider the importance of this technique in structural chemistry it is remarkable that there is still no routine method of predicting ${}^1H$ chemical shifts of organic compounds. Recent advances in *ab initio* calculations are giving promising results (see later) but are not applicable for quick calculations on moderately sized molecules. Also they do not give any breakdown of the different interactions in the ${}^1H$ chemical shift calculations, thus they do not directly assist our understanding of the interactions responsible. We present here a simple mechanistic theory of ${}^1H$ chemical shifts and also detail the methods we have used in this semi-empirical scheme to overcome this challenge. We also include a chapter on HH couplings and the analysis of NMR spectra in order to present a complete picture of ${}^1H$ NMR spectra. In the accompanying CD computer programs are presented which allow the user (a) to draw a molecule on the screen and minimize the conformational energy, (b) to calculate from this file the ${}^1H$ couplings and chemical shifts and (c) to predict and display the resulting ${}^1H$ NMR spectrum. The applications and uses of these programs are discussed in Chapter 9.

1.2   Basic Theory of NMR

The theory of NMR is common to all experiments and all nuclei, but we shall concentrate here on the ${}^1H$ nucleus. This has a nuclear spin (I) of $\frac{1}{2}$ in units of $(h/2\pi)$ and nuclear moment (μ), proportional to I (Equation (1.1)) where γ is called the magnetogyric ratio.

$$\mu = \gamma I h/2\pi \tag{1.1}$$

It is unique for each nucleus: γ for deuterium (${}^2D$) is ca. 1/6th that of ${}^1H$.

In a magnetic field there are $2I + 1$ allowed orientations of the nuclear magnet, thus a ${}^1H$ nucleus has two allowed orientations, defined by the value of the magnetic quantum number m_I. For ${}^1H$ m_I has values of $\pm\frac{1}{2}$.

The energy of interaction of the nucleus and magnetic field is equal to the field times the nuclear moment. Using Equation (1.1) gives Equation (1.2) where **B** is the applied magnetic field.

$$E = -\gamma h/2\pi . m_I\, \mathbf{B} \tag{1.2}$$

The selection rule for NMR transitions is that m_I can only change by one unit, i.e. $\Delta m_I = \pm 1$. Thus the resonance condition for all NMR experiments is given by Equation (1.3) and

$$h\nu = \Delta E = \gamma h\mathbf{B}/2\pi \tag{1.3}$$

eliminating h gives Equation (1.4), the resonance equation for all NMR experiments.

$$\nu = \gamma\mathbf{B}/2\pi \tag{1.4}$$

Note in particular the relationship in Equation (1.4) between field (**B**) and frequency (ν). In older continuous-wave (CW) experiments the frequency or the field was varied and the spectrum obtained. Present day FT experiments remove this dichotomy but we note that all NMR spectra are measured in frequency units (Hz) which increase from right to left.

1.3   The ^{1}H Chemical Shift

Definition. When a molecule containing the ^{1}H nuclei under observation is placed in a magnetic field, the electrons within the molecule shield the nuclei from the external applied field. The s electrons in the molecule are spherically symmetric and circulate in the applied magnetic field (Figure 1.2). A circulating electron is an electric current and this current produces a magnetic field at the nucleus which *opposes* the external field. In order to obtain the resonance condition (Equation (1.4)) it is necessary to increase the applied field over that for the isolated nucleus. If $\mathbf{B}_{ext}$ is the applied field and $\mathbf{B}_0$ the field at the nucleus then the nuclear shielding ($\Delta\mathbf{B}$) is given by Equation (1.5). This increase in the shielding is called the

$$\Delta\mathbf{B} = \mathbf{B}_{ext} - \mathbf{B}_0 \tag{1.5}$$

diamagnetic shift. Diamagnetism is universal as every molecule has s electrons. There is no spherical symmetry for p electrons. These electrons produce large magnetic fields which when averaged over the molecular motions give low-field shifts. This deshielding is called the *paramagnetic* shift.

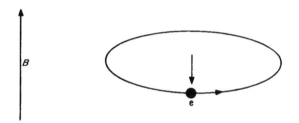

Figure 1.2   *Circulating s-electrons in a magnetic field.*

1.3.1   Nuclear Shielding and Reference Compounds

The **nuclear shielding** $\Delta\mathbf{B}$ is *proportional* to the applied field and the *chemical shift* is defined as the nuclear shielding divided by the applied field. Thus the chemical shift is a molecular quantity. It is a function only of the nucleus and its environment. It is always measured from a suitable reference compound. The standard procedure is to use tetra-methylsilane (TMS) ($Si(CH_3)_4$) as an internal reference compound added to the solution investigated. Sometimes an external reference may be used (e.g. when the solution is very reactive). In this case the external reference could be contained in a capillary tube placed within the sample tube or a coaxial tube outside the sample tube. ^{1}H chemical shifts are usually measured on the δ scale (Equation (1.6)).

$$\delta_H = (\nu_H - \nu_R)/\nu_0 \times 10^6 \text{ ppm} \tag{1.6}$$

Where ν_H is the resonance frequency (Hz) of the proton considered, ν_R the corresponding frequency of the TMS internal reference and ν_0 is the spectrometer frequency. We note that the TMS is at 0 and the scale is from right to left which is the direction of increasing frequency.

In a ^{1}H NMR spectrum at 100 MHz, two peaks with a separation of 100 Hz are 1 ppm apart. The same two peaks when observed in a 500 MHz spectrometer would be 500 Hz apart. It is for this reason that, when the basic data of a spectrum are given, the spectro-meter frequency *must* be recorded. In contrast, the chemical shift δ in ppm is, of course, a molecular parameter dependent only on the sample conditions (solvent, concentration, temperature) and not on the spectrometer frequency. An alternative scale used in early experiments is the τ scale in which TMS is at 10.0 and the scale is from 0 to 10 from left to right. Thus δ = 10 − τ.

As TMS is insoluble in water it is not used for this solvent. The recommended reference is TSP, sodium 3-(trimethylsilyl)propionic-2,2,3,3-d_4 acid ($Me_3SiCD_2CD_2SO_3Na$) in which the reference protons of the methyl groups are defined as 0.0δ. Other useful secondary reference compounds for aqueous solution are given in Table 1.1.

Table 1.1   *Reference compounds for aqueous solution*

TSP	tBuOH	CH_3CN^a	Acetone^a	DMSO^b	$Me_4N^+Br^-$	Dioxane
0.0	1.23	2.06	2.22	2.71	3.18	3.75

^{a} Exchanges in alkaline solution.
^{b} Unsuitable for acid solution.

The hydrogen nucleus is unique as it is the only nucleus, except helium without π-electrons and therefore there is no paramagnetic shielding term from its own valency electrons. The common range of ^{1}H chemical shifts in organic compounds is ca. 0–10δ which contrasts with shift ranges ≥ 200 ppm for all other nuclei. Modern NMR spectro-meters routinely output these shifts to 0.001 ppm. This does not necessarily mean that the absolute accuracy of the chemical shift is to this figure as other interactions may affect these values. Solvent effects in ^{1}H NMR are often appreciable.[9] For example, the chemical shift of acetone in CCl_4, $CDCl_3$, DMSO, methanol and D_2O is 2.09, 2.17, 2.12, 2.15 and 2.22δ, respectively,[10,11] (in the anisotropic benzene solvent it is 1.55δ). To minimize these effects ^{1}H NMR spectra are commonly measured in dilute $CDCl_3$ solution. Even for such 'standardized' conditions variations in sample temperature and/or concentration may affect the chemical shifts and consideration of these factors suggests that routinely measured ^{1}H chemical shifts should be reliable to ca. ± 0.01 ppm.

1.4 ^{1}H Substituent Chemical Shift (SCS)

The influence of any substituent (X) on the chemical shift of any proton is termed the substituent chemical shift (SCS) and defined by Equation (1.7).

$$SCS = \delta(RX) - \delta(RH) \qquad (1.7)$$

It is convenient to divide the SCSs into a one-bond or α effect (i.e. H—X), a two-bond or β effect (i.e. H—C—X), a three-bond or γ effect (i.e. H—C—C—X) and long-range effects (> three bonds). The one-bond or α effect is clearly of considerable theoretical value but is of relatively little practical importance as the great majority of ^{1}H chemical shifts are of hydrogens attached to carbon atoms. Protons attached to almost all other atoms (OH, NH, SH, F, Cl, Br, I, etc.) often show chemical shift changes with solvent and/or concentration of several ppm due to H-bonding interactions.[12] In consequence these chemical shifts have seldom been used for structural identification. We shall show later (Chapters 6 and 8) that given precise experimental conditions many of these protons can give reliable chemical shifts.

1.4.1   Two-bond (H.C.X) Effects

The two-bond or β SCS in methyl derivatives (MeX) was shown.[13,14] to be linearly related to the electronegativity of X and this is shown in Table 1.2.

Table 1.2 ^1H chemical shifts (δ) of CH_3X compounds vs. the electronegativity[a] of X

X	δ_H	E_X	X	δ_H	E_X
SiMe₃	0.0	1.90	SMe	2.08	2.60
H	0.23	2.20	I	2.16	2.65
Me	0.86	—	NH₂	2.46	3.05
Et	0.90	—	Br	2.68	2.95
CCl₃	2.75	2.60	Cl	3.05	3.15
CN	1.98	—	OH	3.38	3.50
CO.Me	2.17	—	F	4.26	3.90

[a] Pauling electronegativity (see Huggins[15]).

The data in Table 1.2 shows the direct influence of the diamagnetic term. As the substituent becomes more electronegative, the electron density round the ^1H nucleus decreases deshielding the nucleus (i.e. increasing δ). The table also shows that for multivalent atoms (e.g. carbon) the chemical shift of the methyl protons is also a function of the γ substituent, e.g. X = Me vs. CCl₃, CN and CO.Me. Originally group electronegativity scales were proposed to take account of this γ effect. In the CHARGE scheme presented the β and γ effects of substituents are considered separately and additive (see later).

A simple and useful extension of the above data is Equation (1.8), originally due to Shoolery.

$$\delta_H = 0.23 + \Sigma \text{ contribution} \tag{1.8}$$

This allows the prediction of any CH_2XY chemical shift by simply adding the substituent shift to the chemical shift of methane (0.23δ). The substituent shifts for some common substituents are given in Table 1.3. The values are from a refined analysis of the SCSs.[16]

We have included the shift for H so that the rules can be extended to methyl compounds. The rules can also be used for methines (CHXYZ) but are much less accurate. For example, the shift for CHCl₃ is calculated as $0.23 + 3 \times 2.48 = 7.67\delta$ compared to the observed value of 7.27.

1.4.2   Three-bond (H.C.C.X) Effects

There are of course many three-bond or γ effects, but we shall consider here the most common one through two saturated carbon atoms. The γ effects of substituents are totally

Table 1.3 *Additive contributions to the chemical shifts of CH$_2$ groups^a*

Group	Shift	Group	Shift	Group	Shift
H	0.17	CO.NR$_2$	1.39	F	3.15
CH$_3$	0.47	CO$_2$R	1.49	Cl	2.48
CH$_2$R	0.67	CO.Ph	2.08	Br	2.29
C=CR	1.33	OR	2.27	I	1.82
C≡CR	1.52	OH	2.46	NH$_2$	1.69
CN	1.73	O.Ph	2.89	NR$_2$	1.41
Ph	1.85	O.CO.R	2.98	NH.CO.R	2.23
CO.R	1.58	SR	1.63	SPh	1.92

^{a} R = alkyl.

Table 1.4 *γ SCS (H.C.C.X) in ethyl derivatives^a*

Substituent	CH$_3$	NH$_2$	OH	SH	F	Cl	Br	I
SCS	0.06	0.25	0.38	0.48	0.51	0.64	0.86	0.99

^{a} From ethane (0.855 ppm).

different from their β effects. Table 1.4 gives a selection of the γ effects of some common substituent groups in ethyl derivatives, i.e. the shift of the methyl protons in ethyl compounds from those in ethane. From Tables 1.2. and 1.4 we note that whereas the β effect of a methyl group is 0.64 ppm (ethane vs. methane) the γ effect of the methyl group is 0.04 ppm (propane vs. ethane). Also the γ effect is not a function of the electronegativity of the substituent. This is clearly demonstrated in Table 1.4 in which the methyl SCSs in the ethyl halides increase from fluorine to iodine, the opposite order of the electronegativity. This demonstrates that ^{1}H chemical shifts are not simply due to the transmission of inductive effects along the carbon–carbon σ bonds. There is a correlation between the methyl chemical shift and the polarizability (i.e. size) of the substituent for the halogen substituents in Table 1.4 and this suggests that the methyl chemical shift is more affected by steric or van der Waals interactions with the substituent rather than inductive effects. This will be considered in more detail subsequently.

1.4.3 ^{1}H SCSs in Olefins and Aromatics

The effects of substituents on ^{1}H chemical shifts in olefins and benzenes have also been determined and Tables 1.5 and 1.6 give the SCSs of some common substituent groups in ethylene and benzene. Table 1.5 is extracted from a more extensive list[17] and Table 1.6 from literature data.[18] Note the chemical shift for ethylene in Table 1.5 is the value in CDCl$_3$[19] and not the value for CCl$_4$ solution[19] (5.25δ).

Although the σ inductive effect of substituents on ^{1}H chemical shifts is only appreciable over one or two bonds this is not true of π-electron shifts. In Table 1.5 large contributions are observed at the β (geminal) and γ (vicinal) protons and often the contributions are reversed for the geminal and vicinal atoms due to competing σ and π effects. For example, the F and OR substituents are π donors and σ acceptors and in consequence the SCSs of

Table 1.5 ^{1}H additive contributions to ethylene chemical shifts ($\Delta\delta$, ppm)
$\delta(C{=}C.H) = 5.405 + \Delta\delta$ gem $+ \Delta\delta$ cis $+ \Delta\delta$ trans

Substituent R	$\Delta\delta$ gem	$\Delta\delta$ cis	$\Delta\delta$ trans
H	0.0	0.0	0.0
Alkyl	0.45	−0.22	−0.28
Alkyl (cyclic)	0.69	−0.25	−0.28
CH$_2$OH	0.64	−0.01	−0.02
CH$_2$SH	0.71	−0.13	−0.22
CH$_2$X(X=F,Cl,Br)	0.70	0.11	−0.04
CH$_2$N	0.58	−0.10	−0.08
C=C	1.00	−0.09	−0.23
C=N	0.27	0.75	0.55
C≡C	0.47	0.38	0.12
C=O	1.10	1.12	0.87
COOH	0.97	1.41	0.71
COOR	0.80	1.18	0.55
CF$_3$	0.66	0.61	0.32
CHO	1.02	0.95	1.17
CO.N	1.37	0.98	0.46
CO.Cl	1.11	1.46	1.01
O.Al	1.22	−1.07	−1.21
O.CO.R	2.11	−0.35	−0.64
CH$_2$.CO;CH$_2$.CN	0.69	−0.08	−0.06
CH$_2$Ar	1.05	−0.29	−0.32
Cl	1.08	0.18	0.13
Br	1.07	0.45	0.55
I	1.14	0.81	0.88
N.Al	0.80	−1.26	−1.21
N.Ar	1.17	−0.53	−0.99
N.CO.	2.08	−0.57	−0.72
Ar	1.38	0.36	0.07
Ar(o-subs.)	1.65	0.19	0.09
S.R	1.11	−0.29	0.13
SO$_2$	1.55	1.16	0.93
F	1.54	−0.40	−1.02

F and OR at the geminal protons are large and positive whereas the corresponding SCSs at both the *cis* and *trans* vicinal protons are large and negative.

Conversely the carbonyl group is both a σ and π acceptor and this gives large positive SCSs at both the geminal and vicinal protons. The benzenes SCSs as may be anticipated lie between those of alkanes and olefins, though there is of course no β proton in substituted benzenes. For example, the γ (H.C.C.X) SCSs of the OH(OR) group is +0.38 ppm for ethanol (Table 1.4), −1.07 for the *cis* proton in ethylene (Table 1.5) and −0.56 for phenol (Table 1.6).

Table 1.6   *¹H substituent chemical shifts (Δδ) in benzenes*

Substituent	Δδ[a]		
	Ortho	Meta	Para
NO_2	0.95	0.26	0.38
$CO.OCH_3$	0.71	0.11	0.21
$CO.CH_3$	0.62	0.14	0.21
CHO	0.56	0.22	0.29
CN	0.36	0.18	0.28
$CH=CH_2$	0.13	0.04	−0.05
F	−0.29	−0.02	−0.23
Cl	0.03	−0.02	−0.09
Br	0.18	−0.08	−0.04
I	0.39	−0.21	0.00
OH	−0.56	−0.12	−0.45
OCH_3	−0.48	−0.09	−0.44
$O.CO.CH_3$	−0.25	0.03	−0.13
CH_3	−0.20	−0.12	−0.22
NH_2	−0.75	−0.25	−0.65
NMe_2	−0.66	−0.18	−0.67
CO.Ph	0.53	0.20	0.31
SO.Me	0.39	0.26	0.24
$SO_2.Ph$	0.69	0.26	0.27

[a] In ppm from benzene (δ_H, 7.341).

There have been many attempts to relate the substituent shifts in benzenes to the electron densities in the molecule, either total or π densities. It can be seen that strongly electron-withdrawing groups (NO_2, CO_2Me) deshield all the protons but the effect is largest at the *ortho* and *para* positions, as expected on simple resonance groups. The converse is true for the strongly electron-donating groups (NH_2, OH), while the halogens, as expected, show less pronounced effects. The general picture agrees with arguments based on electron densities. However, there are other long-range effects which contribute to ¹H chemical shifts and these will now be discussed.

1.5   Long-range Effects on ¹H Chemical Shifts

A pioneering investigation of the effects of substituent groups on distant protons in saturated compounds was given by Zurcher.[20] On this model the influence of a distant group on the chemical shift of a proton may be broken down into a number of separate contributions. These are:

(1) Steric effects due to the proximity of the proton and the substituent ($\Delta\delta_S$).
(2) The electric field produced by the substituent polarizes the C—H bond of the proton considered which affects the proton chemical shift ($\Delta\delta_{EL}$).
(3) Magnetically anisotropic substituents will give rise to magnetic fields at the proton considered which do not average to zero over the molecular tumbling ($\Delta\delta_{AN}$).

In addition for unsaturated compounds there are the effects due to the π densities and in aromatics there are the important ring current shifts. Finally there are the large shifts due to hydrogen bonding.

1.5.1   Steric (van der Waals) Effects

The earliest explanations of long-range substituent effects considered only the electric field and magnetic anisotropy of the substituents. However this explanation becomes questionable when the ^{1}H chemical shifts of saturated hydrocarbons are considered. These range over ca. 2 ppm. which is 20 % of the usual range of ^{1}H chemical shifts, yet these molecules possess neither magnetically anisotropic nor polar substituents. Clearly there are other factors determining the chemical shifts in hydrocarbons. It was then realized that the steric effect due to the proximity of the proton and the substituent was an important factor in ^{1}H chemical shifts. Some examples are shown in Table 1.7. Two contrasting effects can be seen. For each methylene group the proton nearer to the substituent experiences a deshielding effect roughly proportional to the size of the substituent. Also the other proton on the CH_2 group is generally shielded by the substituent. Fluorine is an exception (see below) but this applies even to the methyl group which is neither polar nor magnetically anisotropic. A possible explanation of this effect is as follows. The carbon electrons provide the dominant interaction with the substituent. The repulsion of the two electron clouds causes the electron cloud around the carbon to move away from the substituent. This would cause a deshielding of the closer hydrogen atom and a shielding of the more distant hydrogen on the methylene group as observed. This deshielding and associated shielding is called the 'push–pull' effect and will be considered in more detail in Chapters 4 and 6.

Table 1.7   *SCSs (ppm) of close substituents in cyclohexane and norbornane systems[a]*

X	H_{3ax}	H_{3eq}	H_{7syn}	H_{7ant}	H_{6en}	H_{6exo}
F	0.44	0.07	0.51	0.16	0.65	0.04
OH	0.46	−0.20	0.39	−0.06	0.72	−0.11
Cl	0.65	−0.18	0.59	0.06	0.84	−0.15
Br	0.68	−0.13	0.68	0.11	0.84	−0.07
Me	0.13	−0.15	0.15	−0.15	0.39	−0.20

[a] Data from Abraham.[21]

1.5.2   Electric Field Effects

An important interaction affecting the proton chemical shifts of molecules containing polar substituents is that due to the electric field of the substituent. In an early attempt to calculate the ^{1}H chemical shifts of fluoro- and chloro-substituted alkanes[22] it was noted that there was a pronounced through space effect of the fluorine substituent on ¹H chemical shifts but in contrast to the other substituents investigated there was no push–pull effect for the fluorine substituent (cf. Table 1.7). This is illustrated in Figure 1.3 in which the deshielding due to the axial fluorine substituent in cyclohexane on the 3-axial proton may be compared with the very small effect of the equatorial substituent on the same proton.

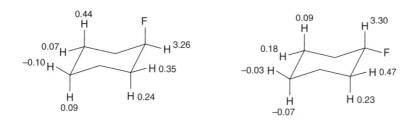

Figure 1.3   *Fluorine SCS (ppm) in axial and equatorial fluorocyclohexane.*

Also the deshielding due to the fluorine atom was better represented by an r^{-3} function than the r^{-6} function used for the other substituents. This suggested that the fluorine SCSs in the compounds examined were primarily due to the electric field produced by the fluorine atom and not due to steric effects. This seemed reasonable in that the fluorine atom was the only substituent atom of comparable size to the hydrogen atom (the hydrogen van der Waals radius is 1.2 Å and the fluorine 1.35 Å). Thus the replacement of a proton by a fluorine atom should not present any large steric perturbations. It will be shown later (Chapter 6) that the long-range (over >3 bonds) SCSs shown in Figure 1.3 can be quantitatively explained by calculating the electric field of the CF bond at the hydrogen atoms considered.

Anisotropic effects. The free circulation of electrons which gives rise to diamagnetic effects in spherically symmetric atoms can also occur around the axis of any linear molecule when the axis is parallel to the applied field. A good example is acetylene (RC≡CH). The electron circulation around the linear axis will give rise to a magnetic effect on neighbouring nuclei in exactly the same manner as any s electron (Figure 1.2). This gives shielding along the molecular axis (e.g. at the acetylenic proton) and deshielding perpendicular to this axis. On hybridization grounds we would expect the ¹H chemical shifts to be in the order of ethane, ethylene and acetylene. The actual shifts are acetylene (1.48δ) compared to ethane (0.88δ) and ethylene (5.31δ). The increased shielding of the acetylene protons is due to this diamagnetic circulation of the π electrons illustrated in Figure 1.4 which gives the sign of Δδ for the shielding contribution.

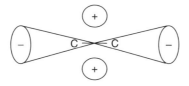

Figure 1.4   *The anisotropic shielding (Δδ) in an axially symmetric molecule such as acetylene.*

This effect occurs in all linear molecules. In the hydrogen halides HCl, HBr and HI, but not HF the same phenomenon gives rise to shielding at the protons and in the gas phase their proton chemical shifts are shielded with respect to TMS.[12] However, hydrogen bonding in solution deshields the protons. The large circulation of the electrons around the C—X bond in the halogens is treated in CHARGE in a similar manner to the acetylene case above (see Chapter 6, Section 6.3).

Most substituents are unsymmetric and therefore in principle magnetically anisotropic. However, in the CHARGE routine only unsaturated groups are regarded as anisotropic. In this case the circulation of the π electrons is less restricted about one molecular axis than the others. This produces a magnetic anisotropy and thus protons near the group will experience both shielding and deshielding effects depending on their position with respect to the anisotropic group.

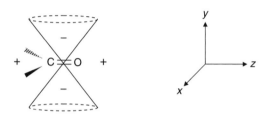

Figure 1.5   *Classical depiction of the shielding ($\Delta\delta$) of the carbonyl anisotropy.*

The most important anisotropic substituent is the carbonyl group. In this case the magnetic anisotropy shields nuclei lying in a cone whose axis is perpendicular to the C=O bond and deshields nuclei outside this cone (Figure 1.5). Thus, an aldehyde proton which lies outside this cone is deshielded due to this anisotropy and resonates at high frequencies (9.5–10.0δ). Figure 1.5 is an over simplification as the carbonyl group has no elements of symmetry and therefore has in principle three different magnetic susceptibilities along the three principal axes (Figure 1.5). This gives two anisotropic susceptibilities which are usually termed the parallel $\Delta\chi_{parl}(\chi_z - \chi_x)$ and perpendicular $\Delta\chi_{perp}(\chi_y - \chi_x)$ anisotropies (see Chapter 3, Section 3.5).

1.5.3 π-Electron Effects

In olefins and aromatic compounds the effects of the π electrons on the chemical shifts of the surrounding protons must be considered. There are two major effects, the direct effect of the π electrons on the carbon atoms on the neighbouring protons and the ring current shifts due to the circulation of the π electrons in the magnetic field. Gunther[23] compared the proton chemical shifts of benzene with similar charged species (cyclopentadiene anion and tropylium cation) and derived the very useful rule that the CH proton is shielded by ca. 10 ppm for a unit increase in the π-electron density at the attached carbon atom. This rule is used in the CHARGE program for both olefinic and aromatic compounds (see Chapters 4 and 5).

An illustrative example of this effect is found in the vinyl compounds above. The calculated excess π-electron density at the β carbon atom is −85.3 me (milli-electrons) for methyl vinyl ether and +51.6 me for acrolein. On Gunther's rule this gives shifts at the attached (vicinal) protons of −0.85 and +0.52 ppm. When compared to the observed SCSs of −1.07 (*cis*) and −1.21 (*trans*) for the ether and 1.12 (*cis*) and 0.87 (*trans*) for the ketone (Table 1.5) it is clear that this is a major contribution to the SCSs. Obviously there are other contributing effects. In the ketone the anisotropy of the carbonyl group makes an appreciable contribution.

Aromatic ring currents. An important contribution to proton chemical shifts in aromatic compounds is due to the aromatic ring current. When a molecule of benzene is oriented perpendicular to the applied magnetic field **B** (Figure 1.6), the π electrons are free to precess in exactly the same way as the s electrons in Figure 1.2. There is now a molecular circulation of the π electrons and the resulting ring current is shown in Figure 1.6. (Remember that the current flows in the opposite direction to the electrons.) Again the induced current gives rise to a magnetic moment which opposes the applied field and the ring current produces the magnetic field shown. Along the sixfold symmetry axis of the benzene ring, the extra magnetic field produced by the ring current *opposes* the applied field, giving a shielding effect. Conversely, at the benzene ring proton the ring current field *adds* to the external field, giving a deshielding effect. The ring current is only induced when the applied magnetic field is perpendicular to the benzene ring. In practice, the benzene molecules are rapidly rotating in solution and the NMR shift is the average over all the orientations. This gives an observed shift equal to one-third of the value in the orientation of Figure 1.6.

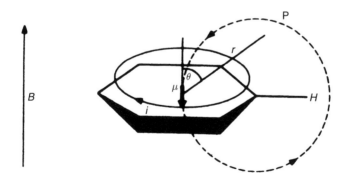

Figure 1.6   *The aromatic ring current of benzene.*

Many calculations of this ring current shift have been attempted. The simplest method is the equivalent dipole model of Pople.[24] In this the ring current shift is given by the field of the equivalent dipole (μ) which is given, for any point P by Equation (1.9) where r and θ are as shown

$$\Delta\delta \text{ (ppm)} = \mu(1 - 3\cos^2\theta)/r^3 \qquad (1.9)$$

in Figure 1.6. Thus, for $\theta = 0°$, i.e. above the benzene ring plane, $\Delta\delta$ is negative, i.e. there is a shielding effect and vice versa for $\theta = 90°$. Also note that when θ is 54.7°, i.e. ($\cos^2\theta = 1/3$) there is zero shift. This is the 'magic angle' in solid state NMR.

Many attempts have been made to estimate the ring current shift in benzene. The simplest is to compare the observed ¹H shifts in benzene (7.34δ) with the 2,3-olefinic protons of cyclohexa-1,3-diene (5.80δ) to give a ring current shift of 1.54 ppm. This ignores other effects such as the different hybridization of the benzene and olefinic protons, any other effects of the aromatic π-electrons, etc. The value of μ was obtained as 26.2 ppm Å³ from a detailed analysis of the ¹H chemical shifts of a series of aromatic hydrocarbons.[25] For benzene ring protons, $r = 2.5$ Å, $\theta = 90°$; this gives a ring current shift of 26.2/2.5³, i.e. 1.67 ppm, in good agreement with the observed data. More refined calculations of the magnetic field due to the two current loops have been performed[26-28] but the results are very similar to those using the simple equivalent dipole.

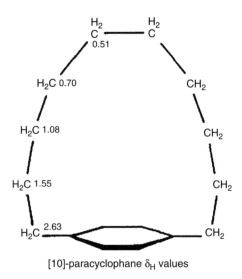

[10]-paracyclophane δ_H values

Many examples of this ring current effect are known. An interesting manifestation of the ring current occurs in [l0]-paracyclophane (above), in which the chemical shifts of the various methylene groups are due to their positions with respect to the aromatic ring, those directly above the ring being most shielded.

Two further examples are shown in Figures 1.7 and 1.8. In Figure 1.7 the olefinic protons in cyclohexene occur at 5.67 ppm, the corresponding protons in 1,4-dihydronapthalene at 5.91 ppm and the aromatic β protons in naphthalene at 7.48 ppm. A ring current shift is observed across the cyclohexene ring in dihydronapthalene but of decreasing intensity. In Figure 1.8 the difference in the chemical shifts of the methyl groups in 1-methylcyclohexene (1.63 ppm) and toluene (2.34 ppm) are again solely due to the ring current.

A more spectacular example is the proton spectrum of *meso* tetra(para-tolyl)porphyrin (Figure 1.8 (bottom)). The large macrocycle of the porphyrin ring is aromatic (it has 18 π electrons) and gives rise to a large ring current. As a consequence, the protons on the periphery of the porphyrin ring are deshielded, the pyrrole protons occurring at 8.84 ppm and the NH protons in the middle of the ring experience a large shielding of several ppm and consequently appear at −2.75 ppm. Indeed, this one spectrum encompasses the entire common ^{1}H NMR region, showing the dramatic effect of the ring current.

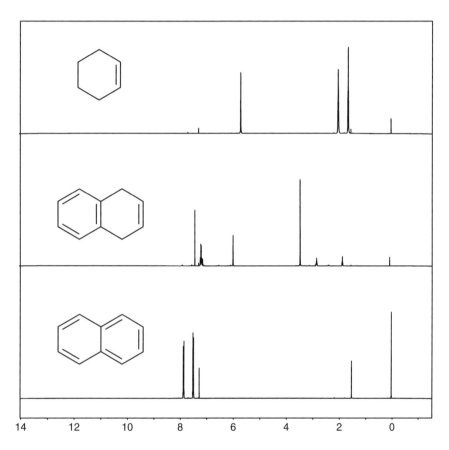

Figure 1.7 *The 400 MHz ^{1}H spectrum of cyclohexene (top), 1,4-dihydronapthalene (middle) and naphthalene (bottom) in CDCl$_3$ solution.*

Because of this large effect, the presence of a ring current is often used as a test for aromaticity. For example, in the annulenes (below), [16]-annulene has ^{1}H chemical shifts of 10.3δ (inner protons) and 5.28δ (outer protons), whereas [18]-annulene has shifts of −4.22δ (inner) and 10.75δ (outer). This shows very clearly that the $4n + 2$ annulene is aromatic, whereas the $4n$ annulene is not. Note both these results are for the low-temperature spectra. At room temperature, ring rotation processes take place, giving an averaged spectrum.

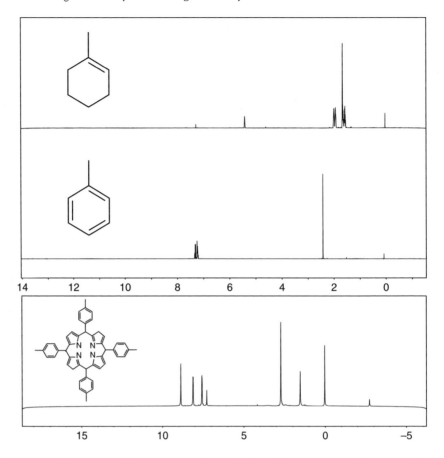

Figure 1.8   *The 400 MHz ^{1}H spectrum of 1-methyl cyclohexene (top), toluene (middle) and meso tetra (para tolyl) porphyrin (bottom) in CDCl$_3$ solution.*

Finally, note that as the ring current is a magnetic effect, the ring current shift will be exactly the same (in ppm) for any nucleus. However, as all other nuclei have chemical shift ranges of greater than 200 ppm compared with ca. 10 ppm for protons, the ring current shifts are much less noticeable for all other nuclei.

[18]- annulene   [16]-annulene

1.5.4   Hydrogen Bonding Shifts

A hydrogen bond (X—H •••Y) is normally formed when both X and Y are electronegative atoms, usually O, N or halides. To a good approximation the interaction may be regarded as electrostatic in character, i.e. the charge distribution X—H •••Y determines the attractive energy of the bond, and, in consequence, when a hydrogen bond is formed, this charge distribution will be slightly enhanced. Thus, the hydrogen becomes more positive and atoms X and Y more negative. Therefore the proton will be deshielded, i.e. moved to increase δ on forming a hydrogen bond. This is precisely what is observed. In compounds capable of forming intermolecular hydrogen bonds (ROH, RNH$_2$), the proportion of H-bonded complexes and therefore the observed chemical shift will depend critically on concentration, solvent, etc. For example, the OH proton in neat ethanol is observed at 5.34δ but on dilution in CDCl$_3$ the OH signal moves to low frequencies until in dilute CDCl$_3$ solution it resonates at 1.1δ.

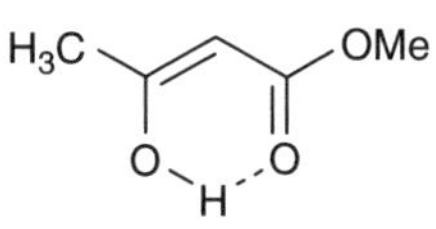

methyl acetoacetate 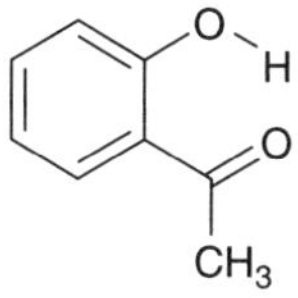   ortho-hydroxyacetophenone

Compounds in which intramolecular H-bonding occurs show, as expected, less dependence of the chemical shifts on dilution, but now the OH chemical shift will be deshielded compared to the analogous compound. For example, in phenol the OH signal moves from 7.45 to 4.60δ on increasing the dilution in CDCl$_3$, but the corresponding proton of o-hydroxyacetophenone occurs at 12.0δ but shows little change on dilution. A particular example of strong intramolecular hydrogen bonding occurs in enols, in which the OH signal is very deshielded. For example, in the enol form of methyl acetoacetate (shown above) the OH signal occurs at 12.1δ.

1.6   Tables of ^{1}H Chemical Shifts of Common Unsaturated and Saturated Cyclic Systems

Table 1.8   *^{1}H chemical shifts of some unsaturated cyclic systems*

Molecule	H-1	H-2	H-3	H-4	H-5	H-6	H-7	H-8	H-9
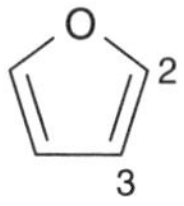	—	7.420	6.380						
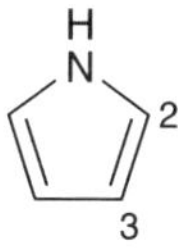	8.00	6.710	6.230						

Table 1.8   (*Continued*)

Molecule	H-1	H-2	H-3	H-4	H-5	H-6	H-7	H-8	H-9
		7.310	7.090						
	—	7.90	—	7.15	7.68				
	—	7.74	—	7.13	7.13				
	—	8.88	—	7.98	7.41				
	—	8.609	7.266	7.657					
	—	—	9.220	7.560					
	—	9.250	—	8.770	7.270				
	—	8.600	8.600						
	—	8.915	7.377	8.139	7.803	7.533	7.709	8.114	
	9.251	—	8.522	7.635	7.808	7.680	7.594	7.955	

—   7.207 6.558 7.647 7.115 7.185 7.396

7.844 7.477

8.009 7.467   8.431

—   —   6.406 7.705 7.502 7.290 7.502 7.290

3.540 —   6.530 7.350 6.170 7.330

Table 1.9   *¹H chemical shifts of some saturated heterocyclic systems*

Molecule	H-1	H-2	H-3	H-4	H-5
	—	3.83	1.85		
	1.60	2.85	1.68		
	—	2.82	1.93		
	—	3.03	2.22		

Table 1.9   (Continued)

Molecule	H-1	H-2	H-3	H-4	H-5
tetrahydropyran (O)		3.67	1.63	1.67	
piperidine (N–H)		1.50	2.80	1.53	1.48
tetrahydrothiopyran (S)		2.60	1.81	1.58	
oxirane (O)	0.3	2.6			
aziridine (N–H)	0.0	1.6			
thiirane (S)		2.38			
2-pyrrolidinone (N–H, O)	6.06	—	2.30	2.14	3.40
γ-butyrolactone (O, O)	—	—	2.490	2.260	4.320

References

1. Bloch, F.; Hansen, W. W.; Packard, M., *Phys. Rev.* 1946, **69**, 127.
2. Purcell, E. M.; Terry, H. C.; Pound, R. V., *Phys. Rev.* 1946, **69**, 37.
3. Knight, W. D., *Phys. Rev.* 1949, **76**, 1259.
4. Dickinson, W. C., *Phys. Rev.* 1950, **77**, 736.
5. Proctor, W. G.; Yu, F. C., *Phys. Rev.* 1951, **81**, 20.
6. Arnold, J. T.; Dharmatti, S. S.; Packard, M. E., *J. Chem. Phys* 1951, **19**, 507.
7. Biospin, B, http://www.bruker-biospin.com/.
8. Nyberg, N. T.; Baumann, H.; Kenne, L., *Magn. Reson. Chem.* 2001, **39**, 236.
9. Lazlo, P., *Prog. Nucl. Magn. Reson. Spectrosc.* 1968, **3**, 203.
10. GlaxoWellcome, *Brochure, NMR Chemical Shifts for Solvents*. GSK: Stevenage, 2002.

11. Tiers, G. V. D., *High Resolution NMR Spectroscopy*. Pergamon Press: Oxford, 1966.
12. Schneider, W. G.; Bernstein, H. J.; Pople, J. A., *J. Chem. Phys.* 1958, **28**, 601.
13. Allred, A. L.; Rochow, E. G., *J. Am. Chem. Soc* 1957, **79**, 5361.
14. Dailey, D. P.; Shoolery, J. N., *J. Am. Chem. Soc* 1955, **77**, 3977.
15. Huggins, M. L., *J. Am. Chem. Soc* 1953, **75**, 4123.
16. Bell, H. M.; Berry, L. K.; Madigan, E. A., *Org. Magn. Reson.* 1984, **22**, 693.
17. Matter, U. E.; Pascual, C.; Pretsch, E.; Pross, A.; Simon, W.; Sternhell, S., *Tetrahedron* 1969, **25**, 691.
18. Abraham, R. J.; Fisher, J.; Loftus, P., *Introduction to NMR Spectroscopy*. 2nd ed.; John Wiley & Sons, Ltd: Chichester, 1988.
19. Abraham, R. J.; Canton, M.; Griffiths, L., *Magn. Reson. Chem* 2001, **39**, 421.
20. Zürcher, R. F., *Prog. Nucl. Magn. Reson. Spectrosc.* 1967, **2**, 205.
21. Abraham, R. J., *Prog. Nucl. Magn. Reson. Spectrosc.* 1999, **35**, 85.
22. Abraham, R. J.; Warne, M.A.; Griffiths, L., *J. Chem. Soc. Perkin Trans. 2* 1997, 2151.
23. Günther, H., *NMR Spectroscopy*. 2nd ed.; John Wiley & Sons, Ltd: Chichester, 1995.
24. Pople, J. A., *J. Chem. Phys.* 1956, **24**, 1111.
25. Abraham, R. J.; Canton, M.; Reid, M.; Griffiths, L., *J. Chem. Soc. Perkin Trans. 2* 2000, 803.
26. Johnson, C.E.,Bovey, F.A., *J.Chem.Phys.* 1958, **29**,1012.
27. Haigh, C. W.; Mallion, R. B., *Mol. Phys.* 1970, **18**, 767.
28. Haigh, C. W.; Mallion, R. B.; Armour, E. A. G., *Mol. Phys.* 1970, **18**, 751.

2

Interpretation of ^{1}H NMR Coupling Patterns

2.1 Fine Structure due to HH Coupling

The fine structure of ^{1}H NMR spectra in solution is due to the nuclear chemical shift, which has been discussed in Chapter 1 and also the spin–spin coupling. The simplest case is two interacting nuclei. For example, the ^{1}H spectrum of 1,1-dichloro-2,2-dibromoethane, $CHCl_2CHBr_2$, consists of two resonances due to the two different H nuclei in the molecule but each of these is split into two peaks (Figure 2.1). This extra splitting is due to the spin–spin coupling interaction between the two nuclei. If we label the ^{1}H nuclei A and B then the resonance frequency of nucleus A is given by the chemical shift plus the spin–spin interaction with nucleus B. Nucleus B has two orientations in the magnetic field, the $m_I = +1/2$ or α orientation, which will produce a small effect at H_A, and the $m_I = -1/2$ or β orientation, which will produce an equal and opposite interaction at H_A. The two orientations are equally populated (to one part in 10^6) and therefore nucleus A gives two equally intense peaks from the two orientations of H_B. Similar reasoning shows that nucleus B gives two equally intense peaks from the interaction with nucleus A *with the same separation.*

This separation or splitting is called the spin–spin coupling constant and has the symbol J. Thus, J_{AB} is the coupling between nuclei A and B. It is a constant, independent of the applied field, and has the units of Hz. Thus, the energy of interaction of nucleus A and nucleus B in this system is simply given by $m_A m_B J_{AB}$ (Hz).

Nuclear coupling constants can be either positive or negative. If the coupling J_{AB} is positive, then the spin states with spins A and B opposed (i.e. $m_A + 1/2(\uparrow)$, $m_B - 1/2(\downarrow)$ and vice versa) will have lower energy than the states with parallel spins. If J_{AB} is negative, the converse is true. In the simple spectrum shown in Figure 2.1 the appearance of the spectrum is independent of the sign of J_{AB}. However, in more complex spectra the relative signs of the couplings do affect the appearance of the spectrum.

Modelling ^{1}H NMR Spectra of Organic Compounds: Theory, Applications and NMR Prediction Software
Raymond Abraham and Mehdi Mobli © 2008 John Wiley & Sons, Ltd

CHCl$_2$CHBr$_2$

Figure 2.1   *Schematic of the ¹H NMR spectrum of 1,1-dichloro-2,2-dibromoethane.*

The spin–spin coupling must *always* be measured in Hz. This may not seem important in the above example, where the splitting of H_A and H_B would be equal if measured in Hz or ppm. Consider, however, the splitting pattern given by the molecule ^{13}CHCl$_3$. This isotopic species is present in 1 % abundance in normal CHCl$_3$. Thus, the ¹H spectrum of pure CHCl$_3$ (Figure 2.2) consists of the large single resonance from the 99 % of C^{12}HCl$_3$ and a doublet of separation $J_{CH} = 210$ Hz from the C^{13}HCl$_3$ molecules. The ^{13}C spectrum will consist of an identical doublet but this coupling is only identical in the two spectra if it is measured in Hz. If the splitting is quoted in ppm, the ¹H spectrum at 500 MHz corresponds to 0.42 ppm, but the ^{13}C spectrum at the same applied field is at 125.5 MHz and at this frequency 210 Hz is 1.67 ppm.

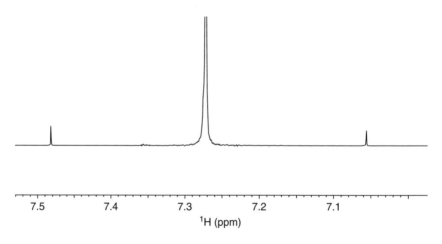

Figure 2.2   *The 500 MHz ¹H spectrum of pure CHCl$_3$, showing the ^{13}C satellites.*

If more than one nucleus is involved all the possible orientations of the nuclei need to be considered. Thus, for three interacting nuclei, if the three nuclei have different chemical shifts and couple to each other as in vinyl chloride (CH$_2$=CHCl) all three nuclei give double doublet patterns due to the four possible orientations of the other two nuclei. If two of the nuclei are equivalent as in CHCl$_2$CH$_2$Cl then the methylene signal is a doublet due to the two orientations of the CH proton, but the CH signal is a 1:2:1 triplet due to the three possible orientations of the CH_2 nuclei. Note that the coupling between the CH_2 protons does **not** appear in the spectrum. This is an example of the general rule that couplings between equivalent protons do not affect the NMR spectrum.

The general rules for splitting patterns are as follows. If one ^{1}H nucleus interacts with n other nuclei with a different coupling to each one, the signal will have 2^n lines all of equal intensity. If one nucleus (A) couples equally to n equivalent nuclei (B) then the pattern of nucleus A is $n + 1$ lines of intensity given by the coefficients of $(1 + x)^n$. Thus, for $n = 3$; $(1 + x)^3 = 1 + 3x + 3x^2 + 1x^3$, i.e. intensities of 1:3:3:1; similarly for $n = 4$, 1:4:6:4:1, $n = 6$, 1:6:15:20:15:6:1, etc. This would make for very complex patterns but fortunately H–H couplings fall off very quickly with increasing number of bonds between the coupling nuclei. For example, for alkanes the coupling over four bonds (H.C.C.C.H) is often too small to be observed. For alkenes and aromatics longer-range couplings are observed but again they are smaller than the two or three bond couplings (see Section 2.3).

These simple rules only hold for certain spectra, which are termed 'first-order' spectra. One condition for these to hold is that all the couplings must be much less ($<$10 %) than the corresponding chemical shift separation (in Hz). If this condition is not met, the spectra show more lines than predicted and the peak intensities will be distorted with a build up of the intensity towards the middle of the coupling patterns. An example of this is shown in Figure 2.3. This shows the ^{1}H spectra of methyl acrylate at 60 and 400 MHz.

The 400 MHz spectrum shows the expected first-order three quartets pattern and the analysis is straightforward. The peak separations give the values of the couplings J_{12}, J_{13}, J_{23} as 1.64, 17.33 and 10.41 Hz directly. The 60 MHz spectrum also shows twelve transitions,

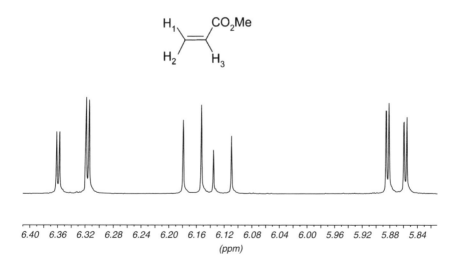

6.40 6.36 6.32 6.28 6.24 6.20 6.16 6.12 6.08 6.04 6.00 5.96 5.92 5.88 5.84
(ppm)

Figure 2.3 *The 60 MHz (below) and 400 MHz (above) ^{1}H spectra of methyl acrylate in acetone solution. The 60 MHz spectrum shows the side-bands from TMS used to measure the spectrum (from Abraham and Castellano[1], reproduced by permission of the Royal Society of Chemistry).*

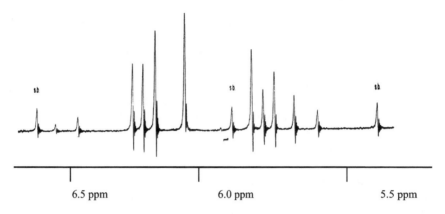

Figure 2.3   (*Continued*).

but the transition intensities are now very different and the analysis of this spectrum proved particularly difficult.[1] It was presented as an example of the uncertainties which can arise in strongly coupled spectra. There are two 'conjugate' solutions which give identical transition energies but which have different chemical shifts and coupling constants. The couplings in the two conjugate solutions were for J_{12}, J_{13}, J_{23}, (a) 2.74, 16.85, 9.87 Hz and (b) 1.56, 17.35, 10.51 Hz. The rms error of the observed vs. calculated spectra was 0.021 Hz; thus these differences are not due to experimental error. It is simply that the spectrum is too closely coupled for an exact well-determined solution. However running the spectrum at 400 MHz resolves these uncertainties and it is clear that solution (b) is the correct answer.

2.2   The Analysis of NMR Spectra

Rigorous spectral analysis is essential if accurate chemical shifts and couplings are to be obtained from the observed spectrum. A knowledge of basic spectral analysis allows one to predict the spectrum from the molecular structure, identify the spectral class by inspection of the spectrum and deduce whether it is first-order or more complex. Also it is important to know when approximate methods of analysis can be used to obtain the couplings and chemical shifts and when the observed spectrum does not provide sufficient information to obtain the molecular parameters.

The principles of spectral analysis have been covered in a number of books and reviews.[2-7] All spectral analysis is derived directly from the quantum mechanical description of NMR and we give this for the simplest possible system, the AB spectrum. For the more complex systems considered, only the results of the quantum mechanical calculations, i.e. the rules for analysing the spectra, are given as these are all that is required to analyse the spectra and no new principles are involved other than those for the AB case.

2.2.1   Nomenclature of the Spin System, Chemical and Magnetic Equivalence

The nomenclature used in the previous section is standard nomenclature for naming the spin systems in NMR. Each different nucleus is given a letter, e.g. A, B, C, etc. If the nuclei have the same chemical shift, i.e. they are *chemically equivalent nuclei,* we use subscripts A_2, B_3, etc. This chemical equivalence can be due to symmetry or rapid rotation or merely by accident; this does not matter here. If the nuclei have very different chemical shifts, then we use X, Y, Z. This differentiation is only strictly valid for different nuclear species but it may be used as an approximation where the couplings are much less than the chemical shift differences. These rules are easier to follow in actual examples. For example, CHCl:CHBr is an AB system, but CHCl:CFBr is an AX system. (Although Cl and Br have nuclear spins, their quadrupole moments relax them so efficiently that there is no interaction with the other spins.)

Similarly, $ClCH_2.CH_3$ is an A_2B_3 system, as the CH_2 and CH_3 protons are chemically equivalent groups, but $ClCH_2CF_3$ is an A_2X_3 system. Also, if we include the ^{13}C isotope, then this is named as well, e.g. $^{13}CH_3I$ is an AX_3 system. The extensions of this process are obvious, e.g. (1) is an AB_2X system; $^{13}CH_3.CH_2Cl$ is an A_3B_2X system etc.

1

A simple extension of the basic nomenclature occurs when there are more than two well-separated groups of nuclei. In this case the middle letters of the alphabet are used. For example, $^{13}CH_3F$ is an AMX_3 system, $CH_3.CH:CHF$ is an $ABXR_3$ system, and so on.

The second, more important, extension is concerned with *magnetically equivalent nuclei.* Magnetically equivalent nuclei are nuclei which possess the same chemical shift and which couple equally to all other groups of nuclei in the molecule. Obviously, therefore, all magnetically equivalent nuclei must be chemically equivalent, but the converse does not apply. This definition is easiest to explain with particular examples. For example, consider CH_2F_2 (below). The two hydrogen nuclei and the two ^{19}F nuclei are chemically equivalent by symmetry. Furthermore, the groups are well separated. Thus we can call them A and X nuclei. The question is: do nuclei 1 and 2 have the same coupling to all the other nuclei in the molecule? That is, is J_{13} equal to J_{23} and J_{14} equal to J_{24}? The answer is Yes. By symmetry there is only one J_{HF} in the molecule and therefore nuclei 1 and 2 *are* magnetically equivalent. By identical arguments the fluorine nuclei are also magnetically equivalent, and the spin system is A_2X_2. We reserve the definition A_2X_2 for magnetically equivalent nuclei.

Now take the similar system $CH_2:CF_2$. The ^{1}H nuclei are again chemically equivalent and well separated from the other nuclei. We ask the same question: is J_{13} equal to J_{23} and J_{14} equal to J_{24}? The answer here is No. J_{13} is a *cis* coupling and J_{23} a *trans* coupling. Thus now H_1 and H_2 are not magnetically equivalent but they are chemically equivalent, and this is also true for the ^{19}F nuclei. A different symbol is used for this spin system. It is an AA'XX' system. By exactly similar reasoning, (A) is ABX$_2$, but (B) is AA'BB'.

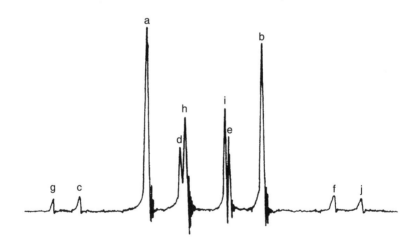

An alternative nomenclature in which square brackets enclose sets of equivalent nuclei proposed by Haigh[8] is also used. In this nomenclature the magnetically equivalent A$_2$X$_2$ case is unchanged but the AA'XX' case is now written as [AX]$_2$. This is useful for more complex systems.

The distinction between the two cases is very important and can be seen immediately in the spectra. The ^{1}H spectrum of CH$_2$F$_2$ is a first-order triplet pattern giving immediately J$_{HF}$, but that of CH$_2$CF$_2$ (Figure 2.4) is a complex pattern of ten lines from which the couplings can only be obtained by analysis. Similarly the spectrum of A above is first-order but that of B complex.

Figure 2.4 *The ^{1}H spectrum of 1,1-difluoroethylene (from Flynn and Baldeschwieler[9], © the American Institute of Physics).*

The importance of this distinction is that there are two conditions to be fulfilled in order to obtain first-order spectra. These are that all chemical shift separations must be much larger than the corresponding couplings and also all groups of nuclei must be magnetically equivalent groups. Thus, the nomenclature describes the complexity of the spectrum. The general rule is that nuclei which are magnetically equivalent *act* as if there is no coupling between them. There is a coupling, but it does not appear in the observed spectrum. It is essential to grasp this principle to understand spectral analysis.

An example of the importance of the nomenclature is the ¹H spectrum of furan (Figure 2.5). The molecular structure clearly classes the spin system as AA′XX′ as there is no reason why J_{23} should equal J_{24}. However, the observed spectrum shows two triplet patterns characteristic of an A_2X_2 spectrum with coupling 1.30 Hz. This spectrum was first interpreted on this basis with $J_{23} = J_{24}$. Later investigations on substituted furans gave J_{23} 1.8 Hz and J_{24} 0.8 Hz and further analyses showed that the observed spectrum is consistent with these values.[11,12] The observed splitting is the *average* of the two couplings. There is insufficient information in the spectrum to obtain all the couplings accurately and only averages can be obtained. This type of spectrum in which there are fewer transitions than expected from the nomenclature was termed a deceptively simple spectrum.[11]

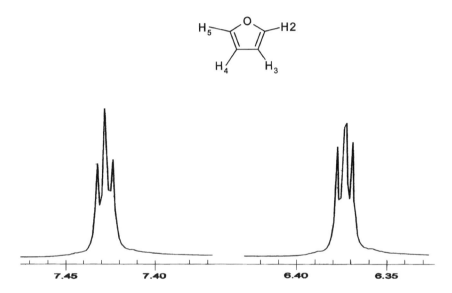

Figure 2.5   *The 300 MHz* ¹*H spectrum of furan in CDCl₃ (reproduced from Puchert and Behnke[10] with permission from the Sigma-Aldrich Company, Ltd, Milwaukee, WI, USA).*

2.2.2   Two Interacting Nuclei, the AB Spectrum

We now wish to consider the detailed analysis of the simplest case, that of two interacting nuclei of spin 1/2, i.e. the AB system. One example of this system was $CH_ACl_2.CH_BBr_2$ (Figure 2.1). The total energy of the system is the sum of the chemical shift term and the coupling interaction.

The **Chemical Shift** term is the interaction between the nuclear moments and the applied field (Equation (1.2)) for the two nuclei H_A and H_B. This gives

$$\mathcal{H}_1 = -\gamma h/2\pi(m_A B_A + m_B B_B) \tag{2.1}$$

where $B_{A,B}$ are the magnetic fields at H_A and H_B, respectively, and $m_{A,B}$ the magnetic quantum numbers for nuclei A and B. The dimensions of Equation (2.1) are energy, and to convert energy to frequency we divide by h. Also, from Equation (1.4), $\nu = \gamma B/2\pi$ and we can substitute ν for B in Equation (2.1). These operations give

$$\mathcal{H}_1 = -(m_A \nu_A + m_B \nu_B)\,\text{Hz} \qquad (2.2)$$

where $\nu_{a,b}$ are the chemical shifts of nuclei A and B *measured in Hz*.

The **Coupling** term is the interaction between the two magnetic moments of nuclei A and B, which we can write in the same way as Equation (2.2) where $\mathbf{I}_{A,B}$ are the nuclear spins of A and B.

$$\mathcal{H}_2 = J_{AB}\mathbf{I}_A\mathbf{I}_B \qquad (2.3)$$

The total energy of the system, expressed in Hz is given by Equation (2.4) and is called the Hamiltonian of the system.

$$\mathcal{H} = \mathcal{H}_1 + \mathcal{H}_2 \qquad (2.4)$$

To find the eigenvalues and wave functions we need to solve the corresponding wave equation Equation (2.5).

$$\mathcal{H}\mathbf{\Psi} = E\mathbf{\Psi} \qquad (2.5)$$

The solutions of this equation define both the energy and wave functions of the system, the latter as combinations of the individual wave functions α and β of the nuclear spins.

The same basic procedure is followed for any NMR system, but in the general case there is not an explicit solution of this equation and iterative computer programs are required. For first-order spectra the equation can be solved exactly. In the AB case the condition for first-order spectra is that the coupling J_{AB} is much less than the chemical shift separation $(\nu_A - \nu_B)$ both measured in Hz. In practise this applies if $J_{AB}/(\nu_A - \nu_B) < 0.1$. In this case the solution of Equation (2.5) is simply Equation (2.6). This gives four energy levels due to the four possible values of m_A ($\pm 1/2$) and $m_B (\pm 1/2)$.

$$E = -(m_A \nu_A + m_B \nu_B) + m_A m_B J_{AB} \qquad (2.6)$$

The **Selection Rule** for single quantum nuclear transitions is a simple extension of that for one nucleus (Chapter 1) in that when one nucleus absorbs energy the other nucleus is unaffected. Thus for nucleus A it is $\Delta m_A = -1$ (i.e. absorption of energy), $\Delta m_B = 0$. This gives the two A transitions shown in Figure 2.1, one for each orientation of B. An analogous rule holds for nucleus B, giving the two peaks due to the two orientations of nucleus A. All these transitions are of equal intensity.

The AB Spectrum. The difference between the general AB spectrum and the first-order case considered above is that the Hamiltonian (Equation (2.5)) has to be solved. This can be done explicitly for this case and the transition energies and intensities are given in Table 2.1.

The appearance of the spectrum is dependant on the values of $\delta\nu(\nu_A - \nu_B)$ and J_{AB}. Using this table calculated AB spectra for different ratios of $J/\delta\nu$ are shown in Figure 2.6

Note that the transitions are only of equal intensity for the first-order (AX) case. As the ratio of $J/\delta\nu$ increases the inner transitions increase in intensity and the outer ones decrease, until for very large values of $J/\delta\nu$ the inner lines merge and the outer lines vanish. Thus for $\delta\nu = 0$ the spectrum is a single resonance at $\frac{1}{2}(\nu_A + \nu_B)$. The system is now an A_2 system and this is another illustration of the general rule that the coupling between magnetically equivalent nuclei does not appear in the NMR spectrum.

Table 2.1 Transition energies and intensities for the AB spectrum[a]

Transition	Origin	Energy	Relative intensity (I_i)
1	$3 \rightarrow 1$ A_1	$\frac{1}{2}(\nu_A + \nu_B) + \frac{1}{2}J + C$	$1 - \sin\theta$
2	$4 \rightarrow 2$ A_2	$\frac{1}{2}(\nu_A + \nu_B) - \frac{1}{2}J + C$	$1 + \sin\theta$
3	$2 \rightarrow 1$ B_1	$\frac{1}{2}(\nu_A + \nu_B) + \frac{1}{2}J - C$	$1 + \sin\theta$
4	$4 \rightarrow 3$ B_2	$\frac{1}{2}(\nu_A + \nu_B) - \frac{1}{2}J - C$	$1 - \sin\theta$

[a] $C\cos\theta = \delta\nu/2 = (\nu_A - \nu_B)/2$; $C\sin\theta = J/2$.

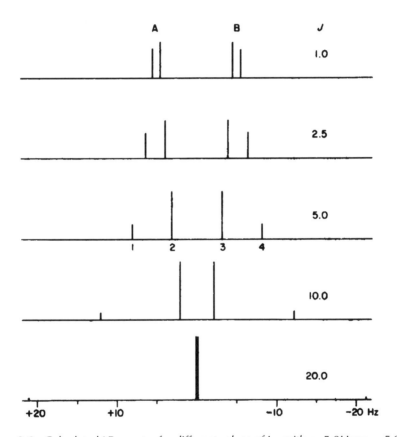

Figure 2.6 Calculated AB spectra for different values of J_{AB} with ν_A 5.0 Hz; ν_B− 5.0 Hz.

The analysis of the general AB spectrum can be performed from the following rules (Equation (2.7)) derived directly from Table 2.1.

$$J = F_1 - F_2 = F_3 - F_4$$

$$\delta\nu = \sqrt{(F_1 - F_4)(F_2 - F_3)} \tag{2.7}$$

$$I_2/I_1 = I_3/I_4 = (F_1 - F_4)/(F_2 - F_3)$$

Here F_i is the frequency and I_i the intensity of peak i. In all analyses the frequency is measured in Hz and increases from right to left in the spectrum. Thus in Figure 2.6 the frequencies of the four lines in the spectrum marked (1, 2, 3, 4) are 8.10, 3.10, -3.10 and -8.10 Hz. from the centre. The above rules give $J = 5.0$ and $\delta\nu = \sqrt{(16.20 \times 6.20)} = 10.0$ and the ratio of the intensities $I_2/I_1 = I_3/I_4 = 16.20/6.20 = 2.6$.

2.2.3   Three Interacting Nuclei, the ABX Spectrum

The general case for three interacting nuclei is the ABC spectrum. This system is however not amenable to direct analysis as the solution of the fundamental Equation (2.5) in this case involves solving a 3×3 determinant. The resulting cubic equation can only be solved by computational methods. It is possible to simplify the determinant by symmetry or other approximations. One very useful method is the X approximation in which one nuclear chemical shift is very different from the other two nuclei. In this case the X nuclear spin can be treated as separate from the other two (AB) nuclei. This allows the breakdown of the determinant into quadratic equations which can be solved explicitly to give a table of transition energies and intensities.[13] It is easier to analyse the spectrum directly as an extension of the AB system as follows.

Because the X spin is independent of the other nuclei the resulting AB spectrum can be broken down into two separate ab sub-spectra,[7] one for each orientation of the X nucleus (Figure 2.7). Each orientation of the X spin (α and β) is equally probable and this gives two equally intense sub-spectra. Each sub-spectrum is an AB-type spectrum which can be analysed immediately from the rules of Equation (2.7). The coupling is J_{AB} but the chemical shifts obtained from these sub-spectra are *not* the true chemical shifts ν_A and ν_B but *effective* chemical shifts given by the chemical shift plus the effect of the X nucleus. The two orientations of the X-nucleus, α and β, give *effective* chemical shifts (ν^*) as shown in Equation (2.8).

$$\alpha(m_x = +1/2) \quad \nu_A^* = \nu_A + 1/2(J_{AX}) \quad \nu_B^* = \nu_B + 1/2(J_{BX}) \qquad (2.8)$$

$$\beta(m_x = -1/2) \quad \nu_A^* = \nu_A - 1/2(J_{AX}) \quad \nu_B^* = \nu_B - 1/2(J_{BX})$$

Thus to analyse the ABX spin system it is necessary to identify the two ab sub-spectra in the AB region and then use the rules of Equations (2.7) and (2.8) to determine the couplings and chemical shifts. This procedure is given in detail elsewhere.[2,13] A typical ABX spectrum is shown in Figure (2.8). The two ab sub-spectra in the AB region can be immediately identified and the spectrum analysed. The observed transition frequencies and parameters obtained from the analysis are given in Section 2.2.5.

2.2.4   Four Interacting Nuclei

The general case of four interacting nuclei requires iterative computer techniques for a complete analysis. However the use of symmetry and the X approximation can simplify the equations to give explicit transition energies and intensities in certain cases. We shall consider two useful examples, the ABRX and AA'XX' systems, which again are most conveniently discussed using sub-spectral analysis.

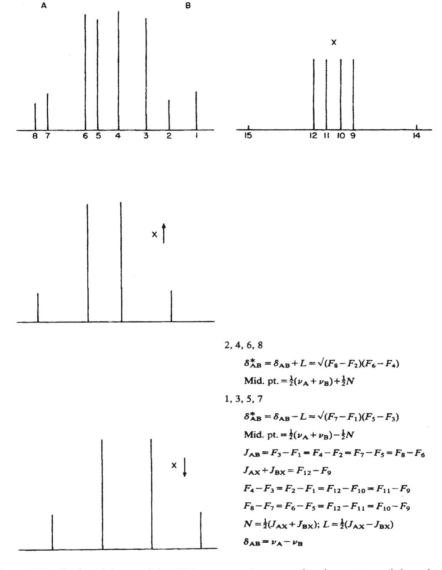

2, 4, 6, 8

$$\delta^*_{AB} = \delta_{AB} + L = \sqrt{(F_8 - F_2)(F_6 - F_4)}$$

Mid. pt. $= \tfrac{1}{2}(\nu_A + \nu_B) + \tfrac{1}{2}N$

1, 3, 5, 7

$$\delta^*_{AB} = \delta_{AB} - L = \sqrt{(F_7 - F_1)(F_5 - F_3)}$$

Mid. pt. $= \tfrac{1}{2}(\nu_A + \nu_B) - \tfrac{1}{2}N$

$$J_{AB} = F_3 - F_1 = F_4 - F_2 = F_7 - F_5 = F_8 - F_6$$

$$J_{AX} + J_{BX} = F_{12} - F_9$$

$$F_4 - F_3 = F_2 - F_1 = F_{12} - F_{10} = F_{11} - F_9$$

$$F_8 - F_7 = F_6 - F_5 = F_{12} - F_{11} = F_{10} - F_9$$

$$N = \tfrac{1}{2}(J_{AX} + J_{BX}); \quad L = \tfrac{1}{2}(J_{AX} - J_{BX})$$

$$\delta_{AB} = \nu_A - \nu_B$$

Figure 2.7 *The breakdown of the ABX spectrum into two ab sub-spectra and the rules for analysing the spectrum.*[13]

The ABRX spectrum. The ABRX analysis is a simple extension of the ABX analysis above. In this case both the R and X chemical shifts are well separated from each other and from the AB resonances. Thus the R and X spins can be treated as separate spins in the equations. The AB spectrum now consists of four ab sub-spectra given by the four combinations of the R and X spins. The four sub-spectra can be identified as they are of equal intensity and all have the same value of J_{AB}. They are then analysed using Equation (2.7) to give J_{AB} and the *effective* chemical shifts ν_A^* and ν_B^* for each sub-spectrum. These effective chemical shifts

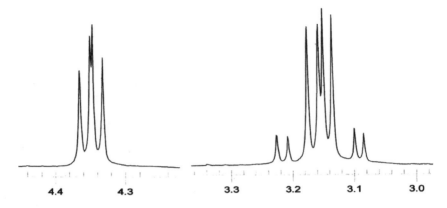

Figure 2.8 *The 300 MHz ¹H spectrum of the CH.CH₂ protons of dl-cysteine HCl (HSCH₂(NH₂).CO₂H) in D₂O (reproduced from Puchert and Behnke[10] with permission from the Sigma-Aldrich Company, Ltd, Milwaukee, WI, USA).*

are given by Equation (2.9) which can be seen to be a simple extension of Equation (2.8). In Equation (2.9) the quantum numbers m_R and m_X have the values $\pm 1/2$ which gives the four possible values of ν_A^* and ν_B^*.

$$\nu_A^* = \nu_A + m_R(J_{AR}) + m_X(J_{AX}) \qquad (2.9)$$

$$\nu_B^* = \nu_B + m_R(J_{BR}) + m_X(J_{BX})$$

An example of such a spectrum is the ¹H spectrum of 1,2-dichlorofluoroethane[14] (Figure 2.9). This shows the AB region of the methylene protons and the R region of the methine proton. The ¹⁹F spectrum of the X (fluorine) nucleus was not observed.

The four ab sub-spectra can be readily identified in the ab region and the analysis follows from this. Further details are given elsewhere.[14]

The AA′XX′ spectrum. A more common spin system which can also be analysed explicitly due to the symmetry of the system and the X approximation is the AA′XX′ spectrum. Again this spectrum is most conveniently discussed using sub-spectral analysis. The couplings are shown in Figure 2.10 and it is convenient to define four new quantities K, L, M, N below.

The analysis of this spectrum is simple in principle. The A and X spectra are identical and are symmetrical about ν_A and ν_X, respectively. The spectra consist of a large doublet of $1/2$ the total intensity of separation $N(J + J')$ and two equally intense ab sub-spectra. The effective coupling (J_{ab}^*) in the two sub-spectra is equal to K and M, respectively and the effective chemical shift δ_{ab}^* equals L in both sub-spectra.

One half of a complete AA′XX′ spectrum is shown in Figure 2.4, the ¹H spectrum of CH₂=CF₂. The ¹⁹F spectrum is identical. The ten possible transitions are well resolved and the large doublet pattern (a,b) and the two ab sub-spectra (cdef and ghij) easily identified. The two sub-spectra are equally intense, thus the two outer transitions and the two inner transitions form one quartet (cdef) and the remaining lines the other quartet (ghij).

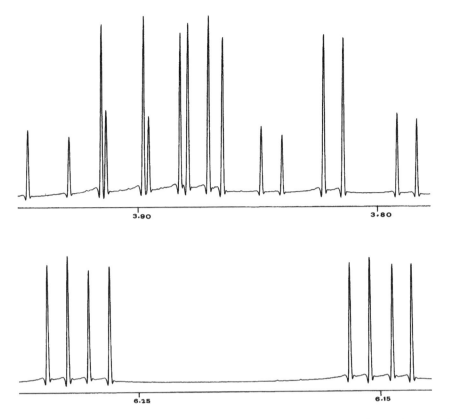

Figure 2.9   *The 400 MHz ^{1}H spectrum of 1,2-dichlorofluoroethane (CHFCl.CH$_2$Cl) in CDCl$_3$.*

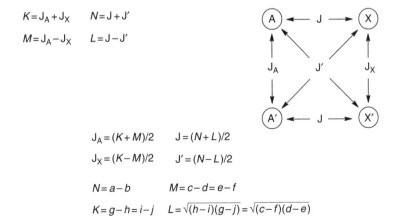

$K = J_A + J_X$ $N = J + J'$

$M = J_A - J_X$ $L = J - J'$

$J_A = (K + M)/2$ $J = (N + L)/2$

$J_X = (K - M)/2$ $J' = (N - L)/2$

$N = a - b$ $M = c - d = e - f$

$K = g - h = i - j$ $L = \sqrt{(h-i)(g-j)} = \sqrt{(c-f)(d-e)}$

Figure 2.10   *Coupling constants for four nuclei AA'XX' (see Figure 2.4 for descriptions of a–j).*

The values of the chemical shifts and couplings are then obtained from the application of Equation (2.7) and the definitions of Figure 2.10.

The problem with this spectrum is that due to the symmetry it is not possible to identify the individual couplings. For example as the A and X spectra are identical then J_A and J_X can be interchanged without affecting the spectra. Furthermore as the ab sub-spectra are independent of the signs of both J_{ab}^* and δ_{ab}^* the parameters K, M and L (Figure 2.10) can have either sign. Thus J and J' may be interchanged without affecting the spectrum. In the case of $CH_2=CF_2$ these couplings are easily assigned by analogy with other data. The cis HC=CF coupling is much smaller than the *trans* and also the *geminal* H.C.H coupling in ethylenes is ca. 0–5 Hz. (Table 2.6). Thus the assignment is obvious. J and J' are 4.8 and 36.4 Hz and J_A and Jx 0.7 and 33.9 Hz.[9] However, both the *geminal* HF and HH couplings may be of either sign in this molecule. The sign of J relative to J' is given from the analysis (N and L are distinct) but the signs of J_A and Jx are not determined by the analysis.

Even more confusing is that the ten allowed transitions in the AA'XX' spectrum are often not observed, due to either low intensity of the transitions or incomplete resolution and in these cases the analysis of the spectrum becomes much more ambiguous. It is clear from Equation (2.7) and Figure 2.10 that if $J = J'$, i.e. $L = 0$ both the ab sub-spectra will collapse to a single line at the centre of the spectrum. This is of course the simple 1:2:1 triplet pattern of the A_2X_2 spectrum as this is the condition for magnetic equivalence of the A and X nuclei. Also if either parameter K or M equals 0, then one ab sub-spectrum will collapse to a single line at the centre but the other will still give the normal pattern.

An illustration of a 'deceptively simple' AA'XX' spectrum of furan was given in Figure 2.5 and some further examples of AA'XX' spectra are shown in Figure 2.11. The deceptively simple spectrum of furan is due to the small value of the parameter L ($J_{23} − J_{24}$) = 1.0 Hz compared to the values of the other couplings, J_{25} and J_{34} (Section 2.3). In this case the outer transitions of the two ab sub-spectra are too small to be observed and the inner transitions coalesce to one broad line, to give the observed spectrum. The small outer transitions were subsequently detected at higher gain.[12] and these authors also noted the increased width of the centre line compared to the outer transitions. This confirmed that the spectrum is indeed a deceptively simple AA'XX' spectrum and not a true A_2X_2 spectrum.

The spectra shown in Figure 2.11 are also all simplified AA'XX' spectra in that they do not show the theoretical ten lines per nucleus. In 1,2-difluoro-4,5-dimethoxybenzene, the triplet pattern is due to similar reasons as in furan. The value of the *ortho* HF coupling in benzenes (ca. 8.0–9.0 Hz) is similar to that of the *meta* HF coupling (ca. 5.0–6.0 Hz) and much less than that of the *ortho* FF coupling (ca. 20 Hz). The value of $L(J_{ortho} − J_{meta})$ is therefore much less than that of K or M and in consequence the ab sub-spectra degenerate to the single line at the centre of the triplet (see next section).

In the ¹H spectrum of 2,6-difluoro-4-bromoaniline (Figure 2.11) the small transitions in the two ab sub-spectra are clearly seen, but the inner transitions of the sub-spectra have coalesced, leaving only four large peaks. However the analysis can proceed using the rules of Equation (2.70) and Figure 2.10.

The ¹H spectrum of para chlorophenol (Figure 2.11) may be considered as an AA'XX' spectrum as the chemical shift separation is large compared to the couplings. Again there is some degeneracy as only six transitions are observed for each nucleus. There is one ab sub-spectrum and the large doublet pattern. The reason for this degeneracy is that the meta HH couplings (J_{26} and J_{35}) in this molecule are very similar (ca. 2–3 Hz) and

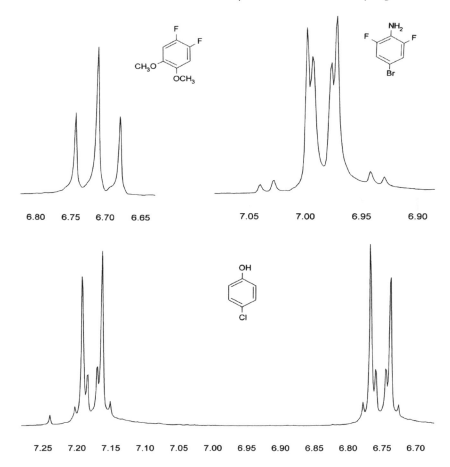

Figure 2.11   *The 300 MHz ¹H spectra of the ring protons of (a) 1,2-difluoro-4,5-dimethoxy benzene, (b) 4-bromo-2,6-difluoroaniline and (c) para chlorophenol, in CDCl₃ (reproduced from Puchert and Behnke[10] with permission from the Sigma-Aldrich Company, Ltd, Milwaukee, WI, USA).*

therefore the value of M $(J_A - J_X)$ is zero. Furthermore the value of the para HH coupling in benzenes is very small (0–1 Hz), thus $J' \approx 0$. One ab sub-spectrum has $J_{ab}^* = M$ which equals 0 and the spectrum is therefore two equally intense lines of separation $L(J - J')$. There is also the large doublet of separation $N(J + J')$ also at the centre of the spectrum. As $J' = 0$ these doublet patterns coalesce to give the resulting six line spectrum.

2.2.5   Iterative Computer Analysis

The only general method of obtaining accurate chemical shifts and couplings from closely coupled spin systems of three or more nuclei is by iterative computer analysis. In this method trial values of the chemical shifts and couplings are input. The program then solves numerically the resulting secular equation to give calculated values of the transition energies and intensities. These are then matched to the observed values and the program then iterates to minimize the observed–calculated differences to give best values of the chemical shifts and couplings.

A number of programs to perform this task have been written. The two most popular early examples were the LAOCOON.[15] and NMRIT/NMREN.[16] programs. These were limited to seven and eight nuclei, respectively. Subsequently useful amendments were added to these programs. Haigh[8] added magnetic equivalence and symmetry factoring to LAOCOON to give the LACX and LAME programs and Ferguson and Marquardt[17] and later Martin and Quirt[18] performed similar revisions on the NMRIT/NMREN program to give the NUMARIT program. These amendments allowed the programs to handle much bigger systems. In both programs the assignment of the observed and calculated transitions was performed manually which for complex spectra with several hundred transitions was a laborious task. Subsequently the commercial Panic program[19] was developed in which the assignment of the observed vs. calculated transitions is performed visually on the computer screen, a much easier task. Very recently the Perch program[20] has been developed which automatically iterates the chemical shifts and couplings to give the best solution using integral-transform iteration techniques.

The CALAC routine.   In order to predict the ¹H NMR spectrum of a compound from the calculated chemical shifts and couplings it is necessary to solve the secular determinant and obtain the transition energies and intensities. Although the LAOCOON program can be easily re-dimensioned to accept larger spin systems, the operation of the program becomes very tedious for larger spin systems and can take minutes or even hours to compute for a large molecule such as a steroid.

It is obvious that for a molecule such as a steroid the couplings between the protons in ring D are not likely to influence the ¹H spectra of the protons in ring A. Thus an amendment to the LAOCOON program was made to reflect this situation. For any proton (H_A) the program searches the CHARGE output for all the nuclei coupling to H_A. Usually only ca. 5 or 6 nuclei have appreciable (> 0.5 Hz) couplings to one proton. These nuclei are identified, their chemical shifts and inter-nuclear couplings obtained from the CHARGE output and the ¹H spectrum is calculated for this sub-system in the usual manner from the secular determinant. However, only the transitions for H_A are retained. The process is repeated for all the other protons in the molecule. Thus for e.g. androsterone ($C_{19}H_{30}O_2$) this replaces one calculation of a spin system of 30 atoms with 30 calculations of spin systems of 6 atoms, an enormous saving in time. The limit of the number of spins in each sub-system can be varied and a calculation was performed to find the optimum size to use. For a steroid molecule, a limit of 8 nuclei took ca. 5 s, 9 nuclei 25 s and 10 nuclei 40 s. There was no difference in the calculated spectra when plotted between the limit of 8 and that of 9 nuclei and thus we have used the limit of 8 nuclei for our calculations. Note that if there are ≤ 8 nuclei coupling with H_A the calculation only uses these nuclei. If there are more than 8 nuclei coupling with H_A the calculation uses those nuclei with the largest couplings to H_A. Equivalent nuclei are always considered together in this calculation.

The CALAC subroutine is incorporated in HNMRSPEC to produce the ¹H NMR spectrum from the chemical shifts and couplings output by CHARGE. It may also be used in a stand-alone capacity where the couplings and chemical shifts are input. For further details and examples see Chapter 9.

We are now in a position to consider computer analysis of ^{1}H NMR spectra and will give in this section a few examples to demonstrate the principles of computer analysis and also to show when even computer analysis, when used unquestioningly can still give incorrect answers. A detailed description and analysis of this subject is given elsewhere.[2] We use the LAOCOON program for these examples. In this case neither symmetry nor magnetic equivalence is included. However the spectrum is still calculated correctly provided the correct data is input.

The ABX spectrum of cysteine is shown in Figure 2.8 and the results of the computer analysis below (Case 1). We use the original nomenclature[15] of $W(x)$ for the chemical shifts and $A(i,j)$ for the couplings. The observed frequencies of the transitions in Figure 2.8 are given in Case 1 under 'EXP. FREQ'. Note that in all spectral analyses the frequencies are measured in Hz increasing from right to left across the spectrum.

Case 1.   LAOCOON III. ABX Spectrum of Cysteine HCl monohydrate in D_2O, 300 MHz.

NN $= 3$ FREQ. RANGE 0.00–1500.00   MIN. INTENSITY 0.010

INPUT PARAMETERS:	$W(1) = 935.00$	$W(2) = 958.00$	$W(3) = 1305.50$
	$A(1,2) = -15.30$	$A(1,3) = 4.40$	$A(2,3) = 5.50$
PARAMETER SETS	1 $W(1)$	2 $W(2)$	3 $W(3)$
	4 $A(1,2)$	5 $A(1,3)$	6 $A(2,3)$

ITERATION 0 R.M.S.ERROR $= 2.140$
ITERATION 1 R.M.S.ERROR $= 0.131$
ITERATION 2 R.M.S.ERROR $= 0.076$
ITERATION 3 R.M.S.ERROR $= 0.076$

BEST VALUES:	$W(1) = 938.64$	$W(2) = 955.08$	$W(3) = 1305.50$
	$A(1,2) = -15.26$	$A(1,3) = 4.18$	$A(2,3) = 5.75$

PROBABLE ERRORS OF PARAMETER SETS

1   0.049   2   0.049   3   0.036
4   0.051   5   0.061   6   0.061

ORDERED LINES

LINE	EXP.FREQ.	CAL.FREQ.	INTEN.	ERROR
11	925.880	925.785	0.299	0.095
1	930.290	930.200	0.340	0.090
15	940.950	941.045	1.677	−0.095
7	945.370	945.460	1.684	−0.090
12	947.570	947.665	1.673	−0.095
2	953.090	953.180	1.688	−0.090
14	963.020	962.925	0.295	0.095
4	968.530	968.440	0.344	0.090
13	1300.570	1300.568	1.028	0.002
9	1304.980	1304.982	1.003	−0.003
5	1306.080	1306.082	0.996	−0.002
3	1310.500	1310.497	0.972	0.003

The trial parameters to input for the iteration are readily obtained in this case from Figure 2.8 and the observed transitions in Case 1 using the first-order rules. These give the input parameters shown. Inputting these into the program provides a spectrum suffi-ciently similar to the observed spectrum that all the calculated transitions can be assigned to the observed lines and this assignment is shown in Case 1. Note that the program does not differentiate between the nuclei. In this case the X approximation is not required. The program then iterates on the observed frequencies to give the best values of the input parameters and continues until the rms error is constant. The progress of the iteration is shown in Case 1 with the best values of the parameters. Also given in the output are error vectors and standard errors (not shown in Case 1) and also the probable errors of the parameter sets. These give the change in the rms error for a given change in the parameter, i.e. they provide a measure of how well each parameter is determined from the spectrum. In this case the probable errors are very similar to the rms error and thus they are determined by the accuracy of the measured transition frequencies. However the probable error is not always the same as the accuracy of the measurements as will be seen later. Note that the values of A(1,3) and A(2,3), which correspond to J_{AX} and J_{BX} in the ABX analysis, are significantly different from the first-order values shown as input parameters. However the **sum** ($J_{AX} + J_{BX}$) remains constant. This is a common feature of this type of spectrum. The first-order splittings come together as the value of J_{AB}/δ_{AB} increases until in the limit they are equal even though the individual couplings may be very different.

The second example is the ¹H spectrum of 1,1-difluoroethylene (Figure 2.4). Trial para-meters can be obtained from the AA'XX' analysis given earlier and are shown in Case 2 with the observed frequencies. Note that the symmetry arises from the input data, i.e. $W(1) = W(2)$, $A(1,3) = A(2,4)$ and $A(1,4) = A(2,3)$. The frequencies were measured from the centre of the spectrum,[9] which is given an arbitrary value of 100.0 Hz. The X approximation is used to calculate the ¹⁹F spectrum using the iso-numbers 1 and 2. This removes some off-diagonal elements and allows the ¹⁹F chemical shift to be input as any number, not the actual shift difference of several million Hz. The ¹⁹F spectrum is identical to the ¹H spectrum and is not shown. The calculated spectrum shows 12 lines for each nucleus, not the 10 lines described earlier for this system. However in Case 2 two pairs of transitions (1,5 and 51,55) are identical in both energy and intensity. This is due to the absence of any symmetry element in the calculations. Including the symmetry element merely coalesces these pairs to give a ten line spectrum, which is identical to the observed spectrum.

Case 2.   LAOCOON III. AA'XX' Spectrum of CH_2=CF_2, pure liquid, 40MHz.[9]

NN = 4   FREQ. RANGE 0.00 – 1000.00   MIN. INTENSITY 0.100

INPUT PARAMETERS:

ISO Numbers	1	1	2	2
	W(1) = 100.000	W(2) = 100.000	W(3) = 500.000	W(4) = 500.000
	A(1,2) = 4.800	A(1,3) = 0.700	A(1,4) = 33.900	
	A(2,3) = 33.900	A(2,4) = 0.700	A(3,4) = 36.400	

PARAMETER SETS

1	W(1), W(2)	2	A(1,2)	3	A(1,3), A(2,4)	
4	A(1,4), A(2,3)	5	A(3,4)			

ITERATION 0 R.M.S.ERROR = 0.031
ITERATION 1 R.M.S.ERROR = 0.011
ITERATION 2 R.M.S.ERROR = 0.011

BEST VALUES

1	W(1), W(2)= 100.000	2	A(1,2)= 4.84	3	A(1,3), A(2,4)= 0.74
4	A(1,4), A(2,3)= 33.86	5	A(3,4)= 36.45

PROBABLE ERRORS OF PARAMETER SETS

1	0.003
2	0.005
3	0.008
4	0.008
5	0.007

ORDERED LINES

LINE	EXP.FREQ.	CAL.FREQ.	INTEN.	ERROR
33	52.900	52.898	0.220	0.002
26	61.300	61.303	0.309	−0.003
51	82.700	82.700	2.000	0.000
55	82.700	82.700	2.000	0.000
46	92.900	92.916	1.691	−0.016
19	94.200	94.180	1.780	0.020
37	105.800	105.820	1.780	−0.020
27	107.100	107.084	1.691	0.016
1	117.300	117.300	2.000	0.000
5	117.300	117.300	2.000	0.000
42	138.700	138.697	0.309	0.003
18	147.100	147.102	0.220	−0.002

The parameters which are equal by symmetry (W(1) and W(2), A(1,3) and A(2,4) etc.) are kept equal in the iteration process. This proceeds as normal to give well-defined solutions. But note that if A(1,3) and A(1,4), and A(1,2) and A(3,4) are interchanged, the only effect is to alter the line assignments, the spectrum is identical. The same happens if the relative signs of A(1,2) and A(3,4), are interchanged. However the spectrum **does** depend on the relative signs of A(1,3) and A(1,4) as the parameters *N* and *L* (Figure 2.10) are distinct. This analysis may be compared with the following one.

The ^1H spectrum of 1,2-difluoro-4,5-dimethoxybenzene (Figure 2.11) may be analysed in a similar manner. Trial values of the chemical shifts and couplings are given in Case 3. The ^1H chemical shift is the centre of the spectrum, measured in Hz from TMS and the ^{19}F chemical shift is an arbitrary value. The values of the couplings are from the literature. The assignment of the transitions to the observed lines is trivial as the calculated ^1H spectrum shows only three peaks. The iteration proceeds as normal to give an acceptable rms error of 0.011 Hz. and the correct chemical shift. However the values of the couplings obtained are unreal and this is confirmed by the large probable errors. The values of A(1,2) and A(3,4) are obviously indeterminate and even those of A(1,3) and A(1,4) are only approximate values. Analysis of the spectrum shows that only the value of *N* (J + J′) (Figure 2.10) which is A(1,3)+A(1,4) in the present nomenclature is well-defined and equals 19.15 Hz.

All the other coupling parameters K, L, M are not defined by the spectrum. The spectrum is another example of a deceptively simple spectrum. Even computer analysis of this spectrum cannot determine the values of the couplings as there is insufficient information in the observed spectrum to allow this. Note that if the probable errors had not been calculated and output there is no indication that the values of A(1,3) and A(1,4) are not accurate. When compared to the input data they are probably in error by ca. 5 Hz. Thus in cases where the couplings are obtained from computer iteration without any output of probable errors the spectra should be examined to determine any possible deceptive simplicity.

We have considered only the explicit ABX and AA'XX' spectra in this section but computer analysis can be used most effectively for more complex spectra which do not give explicit transition energies. The closely coupled analogues of the ABX and AA'XX' spectra are the ABC and AA'BB' spectra which are much more difficult to analyse. These systems have been investigated many times, the ABC spectrum[1,21-23] and the AA'BB' spectrum.[2,12,24-27] The problems of insufficient spectral data to give well-defined solutions can also be present in these systems and have and still are giving rise to considerable misinterpretation.

Case 3.   LAOCOON III. AA'XX' Spectrum of 1,2-Difluoro-4,5-dimethoxybenzene.

NN = 4   FREQ. RANGE 1500.00–2500.00   MIN. INTENSITY 0.001

INPUT PARAMETERS:

ISO NUMBERS   1   1   2   2
              W(1) = 2014.650 W(2) = 2014.650 W(3) = 500.000 W(4) = 500.000
              A(1,2) = 0.500   A(1,3) = 9.500   A(1,4) = 7.500
              A(2,3) = 7.500   A(2,4) = 9.500   A(3,4) = 20.000

PARAMETER SETS

    1   W(1),W(2)   2   A(1,2)   3   A(1,3),A(2,4)
    4   A(1,4),A(2,3) 5   A(3,4)

    ITERATION 0 R.M.S.ERROR = 0.750
    ITERATION 1 R.M.S.ERROR = 0.375
    ITERATION 2 R.M.S.ERROR = 0.011
    ITERATION 3 R.M.S.ERROR = 0.011

BEST VALUES

    1   W(1) =   2014.650
    1   W(2) =   2014.650
    2   W(3) =   500.000
    2   W(4) =   500.000
         A(1,2) =   −78.445
         A(1,3) =   4.037
         A(1,4) =   15.112
         A(2,3) =   15.112
         A(2,4) =   4.037
         A(3,4) =   −5481.534

LARGE RESIDUAL ROTATION
PROBABLE ERRORS OF PARAMETER SETS

1	0.004
2	6198.822
3	7.987
4	7.987
5	13053.200

ORDERED LINES

LINE	EXP.FREQ.	CAL.FREQ.	INTEN.	ERROR
51	2005.090	2005.075	2.000	0.015
55	2005.090	2005.076	2.000	0.014
26	2014.650	2014.644	2.000	0.006
45	2014.650	2014.644	2.000	0.006
42	2014.650	2014.656	2.000	−0.006
21	2014.650	2014.656	2.000	−0.006
1	2024.210	2024.225	2.000	−0.015
5	2024.210	2024.225	2.000	−0.015

2.2.6 Automatic Iteration of Complex Spectra

The above analyses can in cases of complex spectra become very time consuming as the trial parameters must be altered and the result compared to the experimental spectrum by the user. To resolve this tedious process computer programs have been written which can aid in the analysis.[20,28] Such programs use various mathematical representations of the spectrum to automatically compare the experimental spectrum with the calculated one. The program also alters the various parameters in an iterative manner to arrive at the best solution. It should be noted that when using such schemes the result of the analysis is sensitive to the quality of the experimental spectrum, and efforts should be made to collect high quality spectra with little

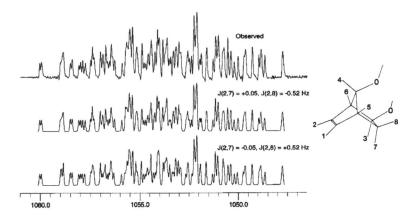

Figure 2.12 *The spectrum of H5 of exo-2-syn-7-norborn-5-enediol, an ABCDEFXY system (400 MHz, 5 % acetone solution, at 310 K). The figure shows the observed (top) and two solutions (bottom), which both yield similar results. The solutions differ in that one coupling constant has reversed sign in the two solutions (from Laatikainen et al.,[20] reproduced with permission of Elsevier).*

noise and artefacts. An example is shown in Figure 2.12 where the Perch program[20] is used to analyse the complex ABCDEFXY spectrum of exo-2-syn-7-norborn-5-enediol, shown with the numbering used by the authors. The figure shows the expanded region for H-5. Although the spectrum is highly complex the procedure only takes minutes on a desktop PC.

The above example illustrates again that in complex cases a unique solution is not always attainable, and alternate solutions may give similar agreement with the experimental data. It is sometimes possible in such cases to resolve the ambiguities if the coupling constants can be calculated accurately by other means. The physical models used to describe the origin of the coupling constant and its calculation based on molecular structure are described in the next section.

2.3   The Mechanism of Spin–Spin Coupling

Before considering the many factors which affect coupling constants and which therefore provide chemical information from the values of these couplings, it is useful to consider briefly the mechanism of the coupling interaction, i.e. how do the nuclear spins interact?

In a stationary molecule in a solid, the magnetic moments of the nuclei interact directly, and this dipolar interaction is very large. For example, for two protons 2 Å apart the interaction energy can be as much as 30 kHz. Also, there is a distance and orientation dependence of the same form as Equation (1.9). Thus, the width of the lines in the ¹H spectrum of a solid will be of this order, and this completely obscures any fine structure due to the chemical shifts. In liquids and gases the rapid molecular motion of the molecules averages these interactions to zero and in consequence we observe very narrow NMR signals.

The internuclear couplings found in high-resolution spectra are transmitted via the bonding electrons, i.e. they are electron coupled interactions. There are three possible mechanisms:[29]

(1) The nuclear moments interact with the electronic currents produced by the orbiting electrons, referred to as either the paramagnetic spin-orbit (PSO) or diamagnetic spin-orbit (DSO).
(2) There is a dipolar interaction between the nuclear and electronic magnetic moments called the spin-dipole (SD) contribution;
(3) There is an interaction between the nuclear moments and the electronic spins in s-orbitals due to the fact that the electron wave function has a finite value at the nucleus. This is called the Fermi-contact (FC) term.

These terms can be computed using quantum mechanical methods; however, such calculations are computationally very demanding and thus far from routine. The computations become heavy as they require very accurate description of the electronic environment, which is achieved by use of very large basis sets (including multiple functions describing each electron orbital). Practically the quantum mechanical methods based on density function theory (DFT) have been shown to require the least amount of computational resources to produce the most accurate results.[30–32] However, even using DFT methods, computing these terms accurately requires large amounts of memory (RAM) and can take several days to compute on a desktop PC. It should be noted that these calculations are performed on a single static structure generally without the inclusion of solvent effects. Although methods exist for applying corrections due to dynamical and solvent effects, these must be used with caution as the accuracy of the corrections are difficult to assess due to the inherent uncertainties associated with the underlying calculations.

For all couplings involving hydrogen the FC term is dominant due to cancellation of the PSO and DSO terms and the minimal contribution of the SD term (see also Figure 2.13). It may thus suffice to simply use the FC term in many cases and as the different terms can be computed separately one may either neglect the non-FC terms or perform two calculations; one using a larger basis set to calculate the FC term and one with a smaller basis set calculating the other terms. Since the FC term is dominant, there is often considerable similarity between H—H couplings and the corresponding H—X coupling. Also, as the size of the contact term is obviously proportional to, among other factors, the percentage of s-character at the coupling nucleus, it is not surprising that there are many correlations of H—X couplings with the percentage s-character of the bonds.

However, for couplings not involving hydrogen the other terms are important, and the couplings are very different from the corresponding H—H and H—X couplings. This occurs to some extent for C—C couplings, and for C—F and F—F couplings the orbital and dipolar contributions are often as large as the contact contribution. For this reason these couplings are fundamentally different from those of the corresponding C—H or H—F couplings.

The mechanism for the ^{13}C—^{1}H coupling in CHCl$_3$ (Figure 2.2) can be visualized as follows:

$$^{13}\text{C} \uparrow\downarrow^e --------- \uparrow^e\downarrow\,^1\text{H}$$

The ^{13}C nucleus interacts with the 2s electron through the contact term and this means that the anti-parallel orientation of the nuclear and electronic spins shown in favoured. By the Pauli exclusion principle, the electrons in the bond will tend to be anti-parallel and the other electron will interact with the proton, giving the favoured orientation of the ^{13}C and ^{1}H nuclei as shown. These spins are opposed which gives a positive coupling constant. Indeed, all directly bonded C—H (and C—C) couplings are positive. For couplings over more than one bond, the mechanism of transmission via the intervening atoms is less direct and, in consequence, both positive and negative couplings are found. Because these couplings are transmitted via the bonding electrons, they fall off rapidly with increasing numbers of bonds between the coupling nuclei, which is in one respect fortunate, as otherwise the resulting spectra would become very complex. This can be seen clearly from Table 2.2, which gives some characteristic H—H couplings. Couplings over more than three bonds are usually quite small (< 3 Hz) and are often not resolved. The division into couplings over one, two and three bonds, i.e. ^{1}J, ^{2}J, ^{3}J couplings, is also a convenient method of grouping these inter-nuclear couplings. We will consider first HH couplings, as these are well understood and well documented.[33–37]

2.3.1 Geminal HH Couplings (^{2}J$_{HH}$)

Selected values of the geminal HH couplings are given in Table 2.3. In contrast to the vicinal coupling (next section) these can be of either sign and they vary much more in magnitude, from -20 to $+40$ Hz. The variations in this coupling shown in Tables 2.2 and 2.3

Table 2.2   *Characteristic* H—H *couplings (Hz)*

System	Coupling range	System	Coupling range
OPEN CHAIN			
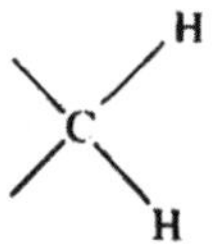 C(H)(H) (geminal, on C)	− 10 to − 18	C=CH—CHO	7 to 9
—N=C(H)(H)	8 to 16	CH—SH	6 to 8
CH—CH / CH—CH=C	5 to 10	CH—OH / CH—NH—	4 to 8
CH₃.CH₂—	7 to 8	CH—C—CH	0 to 1
H...H C=C H	J_{gem} − 3 to + 7	CH—C=CH—	0 to − 2
	J_{cis} 3 to 18	CH—C≡CH	− 2 to − 3
	J_{trans} 12 to 24		
CH—CHO	1 to 3	CH—C=C—CH	0 to 2
		CH—C≡C—CH	2 to 3

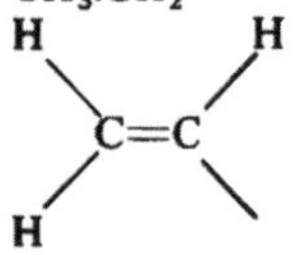

CYCLIC			

Aromatic and olefinic	*ortho*	*meta*	*para*
Benzene derivatives	5–9	2–3	0–1

	J_{23}	J_{34}	J_{24}	J_{25}
Furan	1.8	3.5	0.8	1.6
Pyrrole	2.6	3.4	1.4	2.1
Thiophene	5.2	3.6	1.3	2.7
Cyclopentadiene	5.1	1.9	1.1	1.9

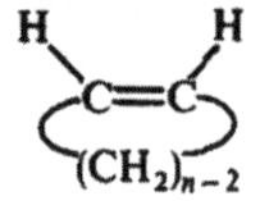

H...H C=C (CH₂)$_{n-2}$

$n = 3$	4	5	6	7	8	9
ca. 1	2.7	5.1	8.8	10.8	10.3	10.7

Saturated

	J_{gem}	J_{cis}	J_{trans}
Cyclopropane	− 4.5	9.2	5.4
Ethylene oxide	+ 5.5	4.5	3.2
Ethylene imine	+ 2.0	6.3	3.8
Ethylene sulphide	*ca.* 0	7.1	5.6
Cyclobutane derivatives	− 11 to − 15	6 to 11	3 to 9
Cyclopentane derivatives	− 11 to − 17	7 to 11	2 to 8
Cyclohexane derivatives	− 12 to − 15	J_{ae}2 to 5	J_{aa}8 to 13 J_{ee}1 to 4

are conveniently discussed in terms of the theoretical predictions of a molecular orbital (MO) theory of coupling[38] which may be summarized as follows:

(i) An increase in the H.C.H angle increases the s-character of the orbitals and makes the coupling more positive (or less negative).
(ii) For both sp^2 and sp^3 CH$_2$ groups, substitution of an electronegative atom in an α-position (i.e. an inductive effect) leads to a positive shift in $^2J_{HH}$.
(iii) Substitution of an electronegative atom in a β-position leads to a negative shift in $^2J_{HH}$.
(iv) A substituent which withdraws electrons from antisymmetric orbitals (i.e. hyper conjugative effects) gives a negative contribution.

From (ii), (iii) and (iv) it follows that substituents which donate electrons inductively or hyperconjugatively give the opposite effects to the above. We note that (ii) and (iv) differ in their orientation dependence, as the inductive effect has no orientation dependence, whereas in the hyperconjugative effect the maximum effect will occur when the 1s orbital of the hydrogen atoms has maximum overlap with the hyperconjugative (i.e. π) substituent.

Table 2.3 *Selected examples of geminal couplings (Hz)*

Molecule	Coupling constant	Molecule	Coupling constant
CH$_4$	− 12.4	CH$_2$:C:C	− 9
CH$_3$.CCl$_3$	− 13.0	CH$_2$(CN)$_2$	− 20.3
CH$_3$.CO.CH$_3$	− 14.9	CH$_2$=O	+ 41
CH$_3$.C$_6$H$_5$	− 14.4	CH$_2$:CH$_2$	+ 2.5
CH$_3$Cl	− 10.8	CH$_2$:CHF	− 3.2
CH$_3$OH	− 10.8	CH$_2$:CHLi	+ 7.1
CH$_3$F	− 9.6	BrCH$_2$.CH$_2$OH	− 10.4
		Br.CH$_2$.CH$_2$.OH	− 12.2
		Br.CH$_2$CH$_2$CN	− 17.5

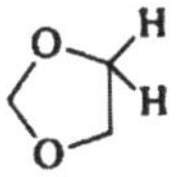

− 8.3   ~0

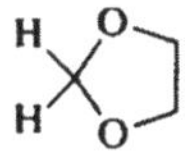

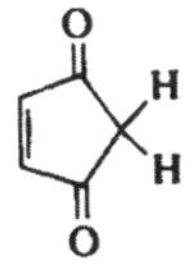

− 21.5   5   6

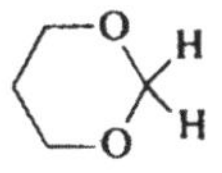

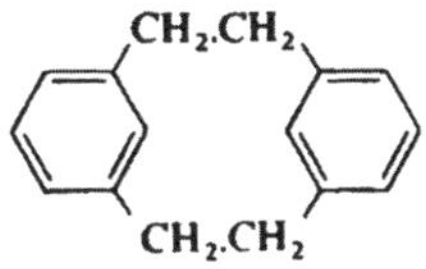

− 12.0

Metacyclophane

Interaction (i) predicts the trends shown in Tables 2.2 and 2.3 for saturated, cyclic and olefinic couplings. For example, in going from methane (−12.4 Hz) to cyclopropane

(−4.5 Hz) to ethylene (+2.5 Hz) a continuous positive change is observed as the H.C.H angle increases.

The influence of electronegative substituents is also clearly observed in Table 2.3, both in saturated and unsaturated systems. For example CH_4 (−12.4 Hz) compared with CH_3OH (−10.8 Hz) and CH_3F (−9.6 Hz) and C_2H_4 (+2.5 Hz) compared with $CH_2{=}O$(+41 Hz), giving as predicted from rule (ii) a positive contribution. In the case of formaldehyde there is also a back donation of the oxygen π orbital in the CH_2 plane and this augments the inductive effect. The effect of β-substituents is clearly seen in substituted vinyl compounds, $CH_2{:}CHX$, and there is a linear dependence of J_{gem} on the electronegativity of the atom X (Equation (2.11)), the *geminal* coupling varying from −3.2 Hz in vinyl fluoride to +7.1 Hz in vinyl-lithium.

The effect of hyperconjugative withdrawal of electrons is also evident in allene, acetone and toluene, in which a negative contribution (rule (iv)) from the unsubstituted compound is observed. Indeed, there is an almost constant change of −1.9 Hz for each adjacent π-bond in a freely rotating CH_2 fragment. This hyperconjugative contribution is proportional to the overlap of the hydrogen 1s orbitals and the π bond, which approximately follows a $\cos^2 \varphi$ dependence, where φ is the dihedral angle between the CH bond and the π orbital. For the two coupled hydrogens this contribution is given by Equation (2.10).

$$\Delta J = -K \cos^2 \varphi . \cos^2(120 + \varphi) \qquad (2.10)$$

As an example consider the coupling in toluene (Figure 2.13). Couplings J_{13} and J_{23} will be the same as in methane, as H_3 lies in the nodal plane of the π-orbital ($\varphi = 90°$; $\cos^2 \varphi = 0$) but J_{12} will experience maximum overlap. The observed coupling (−14.4 Hz) is the rotational average, i.e. $(J_{12} + J_{13} + J_{23})/3$ and taking $J_{13} = J_{23} = -12.5$ Hz gives J_{12} equal to −18.2 Hz. Therefore the constant K in Equation (2.10) for this fragment equals ca. 10 Hz (φ for J_{12} equals 30°).

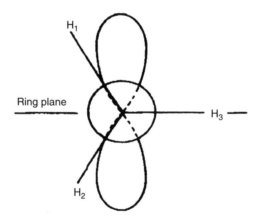

Figure 2.13   *Orientation of methyl protons in toluene.*

In cyclic compounds the value of this coupling can therefore be as large as −20 to −25 Hz if the orientation of the π system and the CH_2 group is favourable. In the almost planar ring of 3,5-dioxocyclopentene (Table 2.3) this favourable orientation occurs, giving a coupling of -21.5 Hz. In contrast, in metacyclophane the orientation of the CH_2 protons with respect to the benzene ring is the same as H_1 and H_3 above, and thus the observed coupling is identical with methane as predicted.

A similar effect occurs in the methylenedioxy group ($O.CH_2.O$), but in this case the oxygen lone-pair is an electron-donating π orbital. In the six-membered 1,3-dioxane ring the CH_2 orientation is similar to H_1 and H_3 above; thus there is no hyperconjugative contribution and the observed coupling (-5 to -6 Hz) reflects only the inductive effect of the two oxygens. In the five-membered dioxolane ring hyperconjugative overlap occurs, giving a positive contribution to the coupling and a resulting coupling of ca. 0 Hz.

2.3.2   Vicinal HH Couplings ($^3J_{HH}$)

The commonest and most useful coupling encountered in NMR is the vicinal proton–proton coupling ($^3J_{HH}$) and some typical values are given in Table 2.2. Our understanding of these couplings stems largely from the theoretical work of Karplus,[39,40] based on the contact term mentioned above. Three main conclusions emerged from this theory, which, in its original form, was concerned with hydrocarbons and neglected the effects of substituents.

(i)   The couplings in both saturated and unsaturated systems are largely transmitted via the σ-electrons, and these are always positive.
(ii)   In olefinic systems J_{trans} is larger than J_{cis}.
(iii)   In a saturated fragment such as H.C.C.H the coupling is proportional to $\cos^2 <\Phi>$, where Φ is the dihedral angle between the two C—H bonds.

It is convenient to consider first olefinic and aromatic couplings. Some typical values are given in Table 2.4. We can see that although the effect of substituents has a considerable effect on olefinic couplings, it is always true that, for the same substituents, J_{trans} is larger than J_{cis}. The effects of substituents in vinyl compounds are simply related to the electronegativity (E_X) of the first atom of the substituent by equations (2.11):[41]

$$J_{trans} = 19.0 - 3.2(E_X - E_H)$$

$$J_{cis} = 11.7 - 4.1(E_X - E_H) \tag{2.11}$$

$$J_{gem} = 8.7 - 2.9(E_X - E_H)$$

where E_X is the electronegativity of the X atom on the Pauling scale (Table 1.2) in which $E_H = 2.2$ and the constants are the observed couplings in ethylene.

These equations enable the couplings in any mono-substituted olefin to be predicted quite well, e.g. these give values of 6.4 and 14.8 for J_{cis} and J_{trans} in vinylether (cf. Table 2.4). Assuming an additivity relationship allows the coupling in any disubstituted olefin to be estimated. These may not be as accurate, particularly for the *cis* couplings, as the two substituents may interact and therefore impair the additivity relationship.

There is also a dependence of the vicinal coupling on the C.C.H angles, and this is reflected in the dependence of the *cis* olefinic coupling on ring size, both in *cis* olefins and in aromatic systems. The dependence on ring size is very clearly seen in the cyclic

Table 2.4 Selected values of vicinal couplings (Hz) in olefinic and aromatic compounds

Molecule	Coupling constant J_{cis}	J_{trans}	Molecule	Coupling constant
$CH_2{:}CH_2$	11.7	19.0	$CH{:}CH$	9.5
$CH_2{:}CH.OR$	6.7	14.2	C_6H_6	7.6
cis- and trans-$CHCl{:}CHCl$	5.3	12.1		J_{12} 8.6
				J_{23} 6.0
	J_{23}	5.5		10–12
	J_{34}	7.5		

olefins (Table 2.2), the *cis* coupling increasing from 2.7 Hz in cyclobutene to the value of 10.8 Hz in cycloheptene, in which the internal angles are relatively unstrained. Increasing the ring size further does not further increase the coupling, as the rings become increasingly buckled.[42]

A similar dependence of the coupling on ring size is observed in aromatic systems. In substituted benzenes the *ortho* coupling is ca. 7–9 Hz, whereas in the seven-membered azulene ring the coupling is 10–12 Hz and in the five-membered aromatic systems of furan, pyrrole and thiophene the couplings are 2–5 Hz. In the latter case there are other factors present which affect the vivinal couplings, in particular the electronegativity of the heteroatom and the partial double bond character of the bonds. The latter effect is clearly seen in the naphthalene couplings (Table 2.4), in which J_{12} is considerably larger than J_{23}. This is not due to any increased contribution of the π electrons to the coupling, but simply due to the shorter bond length in the C_1-C_2 compared to the C_2-C_3 bond, which results in an increased interaction through the σ bonds.

CH.CH couplings In a saturated CH.CH fragment many of the above factors will still influence the coupling, but in addition there is the very important dependence on the CH.CH dihedral angle and some examples of are given in Table 2.5.

In a freely rotating fragment such as $CH_3.CHXY$ there is again a linear dependence on the electronegativity of X and Y. If J_{AV} is the coupling in a $CHR_1R_2.CHR_3R_4$ fragment, then

$$J_{AV} = 8.0 - 0.80 \sum (E_X - E_H) \tag{2.12}$$

where the summation refers to the groups R_1 to R_4 and the constant is the value for ethane.[43] Equation (2.12) shows that the substituent dependence of J_{AV} is only very pronounced for multisubstituted fragments. For the ethyl group ($CH_3.CH_2X$) J_{AV} is ca. 7–8 Hz, and for the $CH_3.CHXY$ group it is ca. 5.5–6.5 Hz. However, for a $C.CH(OH).CH(OH).O$ fragment J_{AV} is calculated as 4.6 Hz. The analogous coupling to olefinic protons is of similar magnitude (cf. the $CH_3.CH$ coupling in propene of 6.4 Hz, Table 2.5), but in aldehydes the coupling is considerably smaller (e.g. $CH_3.CHO$, 2.9 Hz).

Table 2.5 *Selected examples of vicinal couplings across saturated bonds*

Molecule	Coupling constant	Molecule	Coupling constant	
			J_{gauche}	J_{trans}
CH_3CH_2Li	8.4			
CH_3CH_3	8.0	$CH_3.CH\,Br.CH\,Br.CH_3$	2.9	10.3
$CH_3.CH_2.CH_3$	7.3	$CHCl_2.CHF_2$	1.6	8.4
CH_3CH_2OH	7.0	$\diagdown CH.CH{:}C\diagup$	3.5	11.5
$CH_3.CH{:}CH_2$	6.4	$\diagdown CH.CHO$	0.1	8.3
$CH_3.CH(OH)_2$	5.3			
$CH_3.CHO$	2.9			

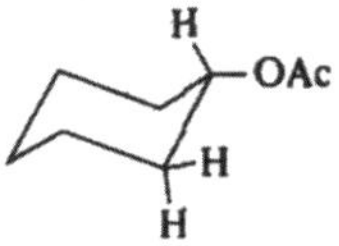

J_{aa} 11.4
J_{ac} 4.2

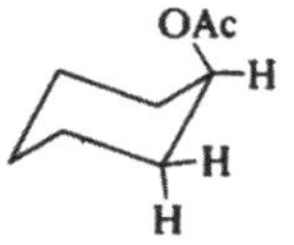

J_{ea} 2.7
J_{ee} 2.7

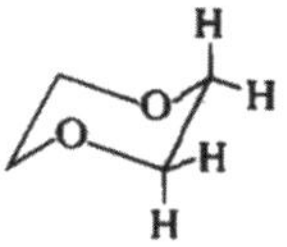

J_{aa} 11.5
J_{ac} 2.7
J_{ee} 0.6

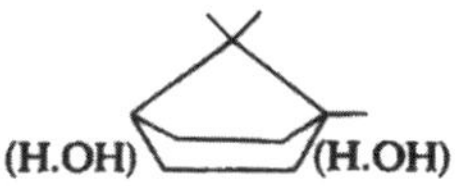

$J_{exo-exo}$ 7.7
$J_{exo-endo}$ 2.3
$J_{endo-endo}$ 8.9

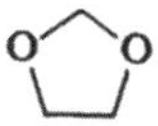

J_{cis} 7.3
J_{trans} 6.0

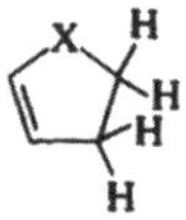

X	CH₂	CO	O	S
J_{cis}	7.4	7.2	10.7	10.0
J_{trans}	4.6	2.2	8.3	7.5

The most important factor influencing these couplings is the dihedral angle between the CH bonds. The original theoretical predictions[39,40] were written in the form

$$J_{(CH.CH)} = 8.5\cos^2\varphi - 0.28 \qquad 0° \le \varphi \le 90° \qquad (2.13a)$$

$$= 9.5\cos^2\varphi - 0.28 \qquad 90° \le \varphi \le 180°$$

$$J_{(CH.CH)} = A\cos^2\varphi + B\cos\varphi + C \qquad (2.13b)$$

$$J_{(CH.CH)} = 11.16\cos^2\varphi - 1.28\cos\varphi + 0.77 \qquad (2.13c)$$

of Equation (2.13a) and the calculated couplings from this equation were in good agreement with the HH couplings in the carbohydrate ring, which were known at that time. Later Karplus[44] suggested Equation (2.13b) noting that the substituent electronegativity also plays an important part. Other equations have been suggested. Altona *et al.* described a complex equation (the Haasnoot equation) in which both the dihedral angle and substituent electronegativity was included[45,46] and Abraham *et al.* derived Equation (2.13c) for alicyclic five and six membered rings.[47] This equation predicts values for the vicinal couplings in cyclohexane (J_{axax}, J_{eqax} and J_{eqeq}) of 13.0, 3.6 and 3.0 Hz, which are in good agreement with the observed values of 13.1, 3.6 and 3.0 Hz. Note also that for $\varphi = 90°$ the coupling is very small (0.77 Hz.), an important result.

Table 2.5 also shows the substituent dependence of these couplings. Thus, J_{aa} varies from 8 to 13 Hz, J_{ae} from 2 to 4 Hz and J_{ee} from 1 to 3 Hz, depending on the number of electronegative substituents present.

In acyclic systems the couplings also follow this pattern, though here the values are extrapolated, not observed values, as the molecules are rapidly rotating in solution Thus, the *gauche* coupling (J_{gauche}) is ca. 2–4 Hz and the *trans* coupling (J_{trans}) ca. 8–12 Hz. Also, in planar (i.e. eclipsed) fragments Equation (2.13c) predicts that J_{cis} (J_0) is ca. 10.5 Hz. and $J_{trans}(J_{120})$ ca. 2.5 Hz and again the results in Table 2.5 support this. In the norbornane ring, J_{exo_exo} and J_{endo_endo} (both J_0) are large (8–9 Hz), whereas $J_{exo\_endo}(J_{120}°)$ is 2.3 Hz. In the planar five-membered ring of cyclopentenone $J_{cis} \gg J_{trans}$ as predicted, but in the other five-membered heterocyclics shown, although J_{cis} is always bigger than J_{trans} the couplings are larger than expected. Other factors such as the non-planarity of the rings and the influence of lone-pairs on the couplings may be affecting the values.

The effect of substituents on these individual couplings is of interest as there is a pronounced orientation dependence. The substituent has a maximum effect when in a planar *trans* orientation with the coupling proton. A good example of this is the 1,4-dioxan ring, where the electronegative oxygens have the largest effect on J_{eqeq}. A simple scheme has been proposed for estimating the substituent contributions to the couplings, as follows.[13] For a *gauche* coupling (J_{gauche}) the substituent X can have one of only two orientations, X_{gauche}, i.e. *gauche* to both protons, or X_{trans}, *trans* to one proton. The substituent contributions in these orientations are given for some common substituents in Table 2.6.

Table 2.6   *Substituent Parameters for ³J^{HH} (gauche) Hz*

$$J^{HH} \ (gauche) = 4.0 + \sum(X_{gauche} + X_{trans})$$

	H	C	Br	N	Cl	O	F
X_{gauche}	0.0	0.2	0.5	0.5	0.7	0.9	1.1
X_{trans}	0.0	−0.6	−1.1	−1.4	−1.6	−1.8	−2.6

2.3.3   *Ab initio* Calculated Couplings

The above examples exclusively use empirical data to determine parameters for the various equations used. However, theoretically calculated data may also be used to obtain such parameters. An example of such a study is that of Mobli *et al.*[48] where the three bond vicinal coupling of the amide group in *N*-acetyl-glucosamine was determined. In that study ab initio methods were used to calculate the various contributions (FC, PSO, DSO and SD) to the vicinal coupling constant at fixed angles of the intervening dihedral angle. The authors found in agreement with previous studies of HH coupling constants[32] that the PSO

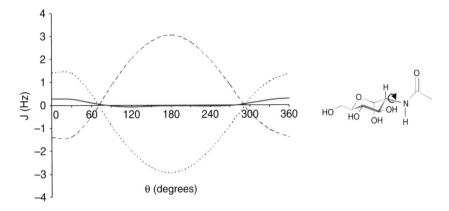

Figure 2.14   *Calculated non-FC ((—) SD; (- - -) PSO; (· · ·) DSO) contributions to the density functional theory (DFT) calculated coupling constants in α-GlcNAc as a function of the intervening dihedral angle (from Mobli and Almond[48], reproduced by permission of the Royal Society of Chemistry).*

and DSO terms nearly cancel one another and the SD term is very small, whereas the FC term is dominant (Figure 2.14). Considering the computational resources required to calculate all terms it is concluded that in these cases the computation of the FC term alone is sufficient.

The summation of the coupling constants (or in this case the FC term) can then be fitted to the empirical Karplus equation (Equation (2.13b)) to yield values of $A = 9.81, B = -1.51$ and $C = 0.62$. It should be noted that when using such an approach one should be aware that the *ab initio* calculations are of static structures *in vacuo* and therefore a correction due to bond libration needs to be applied for non-rigid bonds (solvent effects also affect the values but to a much lesser extent). It should also be noted that for accurate calculations very large basis sets need to be used. In such cases DFT calculations have been found to produce reasonable accuracy albeit with a general overestimation (in some cases up to 10 %) of the exact values.[32]

Although the absolute value of the coupling constants may not be exact the trends are generally good. For this reason it is increasingly popular to use *ab initio* methods to investigate the dependence of a coupling constant on its chemical environment, be it bond rotation as seen above or as in the next example the dependence on substituent groups.

In Table 2.7 we have calculated the various contributions to the coupling constant (as above using a larger basis set for the FC term and a smaller one for all others) for a series of cyclic compounds. The calculated vicinal coupling constants (for atoms connected to the atoms marked with a bold bond) are in very good general agreement with the observed values (where known) and are of considerable interest. In compounds 1, 2 and 3 there is a strong dependence on the electronegativity of the neighbouring atoms, but it is the *reverse* of that expected. The replacement of the carbon atom in cyclopentene by oxygen *increases* the vicinal coupling. In contrast replacing CH_2 by CO *decreases* the same coupling. These trends have not had any previous explanation but the calculations show that the trends are real and due mainly to the Fermi contact term. In addition the remarkable effect of the olefinic bond in norbornene (5) on the vicinal coupling in norbornane (4) has been noted

Table 2.7 *Calculated DFT vs. the observed vicinal couplings in some cyclic compounds*

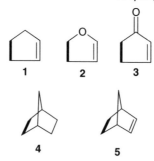

	FC	PSO	DSO	SD	J (TOT)	J(obs)
1						
3Jcis	12.51	−0.41	0.36	0.22	12.68	7.45[a]
3Jtrans	5.21	1.96	−2.03	−0.03	5.11	
2						
3Jcis	14.40	−0.39	0.36	0.21	14.58	10.76[b]
3Jtrans	8.69	2.00	−2.08	−0.03	8.58	8.13
3						
3Jcis	7.84	−0.35	0.31	0.20	8.00	7.20[b]
3Jtrans	1.95	2.01	−2.06	−0.39	1.51	2.24
4						
EN-EN	10.20	−0.52	0.49	0.21	10.38	9.05[c]
EN-EX	4.73	1.82	−1.88	−0.03	4.64	4.62
EX-EX	13.60	−0.61	0.59	0.22	13.80	12.22
5						
EN-EN	9.95	−0.48	0.44	0.22	10.13	9.02[d]
EN-EX	4.08	1.85	−1.92	−0.34	3.67	3.91
EX-EX	10.17	−0.61	0.58	0.23	10.37	9.38

[a] (Jcis + Jtrans)/2; Puchart *et al.*[10]
[b] Abraham *et al.*[50]
[c] Marshall *et al.*[49]
[d] Abraham *et al.*[51]

for some time[49]. The vinylic bond has a large influence on the exo–exo ^{3}J coupling but not on the other vicinal coupling constants. This is due to a change in the Fermi contact term showing the influence of the unsaturated bond on the inner orbitals of the hydrogen atoms involved rather than any direct effect (PSO/DSO) of the small currents produced by the unsaturated bond.

The calculations described here have been performed using the Gaussian[52] software (no. G03, rev. D), and have used the IGLOO-III basis set for the non-FC terms and Huz-IIISu3 as described previously.[32] Examples of input files (i.e. for structure and J-coupling constant calculation including the non-standard basis sets) can be found on the accompanying CD.

2.3.4   Long-range HH Couplings

It is convenient to group all other HH couplings (i.e. over four or more bonds) into one class, although there are a number of different mechanisms involved in these couplings.[53,54] A selection of these couplings is given in Table 2.8.

One well understood mechanism is the σ–π interaction which is involved in allylic (CH:C.CH) and homo-allylic (CH.C:C.CH) couplings. There is a certain amount of overlap of the π-orbital of the double bond and the hydrogen 1s orbital in CH: and CH.C: systems. This can be represented schematically as shown in Figure 2.15. The olefinic proton and adjacent π-electron have opposing spins, whereas the allylic proton and the nearest π-electron prefer the parallel orientation. As a consequence, this mechanism produces a positive contribution to CH:CH couplings, a negative contribution to allylic couplings and a positive contribution to homo-allylic couplings. For freely rotating methyl groups the interaction of the π-electron and methyl group is about the same magnitude, but of opposite sign, to the π-electron olefinic proton interaction. Thus the allylic CH_3.C:CH coupling is about the same size but of opposite sign to the corresponding homo-allylic (CH_3.C:C.CH_3) coupling.

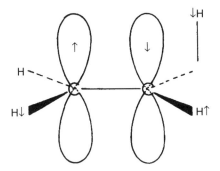

Figure 2.15   *π-Orbitals and proton configuration in olefins.*

Also the interaction in an HC.C: fragment, being proportional to the H(ls), π-orbital overlap is proportional to $\cos^2 \varphi$ where φ is the dihedral angle between the CH bond and the π-orbital. Thus, there will be a $\cos^2 \varphi$ dependence of the allylic couplings, and a $\cos^2 \varphi_1 \cos^2 \varphi_2$ dependence of the homo-allylic couplings, as there are two CH.C: dihedral angles to consider. In particular, the contribution of this mechanism will be zero for $\varphi = 0°$, i.e. when the CH protons are in the plane of the double bond (the nodal plane of the π orbitals).

These general considerations are illustrated in Table 2.8. In propene the *cis* allylic coupling is slightly larger (i.e. more negative) than the *trans,* and this is characteristic of such allylic couplings in acyclic systems (*cis*, 1.5–3.5 Hz; *trans*, 1–3 Hz). The corresponding homo-allylic couplings are similar (1–2 Hz), but of opposite sign, as predicted, but the *trans* coupling is slightly larger than the *cis*, and again this is a general effect in acyclic systems. In acetylenes, and other conjugated systems, the σ–π mechanism can produce larger couplings, which can be transmitted over many conjugated bonds. In cyclic systems the

Table 2.8 Selected examples of long-range HH couplings

Molecule	Coupling (Hz)	Molecule	Coupling (Hz)
CH$_2$:CH.CH$_3$	$^4J_{cis}$ ~ 1.7 $^4J_{trans}$ ~ 1.3	(structure)	1–2
CH$_3$.CH:C(CH$_3$).CO$_2$H	$^5J_{cis}$ 1.2 $^5J_{trans}$ 1.5	(structure)	1–1.5
HC:C.CH$_2$.CH$_3$ CH$_3$.C:C.CH$_2$.CH$_3$ CH$_3$.C:C.C:C.CH$_3$	^4J, −2.4 ^5J, +2.6 ^7J, 1.3	(structure)	7–8
(structure, CH$_3$, Br, Cl)	$^4J_{(Me-H_6)}$, −0.63 $^5J_{(Me·H_3)}$, +0.40 $^6J_{(Me-H_4)}$, −0.58	(structure, Me, H, H)	$^4J_{(Me-H)_{g}}$, 0 − −0.3 $^4J_{(Me-H)_{t}}$, +0.4 − +1.0
(structure O, O, CH$_3$, H, H)	$^4J_{(Me-H_3)}$, −1.6 $^5J_{(Me-H_4)}$, 2.7	(structure, O=C–H, Cl, Cl, H)	$^5J_{(CHO-5)}$, 0.7
(structure, O, O, 7,8,1,2,6,5,4,3)	$^5J_{35}$, +0.32 $^7J_{37}$, +0.34 $^4J_{45}$, −0.31 $^5J_{48}$, +0.65	(structure, H–C=O, O, OH, H)	$^5J_{(CHO-3)}$, 0.6
(structure, H, H, N)	$^5J_{48}$, 0.8		

dihedral angle dependence is noticeable and homo-allylic couplings of up to 3 Hz are not uncommon for the correct orientation of the two protons and the double bond. In aromatic systems a methyl (or CH$_2$X) substituent couples to the *ortho* protons and this is generally considered as a σ–π mechanism, although, as the aromatic ring has less double bond character than a normal olefin, the coupling is smaller (0.5–1.5 Hz). However, the other long-range couplings in aromatic systems, and the very different couplings of aldehyde protons to aromatics, are not examples of σ–π couplings, but include other mechanisms.

The commonest example of a long-range coupling which is not a σ–π coupling is the *meta* (^4J) coupling in aromatics. These are ca. 2–3 Hz and always positive, in contrast to the ^4J allylic couplings; furthermore, the σ–π mechanism has no contribution to these couplings, as both are in the nodal plane of the π system. These couplings are very similar to long-range couplings in saturated systems. ^{4}J couplings in freely rotating fragments, e.g. $CH_3.C.CH_3$, are very small (ca. zero), but for specific orientations of the bonds the coupling may be appreciable. It is largest (1–2 Hz in non-strained systems) if the bonds are in the planar zig-zag (or W) orientation shown. This occurs in the eq–eq coupling in cyclohexanes and the exo–exo couplings in the camphene system (Table 2.8)

In certain strained systems such as the [2,1,l] bicyclohexanes the analogous coupling is very large (7–8 Hz). In acyclic systems with a preferred conformation this coupling is also evident. Methyl groups coupling to *trans-* and *gauche*-oriented protons have couplings of 0.4 to 1.0 and 0 to −0.3 Hz, respectively. An approximate representation of the theoretical calculations for four bond couplings for unstrained systems is given by Equation (2.14) where φ_1 and φ_2 are the two H.C.C.C dihedral angles in the coupling pathway.[55]

$$^4J_{HH} = \cos^2 \varphi_1 + \cos^2 \varphi_2 - 0.7 \qquad (2.14)$$

There are a number of other long-range couplings observed in aromatic systems particularly in which both the above mechanisms may well be present. A well-known coupling is the $^5J_{48}$ coupling, as exemplified in coumarin and quinoline (Table 2.8), which is probably a σ-bond coupling. A similar coupling is that between the aldehyde proton and the ring proton in substituted benzenes, and the example shown illustrates the stereospecificity of this coupling. In 2,4-dichlorobenzaldehyde, in which the aldehyde conformation for steric reasons is as shown in the table, the aldehyde proton couples only to H_5. Conversely, when the aldehyde conformation is reversed, owing to the hydrogen bond with the *ortho* hydroxyl group, the aldehyde proton couples with H_3. Again the planar zig-zag is the preferred coupling path.

2.4 HF Couplings

As would be expected, there are much less data for HF couplings than for the HH couplings but a comprehensive collection of HF couplings has been given.[56] The transmission of HF couplings is similar to that of HH couplings in that the dominant term is again the Fermi contact term. Thus the orientation and substituent dependence of these couplings is generally similar to the corresponding HH couplings (and very different from the FF couplings). However there are considerable differences in practice. HF couplings are more affected by changes in solvent and temperature than the corresponding HH couplings. In π systems the couplings can vary by up to ca. 10 % in different solvents.[56]

2.4.1 Geminal HF Couplings ($^2J_{HF}$)

Selected values of the geminal HF couplings are given in Table 2.9. Inspection of this table shows that, in contrast to the vicinal HF couplings (next section) which are generally similar in outline to the corresponding HH couplings, the geminal HF couplings show little resemblance to the geminal HH couplings. They are all positive, whereas the HH couplings can be of either sign and span a narrow range (45 to 90 Hz) compared to −20 to +40 Hz for the HH couplings. The range is so small for specific fragments that an average coupling is a good approximation. For example, for CH$_2$F.C fragments the coupling is 47.0 ± 1.5 Hz and for CHF$_2$.C fragments it is 55.0 ± 2.0 Hz. It is clear from the results for the fluoromethanes in Table 2.9 that increased electronegativity of the substituents causes a significant increase in the ^{2}J$_{HF}$ coupling. Equation (2.15) relating the coupling to the substituent electronegativity has been given[57] and also rules for predicting this coupling in fluoro sugars.[58]

$$
{}^2J_{HF} = 78.76 + 8.45E_XE_Y - 16.73(E_X + E_Y) \tag{2.15}
$$

In olefins the range of values is larger. For example, for the CHF=C fragments the spread is from ca. 72 to 90 Hz and this implies that π densities on the carbon atom do influence this coupling. There has not been any quantitative analysis of these couplings in terms of the two possible factors responsible, i.e. substituent electronegativity and π densities.

Table 2.9   *Selected examples of geminal ^{2}J$_{HF}$ couplings (Hz)*

Molecule	Coupling constant	Molecule	Coupling constant
CH$_3$F	46.4		49.0
CH$_2$F$_2$	50.2		
CHF$_3$	79.2		
CH$_3$.CH$_2$F	48.5		
CH$_3$.CH$_2$.CH$_2$F	48.5		
CH$_2$F.CO.CH$_3$	47.7		55.1
CH$_2$F.CO$_2$CH$_3$	47.0		
CH$_2$F.CONMe$_2$	47.1		
CF$_3$.CH$_2$F	45.7		
CF$_3$.CHF.CF$_3$	45.0		
CH$_3$.CHF.CO$_2$Me	48.7		
CH$_3$.CH$_2$.CHF$_2$	57		51.7
CHF$_2$.CO.CH$_3$	54.1		
CHF$_2$.CO$_2$Me	53.2		
CH$_2$=CHF	84.7	CHF=CHF	72.7, 74.3
CH$_3$.CH=CHF	89.9, 84.8	CHF=CH.CF(CF$_3$)$_2$	77.3, 77.7

2.4.2   Vicinal HF Couplings

The guidelines given in Section 2.3.2 for vicinal HH couplings apply also to the vicinal HF couplings, namely:

(i) The couplings in both saturated and unsaturated systems are largely transmitted via the σ electrons, and these are nearly always positive.
(ii) In olefinic systems J$_{trans}$ is larger than J$_{cis}$.
(iii) In the saturated fragment (H.C.C.F) the coupling is proportional to cos^2 < φ >, where φ is the H.C.C.F dihedral angle.

As before, we consider first olefinic and aromatic couplings and some typical values are given in Table 2.10. In this table, J$_{cis}$ and J$_{trans}$ can refer to the HC=CF couplings in one molecule such as $CH_2=CHF$ or to the *cis*- and *trans*-couplings in different isomers such as $CHF=CHF$. We note that in any particular olefin J_{trans} is always larger than J_{cis}, but the effect of substituents on the values of these couplings is so large that they overlap. J_{trans} varies from 12.5 to 53.6 Hz and J_{cis} from −4.1 to 20.5 Hz. The values of both these couplings can be correlated reasonably with the sum of the Pauling electronegativity (Ex) of the first atom of the substituents from Equation (2.16):

$$^3J_{HF}(cis) = 84.4 - 6.43 \sum E_X \qquad (2.16)$$

$$^3J_{HF}(trans) = 160.9 - 11.09 \sum E_X$$

Comparison with the analogous Equation (2.11) shows that the affect of substituents on the HF couplings is ca. 1.5 to 4 times greater than in the corresponding HH couplings. The theoretical treatment suggests that these couplings will also depend on the H.C:C and C:C.F angles as in the HH couplings but this has not been verified for these couplings.

Table 2.10 *Examples of vicinal H.C:C.F couplings (Hz) in olefinic and aromatic compounds*

Molecule	Coupling constant		Molecule	Coupling constant
	J_{cis}	J_{trans}		
CH_2:CHF	20.53	53.61	C_6H_5F	9.08
Me.CH:CHF	19.9	41.8		
Me.CF:CH₂	16.6	48.6		J_{16} 9.69
CH_2:CF.CH:CCl₂	16.28	46.45		J_{23} 10.36
Me.CF:CHCl	10.8	24.2		J_{34} 8.17
				J_{45} 7.64
CHF:CHF	4.4	20.4		X $\qquad J_{23}$
CH_2:CF₂	4.8	36.4		CN $\qquad$ 9.1
CHCl:CFCl	4.05	18.1		CO.Me 9.6
CF₂:CHCl	1.0	16.6		NH.Et 11.4
CHF:CF₂	−4.24	12.79		OH $\qquad$ 10.4
				I $\qquad$ 7.4
				F $\qquad$ 10.0

In aromatic compounds the *ortho* H-F coupling also varies, but by much less than the olefins above. The range of the *ortho* HF coupling in substituted benzenes is from ca. 7.5 to 11.5 Hz. Interestingly the four *ortho* HF couplings in 1,2,4-trifluorobenzene (Table 2.10) encompass almost the whole range of these couplings in substituted benzenes. An increase in the substituent electronegativity has been shown to both increase and decrease the HF couplings depending on the position of the substituent in the benzene ring,[59] but there has not been any comprehensive theoretical treatment of the variation of these couplings with substituent. Inspection of the data in Table 2.10 shows clearly that the substituent electronegativity is not the only factor influencing these couplings. For example in the 1-substituted 2,6-difluoro compounds the iodo compound has the smallest coupling but iodine is more electronegative than carbon.

CH.CF couplings. The 3J$_{HF}$ couplings in a saturated fragment show the same general pattern as the corresponding olefinic couplings with the important addition of the dihedral angle dependence. Selected examples of these HF couplings are given in Table 2.11.

Table 2.11   *Selected examples of vicinal H.C.C.F couplings (Hz)*

Molecule	Coupling constant		Molecule	Coupling constant
	J$_{AV}$			
$CH_3.CH_2F$	25.2			43.5
CH_3CHF_2	20.8			
CH_3CF_2Cl	15.0			
$CH_3.CF_3$	12.7			
$CF_3.CH_2.CH_3$	10.5			
$CF_3.CH_2.CF_3$	9.17			
$CF_3.CH_2Cl$	8.41			Jexo-F 12.01
$CF_3.CHF.CF_3$	5.5			Jendo-F 25.10
$CF_3.CH_2F$	7.98			
$CF_3.CF_2H$	2.60			
	Jg	Jt		
$CH_2F.CH_2F$	0	45		
$CHBr_2.CFBr_2$	2.2	22.2		Jexo-H 9.13
$CHCl_2.CHF_2$	2.0	19.6		Jendo-H 0.31
$CHCl_2.CFCl_2$	1.0	18.2		
$CHF_2.CHF_2$	−0.2	13.4		Jcis 17.7
				Jtrans 6.3.

It can be seen that there is again a large dependence of the coupling on the substituent electronegativity. Even in the freely rotating fragments $CF_3.CH$ and $CH_3.CF$ in which only the average coupling (J$_{AV}$) is observed this varies from to 25.2 Hz in $CH_3.CH_2F$ to 2.60 Hz in $CF_3.CHF_2$. There is a linear dependence of this coupling on the sum of the electronegativity of the first atom of the remaining substituents given by Equation 2.17.[60]

$$^3J_{HF}(AV) = 53.0 - 3.38 \sum E_X \qquad (2.17)$$

The data in Table 2.11 also show the large dependence of the coupling on the H.C.C.F dihedral angle. In the cyclic molecules of fixed conformation the value given is the observed coupling. In the acyclic molecules Jt and Jg refer to the *trans* and *gauche* oriented couplings. With the sole exception of $CHBr_2.CFBr_2$ in which the individual rotamer couplings were observed at low temperatures,[61] these have been obtained from measurements of the observed coupling as a function of either temperature or solvent or both and then deducing the individual rotamer couplings, from which the values of Jt and Jg can be derived. These values are therefore not as accurate as the observed couplings but are of considerable interest.

All the couplings generally follow the $\cos^2 \phi$ rule in that Jt (180^0) is much larger than Jg (60^0) and Jexo–exo (0^0) is larger than Jendo-exo(120^0). Note that as in the *cis* CH:CF couplings if there are sufficient electronegative substituents Jg can also be negative. A detailed investigation by Williamson *et al.*[62] using cyclic compounds such as the norbornene derivative (Table 2.11) in which the dihedral angles were varied but the substituents were constant (always carbon) leads to Equation (2.18).

$$^3J_{HF} = 31 \cos^2 \phi \qquad 0 < \phi < 90^0 \qquad (2.18)$$
$$= 44 \cos^2 \phi \qquad 90 < \phi < 180^0$$

A subsequent similar investigation by Ihrig and Smith[63] on related compounds in which the substituents were varied with constant dihedral angles gave the substituent dependence of the individual *cis* (0^0) and *trans* (120^0) couplings. These were shown to have a similar substituent dependence to Equation (2.17). Thus it is possible to combine Equations (2.17) and (2.18) to obtain the general Equation (2.19). This gives the dependence of the $^3J_{HF}$ coupling on both the HC.CF dihedral angle and on the sum of the electronegativity of the first atom of all the substituents.

$$^3J_{HF} = 67.8 \cos^2 \phi (1.0 - 0465 \sum E_X) \qquad 0 < \phi < 90^0 \qquad (2.19)$$
$$= 96.4 \cos^2 \phi (1.0 - 0.0465 \sum E_X) \qquad 90 < \phi < 180^0$$

This equation is derived from Equation (2.17) but differs in that the summation is over all the substituents in the C.C fragment whereas in Equation (2.17) it is over the remaining substituents (i.e. excluding the coupled nuclei). It is simpler computationally to sum over all the substituents. These equations are useful to obtain a reasonable value of the $^3J_{HF}$ coupling in any given C.C fragment, but they should be used with caution. Both Williamson *et al.* and Ihrig and Smith independently emphasized the dependence of the $^3J_{HF}$ coupling on the H.C.C and C.C.F angles. Indeed Williamson *et al.* excluded the HF couplings in the cyclopropane derivative (Table 2.11) from their correlations on the grounds of possible bond length and bond angle changes.

2.4.3   Long-range HF Couplings

There are few examples of long-range HH couplings over more than four bonds, the most common being the para HH coupling in benzenes and the homo-allylic couplings in olefins (Section 2.3.4). There are more examples of HF couplings over more than four bonds and a selection of all the long-range HF couplings is given in Table 2.12. The HF coupling over four saturated bonds has a similar orientation dependence to the corresponding HH coupling. It is large and positive (ca. 4–5 Hz) for the 'planar zig-zag' orientation but much smaller and of either sign for the other orientations. In the fluoro sugars (Table 2.12) Jeq–eq is large and positive (ca. 4 Hz), Jeq-ax small and negative (ca. −0.8 Hz) and Jax-ax not normally resolved, i.e. < 0.5 Hz. These couplings are quite well reproduced by Equation (2.20), which is somewhat different from the analogous HH Equation (2.14).

$$^4J_{HF} = 4.1 \cos \phi_1 \cos \phi_2 + 0.2 \qquad (2.20)$$

The couplings in the rigid fluoro norbornene shown also follow this pattern, except that the Fa–Hb(x) coupling is much larger than the corresponding coupling (Jax-ax) in the six-membered ring. The most likely explanation is that there is a direct 'through space' coupling mechanism in this case as the coupled nuclei are much closer in the norbornene ring. The evidence for the 'through space' mechanism in HF couplings has been reviewed.[64]

The four bond meta HF coupling in benzenes follows the planar zig-zag orientation and is of similar value to the coupling over four saturated bonds. As with other HF couplings, these couplings show a much larger dependence on the substituents than the corresponding HH

Table 2.12   *Selected examples of long-range HF couplings (Hz)* [a]

Molecule	Coupling constant ($^4J_{HF}$)		Molecule	Coupling constant	
				($^4J_{HF}$)	($^5J_{HF}$)
CH$_3$.CH:CHF	CH$_3$–F(*cis*)	2.6	C$_6$H$_5$F	5.69	0.22
	CH$_3$–F(*trans*)	3.3			
CF$_3$.CH=CH$_2$	CF$_3$–H(*cis*)	−2.26	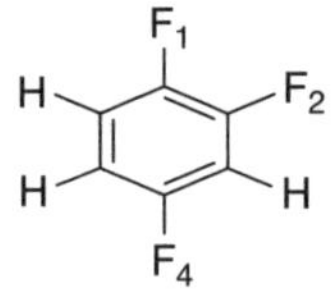	J$_{13}$ 6.24	J$_{25}$ − 2.00
	CF$_3$–H(*trans*)	+0.10		J$_{15}$ 3.26	
				J$_{26}$ 8.71	
				J$_{46}$ 5.03	
CH$_2$F.CH=CH$_2$	CF–H(*cis*)	−4.33	X		J$_{24}$ J$_{25}$
	CF–H(*trans*)	−0.89	CO.Me	6.3	−1.4
CHF$_2$. CH=CH$_2$	CF–H(*cis*)	−3.66	NH.Et	6.0	−1.9
	CF–H(*trans*)	−0.14	OH	6.0	−1.9
(CF$_3$)$_2$.CF.CH=CHF	CF– H(*cis*)	2.1	I	6.4	−1.3
	CF–H(*trans*)	0.0	F	6.9	−2.1

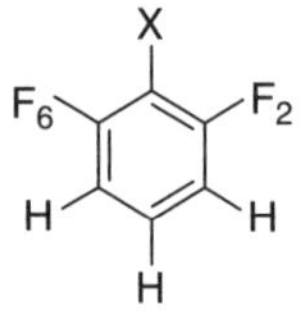

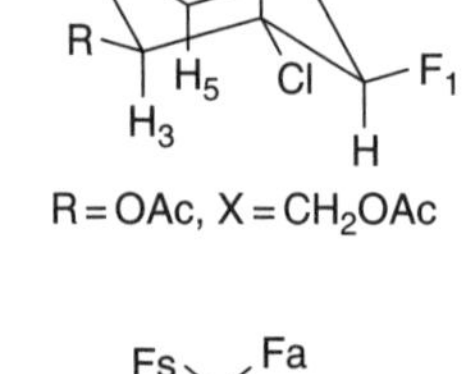

J$_{14}$+4.0

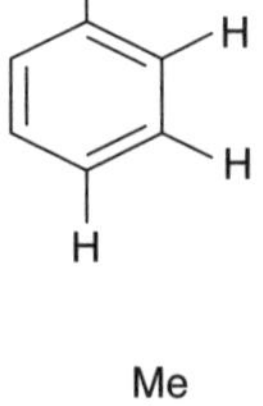

Me–F$_{ortho}$ +2.09
Me–F$_{meta}$ −0.43
Me–F$_{para}$ +1.28

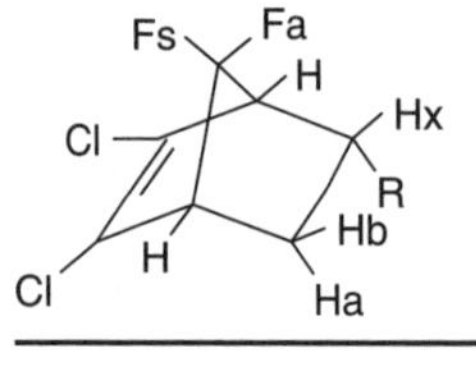

R=OAc, X = CH$_2$OAc   J$_{13}$−0.8   J$_{15}$−0.7

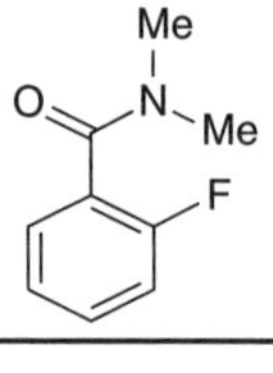

CF$_3$–H$_{ortho}$ −0.75
CF$_3$–H$_{meta}$ +0.82
CF$_3$–H$_{para}$ −0.69

Fs–Ha	+5.0
Fs–Hb(x)	−0.8
Fa–Ha	1.0
Fa–Hb(x)	2.8

F–Me(*cis*)	1.7
F–Me(*trans*)	0.0

[a] The terms '*cis*' and '*trans*' refer to the coupling nuclei.

couplings. In 1,2,4-trifluorobenzene the couplings vary from 3.2 to 8.7 Hz. The coupling is probably influenced by both π and σ transmissions. These couplings are positive, in contrast to the five bond para HF couplings which are much smaller and always negative (Table 2.12).

There is also a significant allylic HF coupling, though here it is necessary to distinguish between HC.C=CF couplings and FC.C=CH couplings. The former would be expected to be similar to the allylic HH couplings with approximately equal values for the *cis*- and *trans*-couplings and a $\cos^2\varphi$ orientation dependence. The values of the *cis*- and *trans*-couplings in the 1-fluoropropenes (2.6 and 3.3 Hz) in Table 2.12 support this suggestion. The Me-ortho F coupling in benzenes is approximately the same value and is positive. The analogous FC.C=CH coupling shows a quite different pattern. In this case the *cis*-coupling is always negative, it is invariably much larger than the *trans* and there is again a large change with substituent. In the *cis* $F_3C.C=CH$ fragment one of the fluorine atoms is eclipsed with the double bond and is therefore near to the coupling proton. A direct H–F mechanism for this coupling has been proposed.[65]

In Table 2.12 there is one example of an HF coupling over six bonds. This is the 1.7 Hz coupling between the *cis* methyl and the *ortho* fluorine in 2-fluoro-dimethylbenzamide. There is no observable coupling to the *trans* methyl group, thus the mechanism is very likely due to a direct 'through space' effect.[64] As the observed coupling is the average over the three methyl protons, only one of which at any time is close to the fluorine atom, the coupling to the near proton could be quite large. If the coupling was zero in the other orientations the value of the near proton-fluorine coupling would be 5.1 Hz.

References

1. Abraham, R. J.; Castellano, S., *J. Chem. Soc. B* 1970, 49.
2. Abraham, R. J., *Analysis of High Resolution NMR Spectra*. Elsevier: Amsterdam, 1971.
3. Pople, J. A.; Schneider, W. G.; Bernstein, H. J., *High Resolution Nuclear Magnetic Resonance*. McGraw-Hill: New York, 1959.
4. Corio, P. L., *Chem. Rev.* 1960, **60**, 363.
5. Corio, P. L., *Structure of High Resolution NMR Spectra*. Academic Press: London, 1966.
6. Emsley, J. W.; Feeney, J.; Sutcliffe, L. H., *High Resolution NMR Spectroscopy*. Pergamon Press: London, 1965.
7. Diehl, P.; Harris, R. K.; Jones, R. G., *Prog. NMR Spectrosc.* 1967, **3**, 1.
8. Haigh, C. W., *Annu. Rep. NMR Spectrosc.* 1971, **4**, 311.
9. Flynn, C. W.; Baldeschwieler, J. D., *J. Chem. Phys.* 1963, **38**, 226.
10. Puchert, C. J.; Behnke, J., *Aldrich Library of 13C and 1H FT NMR Spectra*. Aldrich Chemical Company: Milwaukee, WI, 1993.
11. Abraham, R. J.; Bernstein, H. J., *Can. J. Chem.* 1961, **39**, 216.
12. Grant, D. M.; Hirst, R. C.; Gutowsky, H. S., *J. Chem. Phys.* 1963, **38**, 470.
13. Abraham, R. J.; Fisher, J.; Loftus, P., *Introduction to NMR Spectroscopy*. 2nd ed.; John Wiley & Sons, Ltd: Chichester, 1988.
14. Abraham, R. J.; Leonard, P.; Smith, T. A. D.; Thomas, W. A., *Magn. Reson. Chem.* 1966, **34**, 71.
15. Castellano, S.; Bothner-By, A. A., *J. Chem. Phys.* 1964, **41**, 3863.
16. Swalen, J. D.; Reilly, C. A., *J. Chem. Phys.* 1962, **37**, 21.
17. Ferguson, R. C.; Marquardt, D. W., *J. Chem. Phys.* 1964, **41**, 2087.

18. Martin, J. S.; Quirt, A. R., *J. Magn. Reson.* 1971, **5**, 318.
19. Bruker *Panic 851*. Karlsruhe, 1985.
20. Laatikainen, R.; Niemitz, M.; Weber, U.; Sundelin, J.; Hassinen, T.; J .Vepsalainen, *J. Magn. Reson.* 1996, **120**, 1.
21. Arata, Y.; Shimizu, H.; Fujiwara, S., *J. Chem. Phys.* 1962, **36**, 1951.
22. Castellano, S.; Waugh, J. S., *J. Chem. Phys.* 1961, **34**, 295.
23. Fessenden, R. W.; Waugh, J. S., *J. Chem. Phys.* 1959, **34**, 996.
24. Dischler, B., *Zeit. Fur Naturdor* 1965, **20a**, 888.
25. Gestblom, B.; Hoffman, R. A.; Rodmar, S., *Mol. Phys.* 1964, **8**, 425.
26. Hirst, R. C.; Grant, D. M., *J. Chem. Phys.* 1964, **40**, 1909.
27. Pople, J. A.; Schneider, W. G.; Bernstein, H. J., *Can. J. Chem.* 1957, **35**, 1060.
28. Kolehmainen, E.; Lihia, K.; Laatikainen, R.; Vepsäläinen, J.; Niemitz, M.; Suontamo, R., *Magn. Reson. Chem.* 1997, **35**, 463.
29. Ramsey, N. F., *Phys. Rev.* 1953, **91**, 303.
30. Fukui, H., *Prog. NMR Spectrosc.* 1999, **35**, 267.
31. Deng, W.; Cheeseman, J. R.; Frisch, M. J., *J. Chem. Theory Comput.* 2006, **2**, 1028.
32. Lutnæs, O. B.; Ruden, T. A.; Helgaker, T., *Magn. Reson. Chem.* 2004, **42** (Special issue), 117.
33. Bothner-By, A. A., *Adv. Magn. Reson.* 1965, **1**, 195.
34. Contreras, R. H.; Peralta, J. E., *Prog. NMR Spectrosc.* 2000, **37**, 321.
35. Günther, H.; Jikeli, G., *Chem. Rev.* 1977, **77**, 599.
36. Sternhell, S., *Q. Rev.* 1969, **23**, 236.
37. Thomas, W. A., *Prog. NMR Spectrosc.* 1997, **30**, 183.
38. Pople, J. A.; Bothner-By, A. A., *J. Chem. Phys.* 1965, **42**, 1335.
39. Karplus, M., *J. Chem. Phys.* 1959, **30**, 11.
40. Karplus, M.; Anderson, D. H., *J. Chem. Phys.* 1959, **30**, 6.
41. Schaeffer, T., *Can. J. Chem.* 1962, **40**, 1.
42. Abraham, R. J., Chapter 8, in *NMR for Organic Chemists*, Mathieson, D. W.(Ed.). Academic Press: London, 1967.
43. Abraham, R. J.; Pachler, K. G. R., *Mol. Phys.* 1963, **7**, 165.
44. Karplus, M., *J. Am. Chem. Soc.* 1963, **85**, 2870.
45. Altona, C.; Francke, R.; Haan, R. d.; Ippel, J. H.; Daalmans, G. J.; Hoekzema, A. J. A. W.; Wijk, J. v., *Magn. Reson. Chem.* 1994, **32**, 670.
46. Haasnoot, C. A. G.; Deleeuw, F. A. A. M.; Altona, C., *Tetrahedron* 1980, **36**, 2783.
47. Abraham, R. J.; Koniutou, R., *Magn. Reson. Chem.* 2003, **41**, 1000.
48. Mobli, M.; Almond, A., *Org. Biomol. Chem.* 2007, **5**, 2243.
49. Marshall, J. L.; Walters, S. R.; Barfield, M.; Marchand, A. P.; Marchand, N. W.; Serge, A. L., *Tetrahedron* 1976, **32**, 537.
50. Abraham, R. J.; Parry, K.; Thomas, W. A., *J. Chem. Soc. B* 1971, 446.
51. Abraham, R. J.; Fisher, J., *Magn. Reson. Chem.* 1985, **23**, 856.
52. Frisch, M. J.; Trucks, G. W.; Schlegel, H. B.; Scuseria, G. E.; Robb, M. A.; Cheeseman, J. R.; Montgomery, J. A.; Jr., T. V.; Kudin, K. N.; Burant, J. C.; Millam, J. M.; Iyengar, S. S.; Tomasi, J.; Barone, V.; Mennucci, B.; Cossi, M.; Scalmani, G.; Rega, N.; Petersson, G. A.; Nakatsuji, H.; Hada, M.; Ehara, M.; Toyota, K.; Fukuda, R.; Hasegawa, J.; Ishida, M.; Nakajima, T.; Honda, Y.; Kitao, O.; Nakai, H.; Klene, M.; Li, X.; Knox, J. E.; Hratchian, H. P.; Cross, J. B.; Bakken, V.; Adamo, C.; Jaramillo, J.; Gomperts, R.; Stratmann, R. E.; Yazyev, O.; Austin, A. J.; Cammi, R.; Pomelli, C.; Ochtersk, J. W.; Ayala, P. Y.; Morokuma, K.; Voth, G. A.; Salvador, P.; Dannenberg, J. J.; Zakrzewski, V. G.; Dapprich, S.; Daniels, A. D.; Strain, M. C.; Farkas, O.; Malick, D. K.; Rabuck, A. D.; Raghavachari, K.; Foresman, J. B.; Ortiz, J. V.; Cui, Q.; Baboul, A. G.; Clifford, S.; Cioslowski, J.; Stefanov, B. B.; Liu, G.; Liashenko, A.; Piskorz, P.; Komaromi, I.; Martin, R. L.; Fox, D. J.; Keith, T.; Al-Laham, M. A.; Peng, C. Y.; Nanayakkara, A.; Challacombe, M.;

Gill, P. M. W.; Johnson, B.; Chen, W.; Wong, M. W.; Gonzalez, C.; Pople, J. A., *Gaussian 03*, D; Gaussian Inc.: Wallingford, CT, 2006.
53. Barfield, M., *J. Chem. Phys.* 1964, **41**, 3825.
54. Barfield, M., *J. Am. Chem. Soc.* 1971, **93**, 1066.
55. Abraham, R. J.; Gottschalk, H.; Paulsen, H.; Thomas, W. A., *J. Chem. Soc.* 1965, 6268.
56. Emsley, J. W.; Phillips, L., *Prog. NMR Spectrosc.* 1978, **10**, 83.
57. Olah, G. A.; Pittman, C. V., *J. Am. Chem. Soc.* 1966, **88**, 3310.
58. Emsley, J. W.; Phillips, L.; Wray, V., *Prog. NMR Spectrosc.* 1976, **10**, 83.
59. Hutton, H. M.; Richardson, B.; Schaefer, T., *Can. J. Chem.* 1967, **45**, 1795.
60. Abraham, R. J.; Cavalli, L., *Mol. Phys.* 1965, **9**, 67.
61. Govil, G.; Bernstein, H. J., *J. Chem. Phys.* 1967, **47**, 2818.
62. Williamson, K. L.; Hsu, Y. L.; Hall, F. H.; Swager, S.; Coulter, M. S., *J. Am. Chem. Soc.* 1968, **90**, 6717.
63. Ihrig, A. M.; Smith, S. L., *J. Am. Chem. Soc.* 1970, **92**, 759.
64. Hilton, J.; Sutcliffe, L. H., *Prog. NMR Spectrosc.* 1975, **10**, 27.
65. Myhre, P. C.; Edmonds, J. W.; Kruger, J. D., *J. Am. Chem. Soc.* 1966, **80**, 2459.

3

Chemical Shift Calculations and Molecular Structure

3.1 Introduction

There have been a number of attempts to predict ^{1}H chemical shifts for many years with varying success. They can be loosely divided into three categories. These are quantum mechanical calculations, database methods and semi-empirical calculations, of which the CHARGE calculation is one method. The first two methods will be considered briefly in this chapter and then the theory of the CHARGE procedure will be given in more detail.

The quantum mechanical and semi-empirical methods require as input an accurate molecular geometry. This can be obtained by *ab initio* calculations, molecular mechanics or the experimental geometry if known accurately may be used. The geometries of the compounds studied here were mainly obtained by *ab initio* optimizations using the GAUSSIAN program,[1] originally at the RHF/6-31G* level and later at the B3LYP level with the 6-31G** basis set. For molecules too large to be handled conveniently by GAUSSIAN at this level, either smaller basis sets were used, e.g. 3-21G or the molecular mechanics MMFF94 force field in the PCMODEL program.[2] In some cases even small molecules, e.g. substituted bromobenzenes were minimized much quicker using PCMODEL. For the iodo compounds studied (e.g. iodobenzenes) the 6-31G** basis set could not be used and the recommended basis set (Lanl2DZ)[3] produced bond lengths which were in poor agreement with experiment. In this case a combination of experiment and calculation was used to give satisfactory geometries. In the great majority of the compounds studied the *ab initio* and MM geometries were very similar and the resulting calculated ^{1}H shifts virtually identical. However, in some cases, particularly for sterically crowded molecules (e.g. tri-tert-butylmethane) the two minimized geometries did give different chemical shifts with the CHARGE program. The use of *ab initio* and molecular mechanics methods to obtain molecular geometries is discussed further in Chapter 8.

Modelling ^{1}H NMR Spectra of Organic Compounds: Theory, Applications and NMR Prediction Software
Raymond Abraham and Mehdi Mobli   © 2008 John Wiley & Sons, Ltd

3.2   Quantum Mechanical Calculations of ^{1}H Chemical Shifts

The magnetic shielding of a nucleus is given by the isotropic shielding tensor (σ) of that particular nucleus. It is formally defined[4] as the second derivative of the molecular energy with respect to the external magnetic field and the magnetic dipole moment of the nucleus. The shielding tensor is thus calculated in quantum mechanics by taking the second derivative of the energy, involving the total electronic energy, the magnetic moment of the nucleus and the external applied magnetic field.[5] The quantum mechanical calculation of a chemical shift thus requires the calculation of the molecular energy in a uniform magnetic field. This gives rise to what is termed the gauge problem. In simple terms this is the fact that the results of the quantum mechanical calculation depend on the position of the molecule when magnetic fields are present. This of course is unreal. It is due to the unavoidable truncation of the atomic orbitals used to describe the molecular orbitals in these calculations. Because incomplete basis sets are used the calculated shielding will differ depending on where the origin of the vector potential of the external applied magnetic field is set. Methods have been developed to minimize the errors arising from this. The most common method of performing this calculation is the gauge-including atomic orbital (GIAO) method,[4] but other methods include individual gauge origins for different localized molecular orbitals (IGLO) and individual gauges for atoms in molecules (IGAIM).[5] GIAO works by incorporating the gauge origin into the basis functions themselves, allowing all matrix elements involving the basis functions to be arranged as independent of the origin. These basis functions were first used in 1937 by London[6] to explain magnetic susceptibilities and were introduced into modern electronic theory by Hameka[7] and Pople[8,9] in 1958. The advent of *ab initio* theory gave an additional stimulus to these calculations and this is now the method of choice of calculating nuclear shielding constants. Of the common methods, GIAO is probably the more reliable, but it is possible to obtain good results with all of them.[10] There are comprehensive reviews of the calculation of shielding tensors by *ab initio* methods in the literature.[11–16] It should be noted however that these methods do not routinely include any intermolecular interactions such as solvent or concentration effects.

The use of these calculations to determine chemical shifts for heavy atoms, particularly ^{13}C, is now well established. However there are particular problems with ^{1}H calculations. Pulay *et al.*[4] note that the chemical shift range of ^{1}H is much smaller than any other atom and therefore more accuracy is required for differentiating chemically similar protons. Also protons are usually located at the periphery of the molecule and are sensitive to intermolecular interactions. In the *ab initio* calculations the shielding tensor and therefore the isotropic chemical shift of a nucleus may be obtained once the electronic structure of the molecule has been determined. It is invariably the case that the molecular structure is also obtained from *ab initio* calculations, thus the effect of different basis sets will be seen in both the structure and the chemical shifts.

An illustrative example is due to Lampert *et al.*[17] who compared the experimental and calculated ^{1}H chemical shifts in phenol and 2-hydroxybenzoyl compounds. They used the Gaussian program[1] at three levels of theory (HF, B3LYP and BLYP) and four basis sets from (6-31G(d,p), to 6-311++G(2df,2dp),with four optimized geometries. As usual the calculated isotropic shielding constants were transformed to chemical shifts relative to CH_4 calculated at the same computational level (the chemical shift of an atom 'x' is thus calculated using a reference compound 'ref' according to: $\delta_x = \sigma_{ref} - \sigma_x c + \delta_{ref}$). They found that with the exception of the OH protons, B3LYP and HF optimized geometries gave similar results, but the GIAO-HF calculations were distinctly worse than the GIAO-BLYP calculations. They noted that the calculated shifts vary considerably. For example, in phenol the ortho proton shifts vary from 6.06 (HF) to 6.57(MP2) cf. exp. 6.47δ. The OH protons, as expected, were not as well predicted. In phenol the OH chemical shift ranged from 2.75 to 3.74 (cf. exp. 4.27δ) and in salicylaldehyde the range was 8.7–13.0 (cf. exp. 10.8δ). It is clear that these calculations do reproduce the general trends of the observed data, but the crucial question is whether the basis set dependence can be reduced to the levels of accuracy which are necessary for chemical structure determination.

3.3   The Database Approach

Database methods are popular and very convenient methods of predicting NMR chemical shifts. The target atom is given a chemical shift based on comparisons with atoms in the database having similar environments. These environments are described using a code, and the HOSE code (**H**ierachically **O**rdered **S**pherical description of **E**nvironment) suggested by Bremser in 1978 is the most commonly used example.[18] Each atom of the molecule to be predicted is treated in turn, and is referred to as a *focus*. The environment of the atom is characterized in terms of atom types separated by numbers of bonds. All atom types are given a priority and a ranking. For example, a double bond is ranked higher than a single bond and a carbon is ranked higher than an oxygen. The atoms are also ranked according to the number of substituents they are bound to, giving a higher ranking to atoms with many substituents. This description extends to a chosen number of bonds from the atom considered, or shells, and is given a HOSE code. Figure 3.1 shows how the different shells in the HOSE code are prioritized (more sophisticated code systems can take into consideration bond angles and bond lengths).

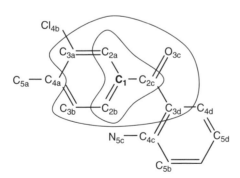

Figure 3.1   *Atom priorities around the C_1 atom according to the HOSE code. Each atom within each shell is ranked alphabetically.*

The *foci* in the database also have codes. The great advantage of the HOSE code is that the codes each have an arithmetic value, and can be ordered by these. The HOSE code value of the target atom *focus* is evaluated and similar environments can be quickly located in the database by finding *foci* with HOSE codes with similar arithmetic values. At one extreme is the case where no similar environments are found, in which case there can be no prediction. At the other extreme, the molecule to be predicted is found in the database and a perfect prediction is made, assuming the experimental conditions are the same. For intermediate cases, a weighted average chemical shift is obtained with atoms in similar environments, i.e. having matching HOSE codes up to a number of shells, contributing more than those in less similar environments.

The prediction of ^{13}C chemical shifts is now almost invariably performed using databases (e.g. ACD Laboratory's ^{13}C Predictor[19], Modgraph's NMRPredict,[20] etc.) and the HOSE code is now the preferred method of ^{13}C prediction although modifications of the code are often used. The major advantage of this approach has been that it provides the user with a systematic way of storing NMR spectra and the user can include their own data to improve the predictions. Another advantage is that the user can retrieve the spectra which have contributed to the prediction and make a more informed choice. A general problem with database approaches is that they require large datasets some of which are acquired from the literature. This can potentially add a further degree of uncertainty to the predictions as literature values may not always be accurate.

The database approach works very well for ^{13}C chemical shift prediction. However, this has not been as successful for ^1H chemical shifts. This is due to the fact that ^1H chemical shifts are very sensitive to geometry changes at large distances (i.e. shells) from the atom considered. Thus a large number of shells are necessary to fully define ^1H shifts and this becomes computationally very demanding.

More recently neural networks have been used for ^{13}C prediction[21] and also investigated for ^1H predictions.[22] Neural networks use a set of descriptors, based on physical or geometric data (e.g. the partial atomic charges or stereochemistry). The descriptors are given a certain weighting and the network is then trained using experimental data. The descriptors will respond to structural features of the training set and the weighting will be altered to better reproduce the training data. This procedure is followed in an iterative manner until the network stabilizes.[22] Neural networks require much less computer time but the user cannot retrieve the data on which the prediction is based.

3.4   Semi-empirical Calculations

As mentioned in Chapter 1 the first comprehensive investigation of the effects of substituent groups on distant protons in saturated compounds was given by Zurcher.[23] All subsequent investigations of long-range effects have been based on Zurcher's analysis. On this model the influence of a distant group on the ^1H chemical shift may be considered as a number of separate contributions. These are:

(1) Steric or van der Waals effects due to the proximity of the proton and the substituent ($\Delta\delta_S$).
(2) The electric field produced by the substituent polarizes the C—H bond which affects the ^1H chemical shift ($\Delta\delta_{EL}$).
(3) Magnetically anisotropic substituents will give rise to magnetic fields at the proton considered which do not average to zero over the molecular tumbling ($\Delta\delta_{AN}$).

The total SCS is given by the sum of these effects. Zurcher also included solvent effects ($\Delta\delta_{SOL}$) explicitly in his treatment. As stated previously only ^{1}H chemical shifts of solutes in dilute solutions in the moderately polar solvent CDCl$_3$ are considered here (the use of DMSO as solvent is considered later in Chapter 8). We assume that under these conditions differential solvent effects may be ignored. Zurcher analysed the substituent effects of a number of common substituents (Cl, OH, CN and CO) using the rigid ring systems of the steroid and bornane rings as templates to obtain the discrete terms listed above. He found that the SCSs for the chloro, hydroxy and cyano groups could be explained satisfactorily in terms of the electric field term only but for the carbonyl group it was necessary to include both the electric field and anisotropic terms. The major limitation of this elegant study was that due to the NMR instrumentation available at that time, the only protons which could be measured and assigned in the spectra were the methyl groups in the steroids and bornanes. These do not always provide sufficient information to quantify the various terms. Zurcher also noted that although this analysis gave reasonable results for the distant protons of the methyl groups when the same approach was used on the vicinal protons in the CH$_2$.CHX fragment of 2-substituted norbornenes no agreement was obtained between the observed and calculated SCSs. Subsequently Apsimon and co-workers[24-26] extended these studies but their studies also did not provide a definitive explanation of the SCSs in ^{1}H NMR.

The explanation of the SCSs in terms of the electric field and magnetic anisotropy of the substituent becomes questionable when the ^{1}H chemical shifts of saturated hydrocarbons are considered (Tables 4.1–4.5). These range over more than 2 ppm., which is 20 % of the usual range of proton chemical shifts, yet these molecules possess neither magnetically anisotropic nor polar substituents. Clearly there are other factors determining the ^{1}H chemical shifts in hydrocarbons. One long standing explanation given for these shifts was that it was due to the magnetic anisotropy of the C—C bonds. This was suggested to explain the chemical shift difference between cyclopentane and cyclohexane[27,28] and also the difference between the axial and equatorial protons in cyclohexane.[29] However, extending this approach[29-32] to larger molecules gave unreasonably large values of the C—C bond anisotropy. Furthermore, attempts to explain the chemical shifts in alkyl derivatives,[33] effects on methyl groups[34] and effects from the methyl group in methylcyclohexanes[35] clearly demonstrated that other factors were important. Subsequent studies[36-42] provide an insight into these factors. Li and Allinger[41] observed a correlation between the steric interaction energy experienced by the hydrogen atoms in a variety of cyclohexanols with the ^{1}H chemical shift and also that the sensitivity of the chemical shift differed for methine, methylene and carbinol hydrogens. Studies of the ^{1}H shifts of methyl substituted cyclohexanes,[39] norbornanes and adamantanes[40] showed that the methyl vicinal SCS was a function of the C.C.C.H. dihedral angle, shielding (−0.5 ppm) at 0° and deshielding (+0.25 ppm) at 180°. Boaz[36] assigned the ^{1}H spectra of some cyclic hydrocarbons and interpreted the observed shieldings as due to electron density changes plus the influence of C—H bonds in a parallel 1,3 arrangement. Curtis *et al.*[37,38] in a study of methylcyclohexanes using ^{2}D NMR obtained good agreement with the observed shifts using an additive scheme with no less than 14 parameters with separate parameters for axial and equatorial hydrogens and four different gauche (C.C.C.H) effects.

A promising approach to the prediction of ^{1}H chemical shifts was the semi-empirical calculations of partial atomic charges in molecules which gave surprisingly good correlations with ^{1}H chemical shifts.[43-45] Gasteiger and Marsili[45] showed that the partial atomic charges calculated by their electronegativity equalization approach gave a good correlation with the ¹H chemical shifts of a variety of substituted simple alkanes. Abraham and Grant[43] also obtained a good correlation of charge versus chemical shifts for a similar set of molecules using the CHARGE3 scheme based on experimental dipole moments.[43,44,46] There were, however, notable deficiencies in both of these schemes. The slope of the chemical shift versus charge differed markedly for different substitution patterns, a serious deficiency in any predictive scheme. Also these schemes were not applied to more complex molecules in which orientational and steric effects were present. We shall now show how the CHARGE scheme has been developed into a semi-empirical method for predicting the ¹H chemical shifts of a wide variety of organic compounds containing diverse functional groups whilst at the same time retaining the ability to give accurate molecular dipole moments.[47]

3.5   Theory of the CHARGE Program

As mentioned in Chapter 1 a basic assumption of the method is the distinction between substituent effects over one, two and three bonds and longer-range effects. The effects over one, two and three bonds are regarded as through bond effects and attributed to the electronic effects of the substituents. The longer-range effects are regarded as through space effects due to the electric fields, steric effects and anisotropy of the substituents.

3.5.1   Through Bond Effects

The CHARGE program calculates the effects of neighbouring atoms on the partial atomic charge of an atom under consideration. The calculations are based on classical concepts of inductive and resonance contributions.[46] If we consider an atom I in a four atom fragment I—J—K—L (Figure 3.2) the partial atomic charge on I is due to three effects, an α effect from atom J, a β effect from atom K and a γ effect from atom L

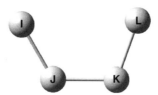

Figure 3.2   *The* I—J—K—L *atom sequence where the effects from the* J, K *and* L *atoms on the* I *(hydrogen) atom are considered.*

The α effect of atom J on atom I to give charge $q_i(\alpha)$ is given by Equation (3.1). E_j and E_i are the electronegativities of atoms I and J and $A(I,J)$ is a constant depending on the exchange and overlap integrals for the bond I—J. In CHARGE there is a set of parameters $A(I,J)$ for all the bonding pairs under consideration.

$$q_i(\alpha) = (E_j - E_i)/A(I, J) \qquad (3.1)$$

The β effect of atom K on atom I is proportional to both the electronegativity of atom K and the polarizability of atom I. Taking the electronegativity of hydrogen as a base, the β effect is defined in Equation (3.2) where $q_i(\beta)$ is the partial atomic charge on I due to atom K and c is a constant.

$$q_i(\beta) = (E_K - E_H)P_i/c \tag{3.2}$$

In order to account for the variation of the polarizability with charge, the β effect calculation is carried out iteratively, according to Equation (3.3), where P_i is the polarizability of atom I with charge q_i, and P_i° and q_i° are the corresponding initial values.

$$P_i = P_i^\circ[1.0 + 3.0(q_i^\circ - q_i)] \tag{3.3}$$

The γ effect of atom L on I is calculated in one of two ways. There is a γ effect given by the product of the atomic polarizabilities of atoms I and L for I = H and L = F, Cl, Br, I and S, given by Equation (3.4).

$$q_i(\gamma) = 0.005P_I P_L^\circ \tag{3.4}$$

where $q_i(\gamma)$ is the partial atomic charge on atom I due to the γ effect and P_I and P_L° are the polarizabilities of atoms I and L, respectively. However, for the second row atoms (C, O, etc.) the gamma substituent effect, GSEF (i.e. X.C.C.H), is parameterized separately and is given by Equation (3.5), where θ is the X.C.C.H dihedral angle and A and B are empirical parameters.

$$GSEF = A + B_1\cos\theta \quad 0° \leq \theta \leq 90° \tag{3.5}$$

$$GSEF = A + B_2\cos\theta \quad 90° \leq \theta \leq 180°$$

The coefficients A and B vary if the proton is in a CH, CH_2 or CH_3 fragment and there are also routines for the methyl γ effect and for the γ effect of the electronegative oxygen and fluorine atoms for CX_2 and CX_3 groups. . The total charge is given by summing these effects (Equation (3.6)).

$$q_i = q_i(\alpha) + q_i(\beta) + q_i(\gamma) \tag{3.6}$$

In order that an element may be included in the scheme, it is necessary to obtain values for the electronegativity and polarizability of that element in the appropriate hybridization state. The electronegativities used were orbital electronegativities based on the Mulliken scale.[48,49] This was used rather than the more common Pauling scale[50] because the orbital electronegativity as opposed to the atomic electronegativity is given. Thus, the electronegativity of $C(sp^3) < C(sp^2) < C(sp)$. The values were from the literature[45] except for Cl, Br and I which were taken directly from the ^1H chemical shifts of the MeX compounds.

For π systems a Huckel subroutine is incorporated to calculate the π charges which are then added to the σ charges (above) to give the total partial atomic charge. The standard coulomb and resonance integrals for the Huckel routine are given by Equation (3.7), where α_0 and β_0 are the coulomb.

$$\alpha_r = \alpha_0 + h_r\beta_0$$

$$\beta_{rs} = k_{rs}\beta_0 \tag{3.7}$$

and resonance integrals for a carbon $2p_Z$ atomic orbital and h_r and k_{rs} the factors modifying these integrals for orbitals other than sp^2 carbon. These factors were obtained from the molecular dipole moments. The standard Huckel routine was further modified by the ω technique to model the very polar π systems of the nucleic acid bases.[46]

3.5.2   ¹H Chemical Shifts of Substituted Methanes and Ethanes

The CHARGE scheme was adapted[51] to give an accurate calculation of the ¹H chemical shifts in substituted methanes and ethanes as follows.

α and β effects. The first calculations did not give an accurate value for the ¹H chemical shift for methane.[43,45] Also the cumulative beta effect of substituents is not a linear function of the number of the substituents as would be predicted from Equations (3.1) and (3.2), but a curved plot. The curvature ranges from a gentle slope for Me and Cl to a sharp bend for F (Figure 3.3(a)). The points on the δ vs. *n* plot (Figure 3.3(a)) are well reproduced by an exponential function (Equation (3.8)) with different values of the curvature parameter *b* for the different substituents and this suggested a possible solution.

$$\delta_i = \delta_0 + A[1 - \exp(-bq)] \tag{3.8}$$

The value of E_H was reduced from 7.17 (the literature value) to 6.9 and Equation (3.3) was replaced with an exponential curve similar to Equation (3.8). One that satisfies the boundary conditions is Equation (3.9).

$$P_i = P_i{}^\circ \exp[10.0(q_i - q_i{}^\circ)] \tag{3.9}$$

The resulting partial atomic charges were plotted against the observed ¹H chemical shifts of methane, ethane, CH_2Me_2, $CHMe_3$, CH_3X and CH_3CH_2X (X = NH_2, OH, F, Cl) to give Equation (3.10) with a correlation coefficient of 0.999 and rms error of 0.059 ppm.

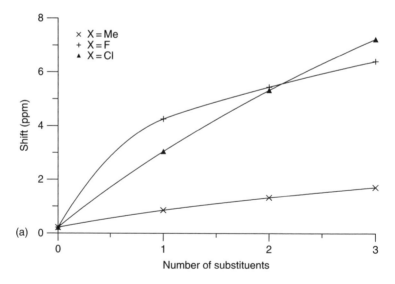

Figure 3.3   *(a)* $\delta(CH_{4-n}X_n)$ *and (b)* $\delta(CH_3CH_{3-n}X_n)$ *versus the number of substituents (n): (x) X = Me; (+) X = F; (▲) X = Cl (reproduced from Abraham* et al.[51] *by permission of the Royal Society of Chemistry).*

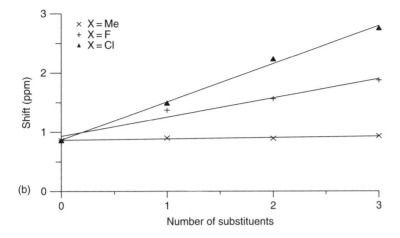

Figure 3.3 (*Continued*).

The observed and calculated ^1H chemical shifts are given in Table 3.1. The value of E_H is equivalent to 2.3 on the Pauling scale (cf. 2.2, Table 1.2) and q_H for methane is 43.2 me corresponding to a C—H bond dipole of 0.22D.

$$\delta = 160.84q - 6.68 \tag{3.10}$$

The coefficient of 160 ppm/electron compares very reasonably with other shift vs. charge equations where values from 130 to 180 have been given.[46,52]

There remains the problem of fluorine (and oxygen) beta substitution. The non-linear effect of multiple fluorine substitution is well known in quantum mechanical calculations of fluoro compounds. The *geminal* fluorine atoms strongly interact with each other, the F.C.F angle is much less than tetrahedral and the CF bond dramatically shortened in the CF_2 and CF_3 groups.[53] Similar effects occur for multiple oxygen substitution.[54] The CHARGE scheme was modified to take explicit account of these effects by reducing the beta fluorine and beta oxygen effects by the appropriate factor. This minor change gave excellent agreement with both the observed ^1H chemical shifts of the fluoro and oxygen substituted methanes (Table 3.1) and the dipole moments.

The γ (H.C.C.X) effect. The γ effect is only a small perturbation ($< 0.1D$) of the dipole moment. However the γ effect of substituents on the ^1H chemical shift is often large and easy to measure and thus it is possible to examine this in more detail. Figure 3.3(b) shows the chemical shift of the methyl protons in substituted ethanes as a function of the number of substituents. There are some similarities to that of beta substitution (Figure 3.3(a)) in that the plot for fluorine is again curved but in contrast those for Cl and Me including the origin are accurately linear. The γ SCS is in the order $I > Br > Cl > F > OH > NH_2$ (Table 3.1) which is proportional to the polarizability of the substituent and thus the γ effect is given by Equation (3.4).

The γ effect for a given substituent is reduced for methylene and methine compared to methyl protons (Table 3.2). Presumably the methylene and methine protons are increasingly shielded from external perturbation by the attached carbon atoms. The γ effect is roughly

Table 3.1 *Observed versus calculated ¹H chemical shifts (δ) of substituted methanes and ethanes*[a]

System				X				
	H	NH₂	OH	F	Cl	Br	I	SH
CH₃X								
obs.	0.22	2.46	3.38	4.26	3.05	2.68	2.16	2.08[b]
calc.	0.27	2.34	3.34	4.26	3.12	2.79	2.21	2.09
CH₃CH₂X								
obs.	0.86	1.11	1.24	1.37	1.49	1.71	1.85	1.33
calc.	0.80	1.30	1.22	1.20	1.49	1.65	1.88	1.31
CH₃CH₂X								
obs.	0.86	2.75	3.71	4.51	3.57	3.47	3.20	2.56
calc.	0.80	2.76	3.71	4.60	3.51	3.19	2.63	2.48
CH₂X₂								
obs.	—	4.90	5.45	5.33	4.94	3.90	—	
calc.	—	4.90	5.52	5.27	4.74	3.79	—	
CHX₃								
obs.	—	4.98	6.41	7.24	6.82	4.91	—	
calc.	—	4.99	6.44	7.00	6.34	5.12	—	
CH₃CHX₂								
obs.	—	5.23[c]	5.94	5.87	5.86	5.24	—	
calc.	—	5.21	5.81	5.57	5.06	5.36	—	
CH₃CHX₂								
obs.	—	1.33[c]	1.56	2.23	2.47	2.96	—	
calc.	—	1.33	1.50	2.12	2.41	2.83	—	
CH₃CX₃								
obs.	—	1.44[d]	1.87	2.75	—	—	—	
calc.	—	1.44	1.84	2.72	—	—	—	

[a] Abraham *et al.*[51].
[b] Me₂S.
[c] In D₂O.
[d] OMe.

Table 3.2 *¹H γ SCS (H.C.C.X) ppm of substituted ethanes and butanes*[a]

System			X				
	NH₂	OH	F	Cl	Br	I	SH
CH₃.CH₂X[b]	0.25	0.38	0.51	0.64	0.86	0.99	0.48
Et.CH₂.CH₂X[c]	0.16	0.30	0.42	0.49	0.59	0.55	0.34
Me₂CH.CH₂X[d]	−0.14	0.06	—	0.26	0.26	0.02	0.03

[a] Abraham *et al.*[51].
[b] From ethane (0.855 ppm).
[c] From butane (CH₂, 1.260 ppm).
[d] From isobutane (CH, 1.715 ppm).

proportional to the number of attached hydrogen atoms, thus for methylene and methine hydrogens Equation (2.10) is multiplied by 2/3 and 1/3, respectively. Finally, as in the case of beta substitution, the γ effects for CX_2 and CX_3 ($X = F, O$) are reduced by the appropriate factors. These simple amendments to the original CHARGE scheme[51] provide a comprehensive calculation of the ^1H chemical shifts of a variety of methyl and ethyl derivatives (Table 3.1) and also give calculated dipole moments in good agreement with the observed values. Also we note that the ^1H chemical shifts in these substituted methanes and ethanes now have a simple chemical explanation. The decrease in shielding on going from $CH_4 \rightarrow CH_3 \rightarrow CH_2 \rightarrow CH$ is simply due to the increased electronegativity of carbon *versus* hydrogen.

Also the effects of the electronegativity of the β substituent are clearly demonstrated in Table 3.1 for a range of substituent groups. The agreement between the observed and calculated shifts in Table 3.1 is very good with an rms error of 0.10 ppm over the shift range of 7 ppm. The major deviations occur for the very polarizable bromine and iodine substituents. In these cases further substitution directly affects the halogen atoms and a more complex iteration procedure would be necessary to accommodate their substituent effects.

3.5.3 Through Space Effects

In CHARGE the effects of more distant atoms on the ^1H chemical shifts are considered to be due to steric, anisotropic and electric field contributions.[47]

H· · ·H steric interactions were found to be shielding in alkanes and deshielding in aromatics. X· · ·H (X = C, O, N, Cl, Br, I, etc.) interactions were found to be deshielding. All these steric interactions are described by a simple r^{-6} dependence, Equation (3.11), where a_s is a constant for any given atom hybridization (e.g. a_s differs for C, C= and C≡).

$$\delta_{steric} = a_s/r^6 \tag{3.11}$$

Furthermore, any X· · ·H steric contribution on a methylene or methyl proton resulted in a push–pull effect (shielding) on the other proton(s) on the attached carbon (cf. Table 1.7).

The electric field for a univalent atom (e.g. chlorine) is calculated as due to the charge on the chlorine atom and an equal and opposite charge on the attached carbon atom (Figure 3.4).

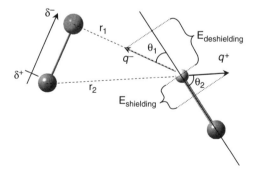

Figure 3.4 *The geometry dependence of the electric field effect on ^1H chemical shifts (see color Plate 1).*

An electric field in the direction of the C—H bond will attract the electrons in the C—H bond towards the carbon atom thereby deshielding the proton. Conversely an electric field along the H—C bond will shield the proton nucleus (see Figure 3.4). Thus the effects of the electric field from a C—X bond (X = H, F, Cl, Br, I) on ¹H chemical shifts is given by Equation (3.12) where A_Z was empirically determined to be 3.67×10^{-12} esu (63 ppm au) (see Section 6.2) and E_Z($E_{deshielding} + E_{shielding}$, in Figure 3.4) is the component of the electric field along the C—H bond according to Equation (3.13).

$$\delta_{el} = A_Z E_Z \qquad (3.12)$$

$$E_Z = \frac{q^- \cos\theta_1}{r_1^3} + \frac{q^+ \cos\theta_2}{r_2^3} \qquad (3.13)$$

The vector sum gives the total electric field at the proton concerned and the component of the electric field along the C—H bond considered is calculated from Equation (3.13).

The magnetic anisotropy of a bond with cylindrical symmetry (e.g. C≡C) is obtained using the McConnell equation[55] (Equation (3.14)), where r is the distance from the perturbing group to the nucleus of interest in Å, θ is the angle between the vector **R** and the symmetry axis and $\Delta\chi$ the anisotropic susceptibility of the C≡C bond. $\Delta\chi = (\chi_{parl} - \chi_{perp})$ where χ_{parl} and χ_{perp} are the susceptibilities parallel and perpendicular to the symmetry axis, respectively.

$$\delta_{an} = \Delta\chi_1(3\cos^2\theta - 1)/3\mathbf{R}^3 \qquad (3.14)$$

For a non-symmetrical group such as the carbonyl group, Equation (3.14) is replaced by the general McConnell equation (Equation (3.15)).[55]

$$\delta_{an} = [\Delta\chi_1(3\cos^2\theta_1 - 1) + \Delta\chi_2(3\cos^2\theta_2 - 1)]/3\mathbf{R}^3 \qquad (3.15)$$

In Equation (3.15), $\Delta\chi_1 = \chi_y - \chi_z$, $\Delta\chi_2 = \chi_x - \chi_z$ where χ_x, χ_y and χ_z are the magnetic susceptibilities along the x-, y- and z-axes (Figure 3.5). θ_1 and θ_2 are the angles between the vector **R** (distance vector to the proton being affected) and the y- and x-axes respectively as shown in Figure 3.5.

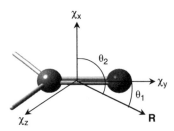

Figure 3.5 *Coordinates and magnetic susceptibilities for the carbonyl group: $\Delta\chi_1$ and $\Delta\chi_2$ are usually termed the parallel and perpendicular anisotropies, respectively.*

3.5.4   Hydrogen Bonding Shifts

The hydrogen bond arises from the overlap of the lone pairs of an electronegative atom with the s-orbital of a hydrogen atom attached to another electronegative atom. This effect is commonly encountered in organic chemistry when the two electronegative atoms are oxygen atoms. In a recent study[56] it was noted that when a $C=O\cdots H-O$ hydrogen bond is formed there is a large change in the OH chemical shift which is not reproduced by the above through space effects. Using *ab initio* calculations it was found that there is a large distance dependence on the chemical shift in such cases and that also the geometrical orientation of the atoms can be important, as described below.

The model system used was that of a phenol OH hydrogen bonding to acetone. The nuclear shielding (σ) of the H-bonded proton was calculated as a function of the distance between the OH and O=atoms along the hydrogen bond axis as shown in Figure 3.6.

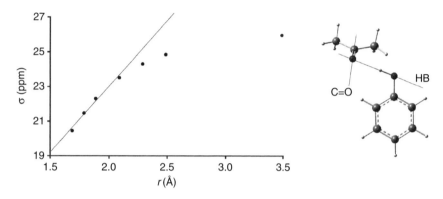

Figure 3.6   *The distance (OH$\cdots$O) dependence of the nuclear shielding of the OH proton of phenol hydrogen bonded to acetone. The best fit to a linear function for distances below 2.1 Å is also plotted. In addition, the minimum energy orientation is shown in the figure; the OH$\cdots$O distance is 1.7 Å (from Abraham and Mobli[56]).*

The figure shows a dramatic increase in nuclear shielding as the OH$\cdots$O distance increases. Beyond 2–2.5 Å the increase in nuclear shielding is comparably moderate. For distances less than 2.1 Å the plot is almost linear. The correlation coefficient (R^2) for acetone using a linear function up to 2.1 Å is 0.98. This demonstrates that the nuclear shielding due to the hydrogen bond for distances <2 Å can be accurately reproduced by a linear equation. It is indeed in this range that ^{1}H chemical shifts of H-bonded protons are poorly predicted using classical models of anisotropy, electric field and steric effects alone (e.g. in the CHARGE program).

The only other geometrical parameter which was found to contribute significantly to the chemical shift was the Me$-C=O\cdots H$ dihedral angle (Figure 3.7). A large change in the nuclear shielding is noted as the dihedral angle is changed. This relationship is true for an sp^2 hybridized oxygen, such as the $C=O$ group and correlates with the position of the lone pairs of the oxygen. In a separate study of the hydrogen bond between phenol and DMSO the same dihedral angle effect occurred. However, the appearance of the plot was consistent with an sp^3 hybridized oxygen atom, as expected for the hybridization of the oxygen atom in DMSO.[57]

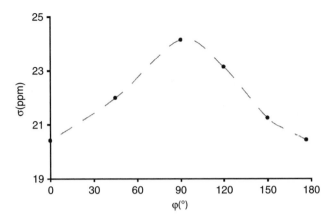

Figure 3.7 *Me—C=O· · ·H dihedral angle (φ) dependence of the nuclear shielding of the OH hydrogen of phenol (from Abraham and Mobli[56]).*

The above findings were formulated into Equation (3.16) reflecting the change with both distance and dihedral angle. Note that this equation only comes into effect for distances < 2.05 Å.

$$\delta_{HB} = [A + B(r - r_0)]\cos^2\phi \qquad (3.16)$$

The constant B reflects the slope of the plot in Figure 3.6 and is given by the *ab initio* calculations as -7.6 ppm Å^{-1}; this result was further confirmed using experimental data. The curve in Figure 3.7 can be described by a $\cos^2\varphi$ function and thus the hydrogen bonding term is maximum when the H· · ·O=C—R entity is planar and $\varphi = 0$ or 180^0 and decreases to zero as the dihedral angle approaches 90^0. The values of A and r_0 were found to be zero and 2.05 Å: the data were confirmed experimentally for strong intramolecular hydrogen bonds involving the phenol OH as the hydrogen bond donor and the nitro, aldehyde, ketone and ester groups (at the *ortho* position of the phenol) as hydrogen bond acceptors. In the CHARGE program this equation is applied to any functional group capable of forming such hydrogen bonds. Note that the study clearly showed that the H-bond energy is not directly related to the nuclear shielding of the H-bonded hydrogen. Although the two parameters are often used synonymously this is largely due to the H· · ·O=C distance and should be used with caution.

3.5.5   Aromatic Compounds

Ring currents. For aromatic compounds it is necessary to include the shifts due to the aromatic ring current and the π-electron densities in the aromatic ring.[58] The equivalent dipole approximation (Equation (3.17)) was used to calculate the ring current shifts. In Equation (3.17), R is the distance of the proton

$$\delta_{rc} = fc\mu(3\cos^2\theta - 1)/R^3 \qquad (3.17)$$

from the benzene ring centre, θ the angle of the R vector with the ring symmetry axis, μ the equivalent dipole of the aromatic ring and fc the π-electron current density for the ring, being 1.0 for benzene.

The π-electron densities are calculated from Hückel Molecular Orbital (HMO) theory.[59] The standard coulomb and resonance integrals for the Huckel routine are given by Equation (3.7), where α_0 and β_0 are the coulomb and resonance integrals for a carbon $2p_z$ atomic orbital and h_r and k_{rs} the factors modifying these integrals for orbitals other than sp^2 carbon. For substituted aromatics the values of the coefficients h_r and k_{rs} in Equation (3.7) for orbitals involving heteroatoms have to be found. These were originally obtained by fitting the molecular dipole moments. In later investigations the factors were obtained so that the π densities calculated from the Huckel routine reproduce the π densities from *ab initio* calculations. It was found that, for non-polar molecules the basis set giving best agreement with the observed dipole moments was the 6-31G* level. For example, propene, observed 0.35D, calculated 0.36D (Chapter 4, Section 4.3.2). However for polar molecules this basis set gave too high values of the dipole moments and the best agreement was with the 3-21G basis set. For example, benzonitrile, observed 4.14D, calculated 4.55D (3-21G), 4.82D (6-31G) (Chapter 6, Section 6.6).

π-Electron density. The effect of the excess π-electron density at a given carbon atom on the ^1H chemical shifts of the neighbouring protons is given by Equation (3.18) where Δq_α and Δq_β are the excess π-electron densities at the carbon atoms α and β to the proton. An experimental determination of the coefficient a_1 is due to Gunther *et al.*[52] The ^1H chemical shifts of some cyclic charged molecules (tropylium cation, cyclopentadienyl anion, etc.) were obtained and compared with benzene. From these data a value of a_1 of 10.0 ppm/el between the ^1H shift $\Delta\delta$ and the excess π charge Δq_α was obtained.

$$\delta_\pi = a_1 \Delta q_\alpha + a_2 \Delta q_\beta \qquad (3.18)$$

It has also been recognized that there is an influence of the excess π charge on the carbon atom β to the proton considered and a related effect gives rise to the phenomenon of negative spin density in ESR spectroscopy.[60] The hyperfine couplings to the α and β protons in alkyl radicals, in which the radical carbon atom is planar and sp^2 hybridized, are quoted as $a^H_\alpha = -22G$ and $a^H_\beta = 4 + 50\cos^2\theta$ where θ is the dihedral angle between the free radical 2p orbital and the C—H bond considered.[60] These considerations suggest that in aromatic compounds in which the C—H bond is orthogonal to the π orbital and θ is 90^0, the value of a_2 in Equation (3.18) is negative and ca. 1/5th of a_1, i.e. -2.0.[58] Thus the values of the coefficients a_1 and a_2 used were 10.0 and -2.0 ppm electron^{-1}.

The above contributions are added to Equation (3.10) to give the calculated shift of Equation (3.19).

$$\delta_{total} = \delta_{charge} + \delta_{steric} + \delta_{anisotropy} + \delta_{el} + \delta_\pi + \delta_{rc} + \delta_{HB} \qquad (3.19)$$

This is the fundamental equation for the CHARGE scheme. The applications of this equation to a wide variety of organic compounds will be considered in the following chapters.

References

1. Frisch, M. J.; Trucks, G. W.; Schlegel, H. B.; Scuseria, G. E.; Robb, M. A.; Cheeseman, J. R.; Montgomery, J. A.; Jr., T. V.; Kudin, K. N.; Burant, J. C.; Millam, J. M.; Iyengar, S. S.; Tomasi, J.; Barone, V.; Mennucci, B.; Cossi, M.; Scalmani, G.; Rega, N.; Petersson, G. A.; Nakatsuji, H.; Hada, M.; Ehara, M.; Toyota, K.; Fukuda, R.; Hasegawa, J.; Ishida, M.; Nakajima, T.; Honda, Y.; Kitao, O.; Nakai, H.; Klene, M.; Li, X.; Knox, J. E.; Hratchian, H. P.; Cross, J. B.; Bakken, V.; Adamo, C.; Jaramillo, J.; Gomperts, R.; Stratmann, R. E.; Yazyev, O.; Austin, A. J.; Cammi, R.; Pomelli, C.; Ochtersk, J. W.; Ayala, P. Y.; Morokuma, K.; Voth, G. A.; Salvador, P.; Dannenberg, J. J.; Zakrzewski, V. G.; Dapprich, S.; Daniels, A. D.; Strain, M. C.; Farkas, O.; Malick, D. K.; Rabuck, A. D.; Raghavachari, K.; Foresman, J. B.; Ortiz, J. V.; Cui, Q.; Baboul, A. G.; Clifford, S.; Cioslowski, J.; Stefanov, B. B.; Liu, G.; Liashenko, A.; Piskorz, P.; Komaromi, I.; Martin, R. L.; Fox, D. J.; Keith, T.; Al-Laham, M. A.; Peng, C. Y.; Nanayakkara, A.; Challacombe, M.; Gill, P. M. W.; Johnson, B.; Chen, W.; Wong, M. W.; Gonzalez, C.; Pople, J. A., *Gaussian 03*. D; Gaussian Inc.: Wallingford, CT, 2006.
2. Gilbert, K., 9.0 ed.; Serena Software: Bloomington, USA, 2005.
3. Foresman, J. B.; Frisch, Æ., *Exploring Chemistry with Electronic Structure Methods*. 2nd ed.; Gaussian, Inc.: Pittsburgh, PA, 1996.
4. Pulay, P.; Hinton, J. F.; Grant, D. M., Harris, R. K. (Eds). *Encyclopedia of NMR*. John Wiley & Sons, Inc.: New York, NY,1995.
5. Bryce, D.; Wasylishen, R. E., *J. Chem. Edu.* 2001, **78**, 124.
6. London, F. J., *J. Phys. Radium* 1937, **8**, 397.
7. Hameka, H. F., *Mol. Phys.* 1958, **1**, 203.
8. Pople, J. A., *Mol. Phys.* 1958, **1**, 175.
9. Pople, J. A., *Disc. Faraday. Soc.* 1962, **34**, 7.
10. Gauss, J., *Chem. Phys.* 1995, **99**, 1001.
11. Cramer, C. J., *Essentials of Computational Chemistry: Theories and Models*. John Wiley & Sons, New York, 2002.
12. Helgaker, T.; Jaszunski, M.; Ruud, K.,*Chem. Rev.* 1999, **99,** 293.
13. Fukui, H., *Prog. NMR Spectrosc.* 1999, **35**, 267.
14. Fukui, H., *Prog. Nucl. Magn. Reson. Spectrosc.* 1997, **31**, 317.
15. Jameson, C. J., *Annu. Rev. Phys. Chem.* 1996, **47**, 135.
16. Dios, A. C. d., *Prog. Nucl. Magn. Reson. Spectrosc.* 1996, **29**, 229.
17. Lampert, H.; Mikenda, W.; Karpfen, A.; Kählig, H., *J. Phys. Chem. A* 1997, **101**, 9610.
18. Bremser, W., *Anal. Chim. Acta* 1978, **103**, 355.
19. Advanced Chemistry Development, Inc.: Toronto ON, Canada, 2003.
20. Modgraph Consultants Ltd: Oaklands, Welwyn, Herts, 2007.
21. Meiler, J.; Maier, W.; Meusinger, R., *J. Magn. Reson.* 2002, **157**, 242.
22. Aires-de-Sousa, J.; Hemmer, M. C.; Gasteiger, J., *Anal. Chem.* 2002, **74**, 80.
23. Zürcher, R. F., *Progr. Nucl. Magn. Reson. Spectrosc.* 1967, *2*, 205.
24. ApSimon, J. W.; Beierbeck, H., *Can. J. Chem.* 1971, **49**, 1328.
25. ApSimon, J. W.; Craig, W. G.; Demarco, P. V.; Mathieson, D. W.; L.Saunders; Whalley, W. B., *Tetrahedron* 1967, **23**, 2399.
26. ApSimon, J. W.; Demarco, P. V.; Mathieson, D. W., *Tetrahedron* 1970, **26**, 119.
27. Bothner-By, A. A.; Naar-Colin, C., *Ann. NY Acad. Sci.* 1958, **70**, 833.
28. Bothner-By, A. A.; Naar-Colin, C., *J. Am. Chem. Soc.* 1961, **83**, 231.
29. Moritz, A. G.; Sheppard, N., *Mol. Phys.* 1962, *5*, 361.
30. Musher, J. I., *J. Chem. Phys.* 1961, **35**, 4.
31. Hall, L. D., *Tetrahedron. Lett.* 1964, **23**, 1457.
32. Musher, J. I., *Mol. Phys.* 1963, **6**, 93.

33. Cavanaugh, J. R.; Dailey, B. P., *J. Chem. Phys.* 1961, **34**, 4.
34. Yamaguchi, I.; Brownstein, S., *J. Chem. Phys.* 1964, **68**, 1572.
35. Bothner-By, A. A., *Disc. Faraday Soc.* 1962, **34**, 66.
36. Boaz, H., *Tetrahedron Lett.* 1973, **14**, 55.
37. Curtis, J.; Dalling, D. K.; Grant, D. M., *J. Org. Chem.* 1986, **51**, 136.
38. Curtis, J.; Grant, D. M.; Pugmire, R. J., *J. Am. Chem. Soc.* 1989, **111**, 7711.
39. Daneels, D.; Anteunis, M., *Org. Magn. Reson.* 1974, **6**, 617.
40. Fisher, J.; Gradwell, M. J., *Magn. Reson. Chem.* 1992, **30**, 338.
41. Li, S.; Allinger, N. L., *Tetrahedron* 1988, **44**, 1339.
42. Pretsch, E.; Simon, W., *Helv. Chim. Acta* 1969, **52**, 2133.
43. Abraham, R. J.; Grant, G. H., *J. Comput. Aid. Mol. Design* 1992, **6**, 273.
44. Abraham, R. J.; Hudson, B. *J. Comput. Chem.* 1985, **3**, 173.
45. Gasteiger, J.; Marsili, M., *Org. Magn. Reson.* 1981, **15**, 353.
46. Abraham, R. J.; Grant, G. H.; Howorth, I. S.; Smith, P. E., *J. Comput. Aid. Mol. Design* 1991, **5**, 21.
47. Abraham, R. J., *Prog. Nucl. Magn. Reson. Spectrosc.* 1999, **35**, 85.
48. Hinze, J.; Jaffee, H. H., *J. Phys. Chem.* 1963, **67**, 1501.
49. Mulliken, R. S., *J. Chem. Phys.* 1934, **2**, 782.
50. Huggins, M. L., *J. Am. Chem. Soc.* 1953, **75**, 4123.
51. Abraham, R. J.; Edgar, M.; Glover, R. P.; Warne, M. A.; Griffiths, L., *J. Chem. Soc. Perkin Trans. 2* 1996, 333.
52. Günther, H., *NMR Spectroscopy.* 2nd ed.; John Wiley & Sons, Ltd: Chichester, 1995.
53. Abraham, R. J.; Chambers, E. J.; Thomas, W. A., *J. Chem. Soc. Perkin Trans. 2* 1994, 949.
54. Mijlhoff, F. C.; Geise, H. J.; Schaick, E. J. M. v., *J. Mol. Struct.* 1974, **20**, 393.
55. McConnell, H. M., *J. Chem. Phys.* 1957, **27**, 226.
56. Abraham, R.J., Mobli, M., *Magn. Reson. Chem.* 2007, **45**, 865.
57. Mobli, M., *Ph.D. Thesis*, University of Liverpool, 2004.
58. Abraham, R. J.; Canton, M.; Reid, M.; Griffiths, L., *J. Chem. Soc. Perkin Trans. 2* 2000, 803.
59. Abraham, R. J.; Smith, P. E., *J. Comput. Chem.* 1987, **9**, 288.
60. Ayscough, P. B., *Electron Spin Resonance in Chemistry.* Methuen: London, 1967.

4

Modelling ^{1}H Chemical Shifts, Hydrocarbons

4.1   Introduction

The theory given in Chapter 3 allows the prediction of the ^{1}H chemical shifts of a variety of organic compounds. Furthermore, Equation (3.19) can be used to provide an insight into the interactions influencing the ^{1}H chemical shifts of any molecule. For example, we can attempt to answer questions such as 'is the effect of a chlorine substituent on the ^{1}H chemical shifts of a given molecule due to the electric field or anisotropy of the C—Cl bond or to the steric effect of the chlorine atom?'. Such questions are difficult to answer by quantum mechanics as these calculations give the components of the total magnetic shielding tensor at any atom and it is not a simple matter to break these numbers down in any rigorous manner.

We commence by considering hydrocarbons in this chapter. Alkanes provide the basic framework on which the various functional groups may be added. Thus we begin this chapter by considering these molecules. Subsequently, alkenes and alkynes are considered. Aromatic hydrocarbons are more conveniently considered together with heteroaromatics in Chapter 5. The effect of various functional groups on ^{1}H chemical shifts can then be analysed. In order to determine the various interactions influencing the ^{1}H chemical shifts it is most convenient to obtain the ^{1}H chemical shifts of molecules in a fixed conformation. The chemical shifts of conformationally mobile molecules can be predicted *provided* the conformational profile of the molecule in solution is known. For many multi-functional molecules of pharmacological interest the conformational profile of the molecule in solution is not known accurately and the ^{1}H chemical shift prediction will be less certain in this case. Thus, most of the molecules considered here exist predominately in one conformation. Note that following the discussions of the geometry optimizations in Chapter 3 unless otherwise

Modelling ^{1}H NMR Spectra of Organic Compounds: Theory, Applications and NMR Prediction Software
Raymond Abraham and Mehdi Mobli   © 2008 John Wiley & Sons, Ltd

stated all *ab initio* minimizations will use the B3LYP/6-31G** theory and all the molecular mechanics minimizations the MMFF94 force field in PCMODEL.

4.2   Alkane Chemical Shifts

As mentioned earlier (Chapter 3, Section 3.4) the ¹H chemical shift range of alkanes extends from 0.3 to 2.2 ppm (Tables 4.1 and 4.2), which is ca. 20 % of the usual range of ¹H shifts in organic compounds (0–10 ppm) and yet these molecules possess neither anisotropic nor polar groups. Thus the accurate prediction of even hydrocarbon chemical shifts is not a trivial task.

4.2.1   H..H and C..H Steric Interactions

In early studies on hydrocarbons,[1] one striking effect observed was the hydrogen *shielding* due to close H...H interactions. Figure 4.1 shows the shielding due to close H..H interactions of the axial H in some condensed hydrocarbons with the unique H atom of perhydrophenalene at 0.32δ. Some of this shielding may be due to the γ *gauche* effect reflecting the number of *gauche* C.C.C.H interactions on the proton considered (see Figure 4.8). In order to further identify the H..H steric contribution the ¹H chemical shifts of the methylene protons of *trans*-butane and the methyl protons in propane were calculated as a function of the rotation of the distant methyl group[1] with the HyperNMR programme.[2–4] In these calculations the only nuclei altering their positions are the methyl protons, thus any γC.C.C.H effect is constant. These calculations showed that the ¹H nuclei on propane and *trans*-butane are *shielded* as the H..H distance decreases (Figure 4.2) and the calculated curves are well reproduced by an r^{-6} function (the curves in Figure 4.2). Also the *trans*

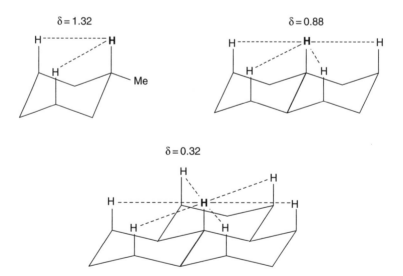

Figure 4.1   *The shielding effect of close H..H interactions.*

(anti) hydrogen atom in propane is *not* affected by the change in the H..H distance, i.e. there is no push–pull effect for H..H interactions (see later).

An alternative explanation of these results is C—H bond anisotropy but trial calculations gave negligible shifts compared to those of Figure 4.2. Thus this interpretation was rejected. Trial calculations for the H..H steric interaction using a non-bonded steric potential[5] and an r^{-6} function with a cut-off at the van der Waals minimum[6] were subsequently replaced with the simple Equation (3.11) with a global cut-off at 6 Å. In Equation (3.11) the coefficient a_s is obtained from the observed data. In contrast to the H..H steric interaction, the steric effects of other substituents on ^{1}H chemical shifts can be observed experimentally (cf. Chapter 1, Table 1.7). The C..H interactions are deshielding as expected and were also reproduced from Equation (3.11) with the value of the coefficient (a_s) for C..H interactions again fitted to the observed data. Also the ratio of the shift of the distant hydrogen in the methylene group to that of the hydrogen experiencing the steric effect (the push–pull coefficient) has to be found. The push–pull effect was identified originally in a very sterically hindered alcohol[7] but was not considered as a general effect. From the data in Table 1.7 it can be seen that the ratio of the two effects on the hydrogens of any given methylene group for the methyl interaction is ca. 0.5–1.0 and this ratio was subsequently refined using the whole data set.

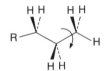

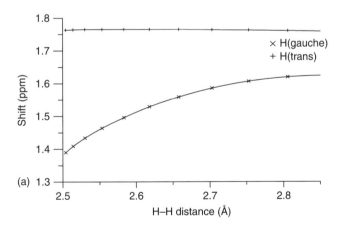

Figure 4.2 *The effect of rotating a methyl group on (a) the ^{1}H chemical shifts of the other methyl hydrogens in propane and (b) the delta CH$_2$ hydrogens of trans-butane: (a) x, H(gauche); +, H(trans): (b) ▲, CH$_2$ (reproduced from Abraham et al.,[1] with permission of the Royal Society of Chemistry).*

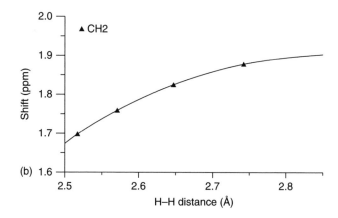

Figure 4.2 (*Continued*).

The ^{1}H chemical shifts of a variety of alkanes are given in Tables 4.1–4.5 and a number of attempts to calculate these shifts were made using various CHARGE models. In CHARGE3[1] the orientation dependence of the methyl γ SCS was addressed using two possible approaches (A and B) for this effect to give the following interactions.

(1) A shielding H..H steric interaction (Equation (3.11)) with *different* coefficients depending on the types of the two protons involved: CH, CH$_2$ and CH$_3$.
(2) A deshielding C..H steric interaction again using Equation (3.11).
(3) A push–pull routine for the proton of a methylene or methyl group other than the proton which is experiencing the C..H steric shift.
(4) An explicit carbon gamma effect (H.C.C.C) given by (A) a simple through bond shift and (B) an orientation effect proportional to $\cos\theta \times |\cos\theta|$ where θ is the C.C.C.H dihedral angle.

This scheme did not give any steric shifts for methyl hydrogens, as the push–pull effect on the methyl protons combined with the averaging of their shifts due to rapid rotation of the methyl group meant that all steric effects average to zero. It is possible that this is the reason for the remarkable consistency of the methyl group chemical shift. In all the hydrocarbons examined except methane and t-butyl compounds, the methyl chemical shift was 0.85–0.95δ. This explanation is strongly supported by the low temperature measurements on tri t-butylmethane in which one methyl group was observed in the low exchange limit.[8] In this case the observed shifts were 0.63, 1.25 and 1.67δ, a spread of more than 1 ppm over one methyl group.

The CHARGE3 scheme gave calculated ^{1}H chemical shifts for a wide range of molecules in reasonable agreement with the observed shifts. Using the available hydrocarbon shifts (Tables 4.1–4.5) both options A and B gave identical rms errors (obs. vs. calc. shifts) of 0.17 ppm. Thus it was not possible to distinguish between the two options A and B, i.e. whether carbon has an intrinsic orientation dependent γ effect or not. Even though the two schemes give very different results in certain cases and the calculated shifts for some protons were in error by ca. 0.5 ppm.

These calculations did not include any explicit magnetic anisotropy or electric field effects. It is possible that option B is implicitly including in the dihedral angle term the effect of C—C bond anisotropy and this possibility was examined subsequently by including the C—C anisotropy in the charge calculations (see below).

Also of note are the different interpretations of the cyclohexane chemical shifts in the two schemes. In option B the orientation dependence of the carbon γ effect produces a deshielding of the equatorial protons relative to the axial protons due to the H.C.C.C dihedral in a *trans* (anti) vs. *gauche* orientation. In option A the difference between the axial and equatorial proton chemical shifts is solely due to the H..H steric shifts (cf. Figure 4.8).

4.2.2   The Methyl Effect

A further advance in this work was to include a specific methyl effect.[9] The SCS of the methyl group in cyclohexanes[10,11] is shown in Figure 4.3. The SCS of an equatorial methyl on H_{2e} is −0.03 ppm but on H_{2a} is −0.31 ppm yet the orientation of the methyl group is symmetrical with respect to both protons and the H...H distances virtually identical. These SCSs are well documented, reproducible and additive.[10] This was not explained by any of the above mechanisms nor by C—C anisotropy (see later).[10,11] Note that the methyl SCSs in $CH_3.CH.CH$ and $CH_3.CH.CH_2$ fragments are very similar. For example, the SCS for the CH proton in *trans* 1,2-dimethylcyclohexane vs. methylcyclohexane is −0.38 ppm (Table 4.3), compared to the 2a proton in methylcyclohexane of −0.31 ppm (Figure 4.3).

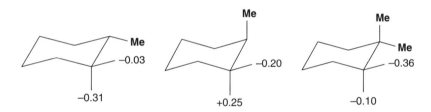

Figure 4.3   *Experimental SCSs of the methyl group on the gamma protons in methylcyclo-hexanes.*

The methyl effect can be visualized somewhat differently as follows. In Figure 4.4 the methyl SCSs on the γ protons in methylcyclohexanes and norbornanes are plotted against the Me—C—C—H dihedral angle. The fragments considered are $CH_3.CH(C).CH(C)$ and $CH_3.CH(C).CH_2(C)$.

It can be seen that for a dihedral angle of ca. 60° there is the anomaly noted above and thus the data cannot be fitted by any curve which is simply a function of the Me—C—C—H dihedral angle. The data in Figure 4.4 was fitted with a carbon gamma effect for the $CH_3.CH.CH$ and $CH_3.CH.CH_2$ fragments which is a function of the two dihedral angles (θ and ϕ Figure 4.5). The function chosen was a simple $\cos\theta \sin\phi$ function (Equation (4.1)).

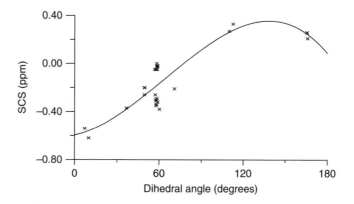

Figure 4.4   *Methyl SCSs in cyclohexanes and bicyclo[2.2.1] heptanes vs. the Me—C—C—H angle. The solid curve is a computer generated best fit curve, a polynomial function of order 3 (reproduced from Abraham et al.,^{9} with permission of the Royal Society of Chemistry).*

The 2a and 2e protons are both *gauche* to the methyl carbon (Figure 4.5) but the 2a proton is also *gauche* to the ring carbon attached to the beta carbon (see (a)), but the 2e proton is *trans* (see (b)). With this distinction noted, all the data in Figure 4.4 can be fitted.

$$q_H = A_1 \cos \theta \sin \phi + k \qquad 0 < \theta < 90° \tag{4.1}$$

$$q_H = A_2 \cos \theta \sin \phi + k \qquad 90 < \theta < 180°$$

This function was not applied to the $CH_3.C_q.CH$ or $CH_3.C_q.CH_2$ fragments where C_q is a quaternary carbon as the β carbon no longer possesses two different substituent atoms, hence the standard function of θ (Equations (3.5) and (3.6)) was used.

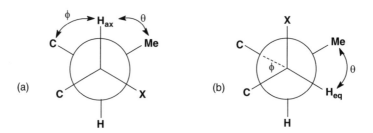

Figure 4.5   *Definition of the dihedral angles chosen to distinguish equatorial and axial gamma protons relative to an equatorial methyl substituent (reproduced from Abraham et al.,^{9} with permission of the Royal Society of Chemistry).*

4.2.3   C—C Bond Anisotropy

Another mechanism suggested for many years to explain the ^{1}H shifts of hydrocarbons is the C—C bond anisotropy. In hydrocarbons the magnetic anisotropy effects were usually ascribed to the C—C bonds and the simple McConnell equation (Equation (3.14)) may be used where $\Delta\chi$ is $\Delta\chi^{C-C}$, the molar anisotropy of the C—C bond. A value of the C—C bond anisotropy of about 3.3×10^{-6} cm^3 mol^{-1} would explain the observed chemical shift difference between cyclopentane and cyclohexane and between the axial and equatorial protons in cyclohexane.[12–14] However extending this approach[12,15–17] to larger molecules gave much larger values of $\Delta\chi^{C-C}$[18] and attempts to explain the chemical shifts in alkyl derivatives,[19] effects on methyl groups[20,21] and effects from the methyl group in methylcyclohexanes[22] clearly demonstrated that other factors were important.

In a seminal paper, Bothner-By and Pople[23] reviewed this early work and also obtained a limiting value of the C—C anisotropy since:

$$\Delta\chi^{C-C} = \chi^{C-C}_{\parallel} - \chi^{C-C}_{\perp} \tag{4.2}$$

and

$$\chi^{C-C} = (\chi^{C-C}_{\parallel} + 2\chi^{C-C}_{\perp})/3 \tag{4.3}$$

where χ^{C-C} is the mean molar susceptibility and $\chi^{C-C}_{\parallel}$ and $\chi^{C-C}_{\perp}$ are the susceptibilities parallel and perpendicular to the C—C bond. To avoid the bond being paramagnetic in the longitudinal direction, the C—C anisotropy must be less than one and half times the mean susceptibility. Using a value of 3.0×10^{-6} cm^3 mol^{-1} for the mean susceptibility from crystal data a limiting value of 4.5×10^{-6} cm^3 mol^{-1} for $\Delta\chi^{C-C}$ was obtained.

The C—C anisotropy contribution was calculated from Equation (3.14) and included with the above refined treatment of the methyl effect in the CHARGE3 routine which was then parameterized and tested on the available hydrocarbon data.[9] Equation (4.1) gave good results for the CH$_3$.CH(C).CH/CH$_2$ fragments in cyclic molecules but for isopropyl groups the carbon γ effect was left unchanged.

Including the anisotropy within the CHARGE3A scheme gave a ca. 15 % improvement in the overall fit with a value of $\Delta\chi^{C-C}$ of 4.98 ppm Å^3 molecule^{-1} (3.0×10^{-6} cm^3 mol^{-1}) which is within the limit specified above. In contrast the CHARGE3B scheme showed no improvement with any value of $\Delta\chi^{C-C}$. On closer examination the improvement with CHARGE3A was found to be due mainly to the C$_\beta$—C$_\gamma$ bond contributions, with no change in the fit if the more distant C—C bonds were excluded. This explains why the CHARGE3B scheme shows no improvement since a carbon cos θ γ effect is already included.

In order to further examine this result and provide a direct contrast with the refined CHARGE3A routine (henceforth CHARGE4) the CHARGE3B routine was further refined. It was found that the simple cos θ function (Equation (3.5)) gave values for the carbon γ effect identical to the cos θ x| cos θ| function used previously and this was incorporated into CHARGE3B with the above refinements but without any C—C anisotropy contribution. This model (CHARGE5)[24] was further improved by extending the methyl γ effect to all CH$_2$ groups to give CHARGE6, i.e. the same formula was used to represent the influence of a methylene group adjacent to a methine carbon. This treatment may be considered as reflecting the intrinsic chirality of the central methine carbon atom.

The CHARGE program also now included the electric field calculation (Equation (3.12)). This is a small term for hydrocarbons and does not affect the fit of the data, but reduces the

H...H steric contribution. The calculated shifts from this model (CHARGE6) were compared with the observed shifts for all the hydrocarbon data (Tables 4.1–4.5). The rms error was 0.09 ppm, slightly better than the previous scheme. This was achieved also with a reduction in the number of variable parameters from thirteen to ten as it was found that in CHARGE6 only one carbon steric coefficient was required instead of the four used in CHARGE4 for the different types of carbon atom (quaternary, CH, CH_2, CH_3). In general both schemes gave good agreement with the observed shifts thus providing the first definitive quantitative explanation of the ¹H chemical shifts in these molecules. A full discussion of these results is given elsewhere.[9,25]

The comparison of CHARGE4 and CHARGE6 showed that the introduction of an orientation dependant carbon γ effect produces as good, if not better agreement with the observed shifts of hydrocarbons than the scheme including the C—C anisotropy term. It was noted previously that the anisotropy term was mainly due to the contributions from the C_β—C_γ bond with only minor contributions from the more distant C—C bonds and this result confirms and extends this conclusion. Indeed, iteration of the parameterized scheme including only the more distant C—C bond anisotropic term gave an almost zero value for the anisotropic coefficient. Thus there is no evidence for a general long range (further than C_β—C_γ) C—C anisotropic contribution to ¹H chemical shifts in hydrocarbons.

This is further supported by comparison of the anisotropic contribution of the C_β—C_γ bonds with the simple cos θ dependence (Equation (3.5)) used here. The functional dependence of the chemical shift in a H.C.C.C fragment is almost identical to the anisotropic contribution, which varied from -0.12 ppm at 0^0 to $+0.09$ ppm at $180°$. The anisotropic contribution depends also on the bond lengths and angles in the H.C.C.C fragment, which is not the case for Equation (3.5). But the differences were very small even for the variety of compounds considered. This implies that the C—C anisotropy contribution is essentially a C_β—C_γ contribution to the chemical shifts. However this is theoretically untenable as an anisotropy contribution is by definition a long-range through space effect.

There are however certain situations in which the more distant anisotropic contribution to the ¹H chemical shifts may be significant. One of these is the bridging protons in norbornane and it is possible that there may be a shielding or anisotropic contribution from the eclipsed CH_2—CH_2 bonds in the norbornane molecule. The planar C.CH_2—CH_2.C fragment could have a larger directional anisotropy than an equivalent staggered fragment. Thus an extra shielding term proportional to r^{-6} was introduced for this fragment.

4.2.4   Observed versus Calculated Shifts

All the hydrocarbon shifts examined are given in Tables 4.1–4.5 with the calculated shifts from the latest version of CHARGE, CHARGE8d. This is based on CHARGE6 with some additional routines. Cyclopropane and cyclobutane are special cases and their shifts were modelled separately. In particular an additional steric interaction across the four-membered ring was included. Also it was found that better agreement was obtained if the HH steric coefficient for acyclic hydrocarbons differed from that for the cyclic compounds.

Acyclic alkanes. The results for the acyclic alkanes (Table 4.1) are very good with the largest error 0.22 ppm. In most cases the calculated shifts are not very dependent on the

Table 4.1   *Observed versus calculated ¹H chemical shifts (δ) of acyclic alkanes*

Molecule	Proton	Observed[a]	Calculated
Methane	CH_4	0.22	0.27
Ethane	Me	0.86	0.80
Propane	CH_2	1.30	1.30
	Me	0.90	0.87
n-Butane	CH_2	1.29	1.25(t), 1.38(g)
	Me	0.89	0.84(t), 0.87(g)
n-Pentane	2,4 CH_2	1.27	1.27
	3,5 CH_2	1.31	1.19
	Me	0.88	0.84
n-Hexane	2,5 CH_2	1.27	1.28
	3,4 CH_2	1.29	1.21
	Me	0.88	0.84
n-Heptane	2,6 CH_2	1.27	1.28
	3,5 CH_2	1.29	1.22
	4 CH_2	1.27	1.24
	Me	0.88	0.84
n-Octane	2,7 CH_2	1.27	1.27
	3,6 CH_2	1.27	1.22
	4,5 CH_2	1.27	1.24
	Me	0.88	0.84
Iso-butane	CH	1.74	1.77
	CH_3	0.89	0.93
2-Methylbutane	CH	1.45	1.82(t), 1.58(g)
	CH_2	1.20	1.45(t), 1.32(g)
	CH_3 (Et)	0.86	0.88(t), 0.89(g)
	CH_3 (iPr)	0.87	0.92(t), 0.92(g)
2,2-Dimethylbutane	CH_2	1.21	1.27
	CH_3	0.82	0.79
	ᵗBu	0.86	0.88
2,3-Dimethylbutane	CH	1.41	1.42(t), 1.60(g)
	CH_3	0.83	0.94(t), 0.93(g)
2,2,3-Trimethylbutane	CH	1.38	1.60
	CH_3	0.83	0.87
	ᵗBu	0.83	0.92
2,2,3,3-Tetramethylbutane	ᵗBu	0.87	0.92
Neo-pentane	CH_3	0.93	0.89
2,2-Dimethylpentane	Me	0.88	0.85
	ᵗBu	0.87	0.86
	3 CH_2	1.15	1.20
	4 CH_2	1.25	1.42
Di-t-butyl-methane	CH_2	1.23	1.18
	ᵗBu	0.97	0.91
1,1-Di-t-butyl-ethane	CH	1.18	1.40
	CH_3	0.86	0.84
	ᵗBu	0.97	0.98
2,2-Di-t-butyl-propane	CH_3	0.83	0.99
	ᵗBu	0.99	0.96
Tri-t-butyl-methane	CH	1.38	1.23
	ᵗBu	1.22	1.06

[a] Data from Abraham *et al.*[24].

molecular geometry used, but in molecules where there is severe steric strain the calculated shifts are sensitive to the molecular geometry used. For example in tri-tert butylmethane the calculated shifts in Table 4.1 using *ab initio* geometry (CH 1.23, t-Bu 1.00) may be compared with those using a molecular mechanics geometry (PCMODEL with the MMFF94 force field)[26] of 1.13 (CH) and 0.90 (Me).

Cyclic alkanes. The extensive data set for the cyclic compounds (Tables 4.2–4.5) also includes some very strained and sterically hindered molecules and with the ^{1}H chemical shifts ranging from 0.3 to 2.2 ppm this provides a stringent test of any theoretical calculation. The chemical shifts in the highly strained rings of cyclopropane, cyclobutane and bicycle-1,1,2-hexane are well reproduced. Also the ^{1}H shifts in *cis*-decalin are in good agreement with the observed shifts. The numbering is given in Figure 4.6 and the corresponding assignment in Table 4.2.[27] There is still some ambiguity concerning the assignment of this spectrum. The ^{1}H shifts can only be resolved at temperatures of $-70\,^\circ$C where the molecule is not interconverting between the two enantiomeric chair forms.

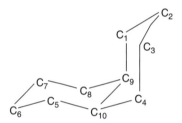

Figure 4.6   *Nomenclature used for cis-decalin.*

This experiment was performed and the assignment in Table 4.2 obtained from an HMQC correlation from the C-13 shifts together with consideration of the axial and equatorial splitting patterns.[9] This experiment was repeated recently[28] to give very close agreement with the first assignment (see Table 4.2) though there is still some uncertainty about the assignment of H3,7ax and H3,7eq. The latter authors also calculated the ^{1}H shifts by *ab initio* GAIO calculations at the DFT/B3LYP level with a large basis set and these are also given in Table 4.2. There is good agreement with both sets of calculations though it should be noted that the shifts from the *ab initio* calculations are only relative shifts as the value for one hydrogen (starred in Table 4.2) in both *trans-* and *cis*-decalin has been fitted in the calculations.

Other molecules for which the present scheme fits well include adamantane and the strained rings of norbornane and bornane. In the first attempts to fit norbornane[9] the calculated shifts of both the bridge and endo protons of norbornane were not in good agreement with the observed shifts. A possible explanation of this came from the work of Marshall *et al.*[29] in their observation of the anomalous values of the vicinal H–H couplings in the $CH_2.CH_2$ group in this molecule. The exo–exo coupling (ca. 12 Hz) is much larger than the endo–endo coupling (ca. 9 Hz.) though both the dihedral angles are 0°. Theoretical calculations showed that there was a significant interaction between the orbitals of the methylene bridge and the $CH_2.CH_2$ group and this interaction affected only the endo–endo coupling (see Chapter 2, Section 2.3.3). It is possible that this interaction could affect the corresponding ^{1}H chemical shifts and this was modelled by an r^{-6} shielding function from the eclipsed C.C.C.C fragment to give additional shielding to the bridge and endo hydrogens at these close distances (< 2.7 Å). With this small addition the data for both norbornane and the methyl substituted norbornanes (Tables 4.2 and 4.4) are very well reproduced with no error > 0.2 ppm.

Table 4.2 *Observed versus calculated ^{1}H chemical shifts (δ) of cyclic alkanes*

Molecule	Proton	Observed[a]	Calculated
Cyclopropane	CH$_2$	0.22	0.22
Cyclobutane	CH$_2$	1.96	1.88
Cyclopentane	CH$_2$	1.51	1.50
Cyclohexane	ax	1.19	1.19
	eq	1.68	1.65
Norbornane	1,4 (CH)	2.19	2.17
	endo	1.16	1.19
	exo	1.47	1.51
	7a,s	1.18	1.22
Bicyclo[1.1.2] hexane[b]			
	1,4 (CH)	2.49	2.61
	2,3 CH2	1.58	1.57
	5,6 endo	1.56	1.70
	5,6 exo	0.87	0.77
Bicyclo[2.2.2]octane			
	CH	1.64	1.92
	CH$_2$	1.50	1.60
Trans-decalin			
	1,4,5,8a	0.93 (0.93)[c]	0.97 (0.93)[c*]
	1,4,5,8e	1.54 (1.53)	1.57 (1.52)
	2,3,6,7a	1.25 (1.24)	1.26 (1.28)
	2,3,6,7e	1.67 (1.68)	1.67 (1.68)
	9,10 (CH)	0.88 (0.87)	0.88 (0.80)
Cis-decalin			
	1,5a	1.59 (1.54)[c]	1.36 (1.59)[c]
	1,5e	1.18 (1.12)	1.23 (1.12)*
	2,6a	1.19 (1.13)	1.22 (1.21)
	2,6e	1.70 (1.65)	1.67 (1.67)
	3,7a	1.36 (1.35)	1.53 (1.34)
	3,7e	1.33 (1.38)	1.38 (1.30)
	4,8a	1.45 (1.40)	1.44 (1.47)
	4,8e	1.45 (1.40)	1.51 (1.43)
	9,10 (CH)	1.63 (1.58)	1.69 (1.56)
Perhydro-phenalene	1,3,4,6,7,9a	0.95	1.02
	1,3,4,6,7,9e	1.57	1.59
	2,5,8a	1.29	1.31
	2,5,8e	1.65	1.69
	10–12 (CH)	0.96	0.98
	13 (CH)	0.32	0.24

Table 4.2   (*Continued*)

Molecule	Proton	Observed[a]	Calculated
Perhydro-anthracene			
	1,4,5,8a	0.95	0.99
	1,4,5,8e	1.56	1.59
	2,3,6,7a	1.23	1.27
	2,3,6,7e	1.67	1.67
	9,10a	0.72	0.75
	9,10e	1.43	1.49
	11–14 (CH)	0.91	0.96
Adamantane	CH	1.87	1.98
	CH_2	1.75	1.62
Bornane	2n	1.23	1.16
	2x	1.49	1.60
	3n	1.13	0.94
	3x	1.71	1.76
	4 (CH)	1.60	1.72
	7,8-CH_3	0.83	0.78
	10-CH_3	0.83	0.86
Tertiary-butylcyclohexane			
	1a (CH)	0.94	0.87
	1-ᵗBu	0.84	0.87
	2,6a	0.91	0.92
	2,6e	1.75	1.68
	3,5a	1.19	1.23
	3,5e	1.75	1.66
	4a	1.08	1.23
	4e	1.64	1.66
Cis-4-ᵗbutyl-methylcyclohexane			
	1a-CH_3	0.86	0.87
	1e (CH)	1.90	1.89
	2,6a	1.45	1.44
	2,6e	1.48	1.55
	3,5a	1.17	1.08
	3,5e	1.49	1.59
	4a (CH)	0.93	0.87
	4e-ᵗBu	0.84	0.88
Trans-4-ᵗbutyl-methylcyclohexane			
	1a (CH)	1.24	1.43
	1e-CH_3	0.86	0.87
	2,6a	0.93	0.95
	2,6e	1.73	1.69
	3,5a	0.93	0.86
	3,5e	1.73	1.62
	4a (CH)	0.95	0.90
	4e-ᵗBu	0.84	0.87

[a] Data from Abraham *et al.*[2].
[b] Data from Wiberg *et al.*[30].
[c] Data from Dodziuk *et al.*[28].
* Referred to this proton.

Table 4.3   *Observed versus calculated ¹H chemical shifts (δ) of methyl cyclohexanes*

Molecule	Proton	Observed[a]	Calculated
eq-Methylcyclohexane			
	1a	1.32	1.39
	1e-CH₃	0.86	0.87
	2,6a	0.88	0.83
	2,6e	1.68	1.61
	3,5a	1.20	1.22
	3,5e	1.68	1.66
	4a	1.11	1.22
	4e	1.68	1.66
ax-Methylcyclohexane			
	1a-CH₃	0.92	0.85
	1e	1.98	1.87
	2,6a	1.40	1.41
	2,6e	1.48	1.58
	3,5a	1.32	1.35
	3,5e	1.53	1.52
	4a	1.19	1.18
	4e	1.68	1.65
1,1-Dimethylcyclohexane			
	1a-CH₃	0.87	0.82
	1e-CH₃	0.87	0.81
	2,6a	1.09	1.04
	2,6e	1.32	1.38
	3,5a	1.36	1.36
	3,5e	1.48	1.51
	4a	1.04	1.21
	4e	1.65	1.66
Trans-1,2-dimethylcyclohexane			
	1,2a (CH)	0.94	1.07
	1,2e-CH₃	0.88	0.91
	3,6a	0.88	0.87
	3,6e	1.63	1.57
	4,5a	1.21	1.24
	4,5e	1.66	1.67
Cis-1,3-dimethylcyclohexane			
	1,3a (CH)	1.34	1.43
	1,3e-CH₃	0.86	0.88
	2a	0.54	0.47
	2e	1.63	1.57
	4,6a	0.76	0.85
	4,6e	1.63	1.63
	5a	1.25	1.25
	5e	1.69	1.67
Trans-1,4-dimethylcyclohexane			
	1,4a (CH)	1.26	1.33
	1,4e-CH₃	0.86	0.88
	2,3,5,6a	0.90	0.89
	2,3,5,6e	1.65	1.60

Table 4.3   (*Continued*)

Molecule	Proton	Observed[a]	Calculated
Cis, cis-1,3,5-trimethylcyclohexane			
	1,3,5a (CH)	1.39	1.45
	1,3,5e-CH₃	0.86	0.88
	2,4,6a	0.47	0.53
	2,4,6e	1.61	1.57
Trans, cis-1,3,5-trimethylcyclohexane			
	1-CH₃	0.97	0.88
	1e (CH)	2.02	1.93
	2,6a	1.03	1.05
	2,6e	1.43	1.55
	3,5a (CH)	1.61	1.62
	3,5-CH₃	0.83	0.88
	4a	0.48	0.44
	4e	1.60	1.58

[a] Data from Abraham *et al.*[24].

Table 4.4   *Observed[a] versus calculated ¹H chemical shifts (δ) of methyl norbornanes*

Proton	2-Exo-methyl		2-Endo-methyl	
	Obs.	Calc.	Obs.	Calc.
1	1.82	1.96	1.98	2.18
2n	1.49	1.65	0.93[b]	0.83
2x	0.86[b]	0.88	1.90	1.80
3n	1.43	1.41	0.53	0.45
3x	0.93	0.80	1.74	1.65
4	2.16	2.20	2.11	2.19
5n	1.11	1.19	1.08	1.22
5x	1.44	1.52	1.47	1.49
6n	1.14	1.23	1.55	1.49
6x	1.46	1.52	1.27	1.30
7a	1.04	1.08	1.25	1.22
7s	1.33	1.38	1.33	1.21

[a] Data from Abraham *et al.*[24].
[b] Methyl shift.

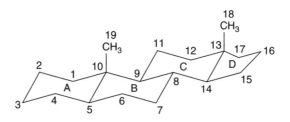

Figure 4.7   *Nomenclature used for 5α-androstane.*

Table 4.5   *Observed versus calculated* ^{1}*H chemical shifts for 5α-androstane*

Proton	Observed A[31]	Observed B[9]	Calculated
1α	0.89	0.87	0.95
1β	1.66	1.67	1.55
2α	1.50	1.48	1.51
2β	1.41	1.41	1.44
3α	1.23	1.21	1.25
3β	1.67	1.67	1.68
4α	1.22*	1.22*	1.24
4β	1.22*	1.22*	1.19
5 (CH)	1.06	1.02	1.14
6α	1.22*	1.22*	1.22
6β	1.22*	1.22*	1.37
7α	0.93	0.91	0.86
7β	1.69	1.68	1.89
8 (CH)	1.29	1.28	1.20
9 (CH)	0.69	0.68	0.76
11α	1.55	1.53	1.53
11β	1.26	1.26	1.23
12α	1.10	1.09	1.12
12β	1.71	1.70	1.60
14 (CH)	0.90	0.89	0.75
15α	1.65	1.63	1.50
15β	1.15	1.14	1.27
16α	1.56*	1.58*	1.55
16β	1.56*	1.61*	1.63
17α	1.13	1.12	1.17
17β	1.42	1.39	1.62
18-Me	0.69	0.69	0.75
19-Me	0.79	0.78	0.81

* Unresolved.

Methyl cycloalkanes.   The extensive data set for the methyl cyclohexanes (Table 4.3) shows excellent agreement between the observed and calculated shifts. The treatment of the methyl group γ effect given in Section 4.1.2 is included in the present scheme and the good agreement shown supports this treatment of the methyl group effect.

Androstane.   5α-Androstane was originally included as a test of the general applicability of the scheme to the important class of steroids and also to determine the importance of long-range effects, e.g. whether the C ring effects the ^{1}H chemical shifts in the A ring.

It can be seen from Table 4.5 that the calculated shifts are in general very good agreement with the experimental data. Indeed out of the 28 recorded shifts only two are 0.20 ppm in error, a pleasing result when it is considered that two possible geometries of 5α-androstane, the *ab initio* one considered here and a derived crystal geometry give differences in the calculated shifts of the 11β and 17α protons of −0.18 and 0.30 ppm respectively.[1] The *ab initio* calculations gave the geometry of the flexible 5-membered D ring as a 13-envelope (C14, C15, C16 and C17 are more or less in a plane with only a 9.5° twist). However, the exact conformation in solution of the unsubstituted ring D has not been determined and this may affect the calculated shifts of these protons.

Chemical shift contributions in cyclohexane. Finally it is of interest to consider the contributions to the ^{1}H chemical shifts in cyclohexane, shown in Figure 4.8. The difference between the axial and equatorial protons is multi-functional. There are contributions due to the different charges resulting from the orientation dependent carbon γ effect as well as the H..H and C..H steric terms and C—H electric field. Both H_{1ax} and H_{1eq} are shielded by $H_{3,5\,ax}$ but the steric effect at H_{1ax} is much greater. The C_δ carbon atom provides the only deshielding steric term and again the effect is greater at the axial proton. The electric field is shielding and is of similar size at the equatorial and axial protons.

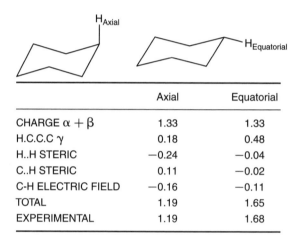

	Axial	Equatorial
CHARGE α + β	1.33	1.33
H.C.C.C γ	0.18	0.48
H..H STERIC	−0.24	−0.04
C..H STERIC	0.11	−0.02
C-H ELECTRIC FIELD	−0.16	−0.11
TOTAL	1.19	1.65
EXPERIMENTAL	1.19	1.68

Figure 4.8   *Contributions to the calculated shifts of the protons in cyclohexane.*

4.3   Alkene Chemical Shifts

4.3.1   Introduction

The ^{1}H spectra of alkenes have been investigated for many years but there has been continuing controversy over the shielding effect of the double bond. Jackman[32] first suggested the anisotropic shielding of the olefinic bond from the enhanced shielding of one of the CMe_2 groups in α-pinene which was situated over the double bond. This lead to the well-known shielding cone (Figure 4.9) in which any nucleus situated above the double bond is shielded whilst any nucleus in the plane of the double bond is deshielded. In an authoritative review of this field, Bothner-By and Pople[23] noted that whereas Jackman's model is due to a large diamagnetism along the x−axis (Figure 4.9), Conroy[33] had suggested a large diamagnetism in the y-direction and Pople from theoretical calculations[34] a paramagnetism in the y-direction centred on the carbon atoms rather than the centre of the C=C bond. Both Jackman's and Pople's theories give increased shielding in the x-axis and deshielding in

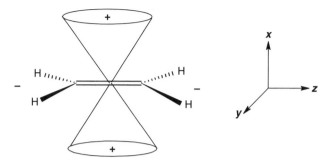

Figure 4.9   *Classical shielding cone for ethylene.*

the y-axis. They differ only in their predictions for shielding along the z- (i.e. C=C) axis, which is not easy to observe.

The shielding cone hypothesis was implicated in an early controversy over the assignment of the bridge methylene protons in norbornene. Deuteration studies[35] unambiguously assigned the 7-*syn* protons in norbornene to higher frequency than the 7-*anti* proton, contrary to Jackman's theory. A later investigation of olefinic shielding was due to ApSimon *et al.*[36]. They derived comparable values for the parallel $(\chi_z - \chi_y)$ and perpendicular $(\chi_x - \chi_y)$ anisotropies of the double bond but concluded: 'the conventional picture of a shielding cone around the C=C bond appears to require substantial modification. It would appear that deshielding is confined to a restricted region at the ends of the double bond: outside this region a nucleus is shielded whether it lies in the plane of the double bond or above it'.

The central problem of this early work was that the NMR instrumentation at this time was inadequate to analyse the complex ¹H spectra of the rigid molecules needed to examine olefinic shielding. ApSimon *et al.*[36] could use only the C-18 and C-19 methyl groups of unsaturated steroids as probes, which was a major limitation in this investigation.

More recently *ab initio* DFT-GIAO (density functional theory-gauge including atomic orbitals) calculations have been applied to calculate the shielding effects of a double bond. Alkorta and Elguero[37] using a probe methane molecule situated near to an ethylene molecule calculated that the methane proton nearest the ethylene molecule was *de*shielded in every direction with the largest deshielding above the C=C bond. At 2.5 Å in the x-direction (Figure 4.9) the deshielding was 1.27 ppm and at 3.7 Å from the C=C bond in the y- and z- directions the deshielding was 0.11 and 0.06 ppm, respectively.

Martin and co-workers[38–40] in a number of publications using the same DFT-GIAO technique again with a methane probe molecule but a different basis set, obtained more detailed information. They varied the orientation of the methane protons and averaged the results for the methane protons. They calculated the shielding over a box with $x = 2.5, 3.0$ and 3.5 Å and y and z varying from 0 to 2 Å in 0.5 Å steps from the centre of the C=C bond (Figure 4.9). The resulting shielding increments were fitted by a quadratic equation in (x, y, z), which was however only valid over the box dimensions. For $x = 3.5$ Å the methane protons were *shielded* by the double bond for all values of y and z, but for $x = 2.5$ Å the methane protons were *de*shielded. At $x = 3.0$ Å the shielding was positive or negative depending on the values of the other coordinates.

These authors also calculated the shielding increments of protons over a C=C bond in some rigid molecules. In norbornene the calculations reproduced the experimental result

(δ 7-*syn* > δ 7-*anti*) but in α-pinene the calculations predicted that the *syn* methyl group is deshielded compared to the *anti* methyl group. Although the authors regarded this as agreeing with the experimental data, this is the reverse of the correct experimental value (see later).

It should be stressed that all such *ab initio* calculations are basis set dependent and also they do not give direct information on the mechanism responsible for the shielding. Thus in this case it is not possible to tell whether the results are due to C=C bond anisotropy or some other mechanism (e.g. van der Waals interactions). This is of importance as whereas anisotropy is independent of the probe nucleus, the H—H van der Waals interactions are a function of both interacting atoms. In alkanes H$\cdots$H interactions are shielding (Section 4.2) but in aromatics deshielding (Section 5.1). The *ab initio* calculations are very useful in visualizing the spatial dependence of the olefinic shielding. It is clear from these results that this must be a complex function of the distance as a simple $1/r^n$ term would not give both positive and negative shielding along one axis. This important aspect will be considered further subsequently.

A systematic investigation of ^{1}H chemical shifts in alkenes was presented by Abraham *et al.*[41]. The complete assignment of the ^{1}H spectra of a variety of aliphatic and aromatic alkenes was obtained and this provided a sufficient amount of data for a quantitative analysis of alkene shielding using the CHARGE model. This showed that the ^{1}H chemical shifts of olefins are influenced by both the magnetic anisotropy and steric effects of the double bond.

4.3.2   C=C Bond Anisotropy and Shielding

The theory has been given in Chapter 3 but it is useful to note here some specific aspects relating to the C=C bond. The magnetic anisotropy of the C=C bond was obtained from the general McConnell equation[16] (Equation (3.15)). R is the distance from the C=C bond to the H nucleus of interest in Å and $\Delta\chi$ is the molar anisotropy. $\Delta\chi_1 = \chi_x - \chi_y$ and $\Delta\chi_2 = \chi_z - \chi_y$ where χ_x, χ_y and χ_z are the susceptibilities along the *x*-, *y*- and *z*-axes and the angles θ_1 and θ_2 are defined as shown in Figure 4.10.

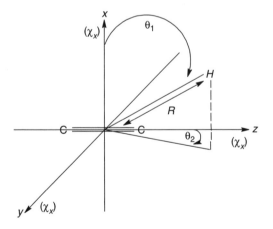

Figure 4.10 *Principal axes of the C=C bond.*

Note that the Jackman model (Figure (4.9)0 is given by the first term in Equation (3.15). This will be referred to henceforth as the perpendicular anisotropy and the second term as the parallel anisotropy.

For the aromatic olefins investigated it is necessary to include the shifts due to the aromatic ring current and the π-electron densities in the aromatic ring. The treatment given in Chapter 3 is incorporated unchanged here. The π-electron densities are calculated from Huckel theory.[42,43] The standard coulomb and resonance integrals for the Huckel routine are given by Equation (3.7) where α_0 and β_0 are the coulomb and resonance integrals for a carbon $2p_z$ atomic orbital and h_r and k_{rs} the factors modifying these integrals for orbitals other than sp^2 carbon. For substituted aromatics the appropriate values of the coefficients h_r and k_{rs} in Equation (3.7) for the orbitals involving hetero atoms were obtained so that the π densities calculated from the Huckel routine reproduce the π densities from *ab initio* calculations.

The effect of the excess π-electron density at a given carbon atom on the proton chemical shifts of the neighbouring protons is given by Equation (3.18) where Δq_α and Δq_β are the excess π-electron densities at the α and β carbon atoms. The above contributions are added to the shifts to give the calculated shift of Equation (3.19).

Application to alkenes. The olefinic group has γ effects on protons three bonds away and in principal steric, anisotropic and electric field effects on protons more than three bonds removed. All these need to be considered. There are a number of different γ effects as there are many different pathways in olefines. For example, for the alkene protons there are C.C=CH, C.C.C(sp^2)H etc. and for the alkane protons C=C.CH, C.C(sp^2).CH etc. For the saturated protons, the γ effects vary if the proton is in a CH, CH_2 or CH_3 fragment. The coefficients A and B (Equation (3.50) for each γ effect need to be obtained to give the best fit with the observed data.

The π densities were obtained from *ab initio* calculations, using the GAUSSIAN software at the RHF/6-31G* level.[44] This basis set gave the best agreement with the observed dipole moments (e.g. propene, observed 0.35D, calculated 0.36D). The h_r and k_{rs} parameters in Equation (3.7) were then varied in order to obtain the same π densities as the *ab initio* calculations. Simple Huckel theory gives the same π densities ($= 1.0$) for the olefine carbon atoms in propene and butadiene. In order to obtain more realistic π densities in these cases two modifications were introduced. The hyperconjugative effect of a saturated substituent (e.g. CH_3) on the π-electron densities was modelled by Equation (4.4). The coulomb integral (α_r) of the sp^2 carbon connected to an sp^3 carbon is modified in order to reproduce the increased charge on the attached sp^2 carbon; q_r is the charge on the attached sp^3 carbon atom.

$$\alpha_r = \alpha_r{}^0 + 0.06 - 0.13 \, q_r \tag{4.4}$$

This gave excess π densities on the olefine atoms of propene as $+/-0.037$ electrons which compares reasonably with the *ab initio* calculated values of -0.104 (C_1) and $+0.029$ (C_2).

A similar modification was made to the coulomb integral of an alkene carbon attached to another alkene carbon via a single bond (e.g. C_2-C_3 in butadiene). In this case the coulomb integral was altered from 0.0 to 0.043. Again this gave reasonable agreement with the *ab initio* calculations. For butadiene the excess π densities on the olefine atoms were $+/-0.0154$ which compare well with the *ab initio* calculated value of $+/-0.0157$.

The shielding or steric effect due to the carbons in a C=C bond has to be calculated as well as the C=C bond anisotropy as they are both an integral part of the total shielding. The C=C bond anisotropy is a complex function depending on the values of the perpendicular and parallel anisotropies (Equation (3.15)). If only the perpendicular anisotropy is present this gives the shielding cone of Figure 4.9, i.e. shielding above the double bond deshielding in the olefine plane. The steric effects of all non hydrogen atoms are deshielding and given by Equation (3.11) with the appropriate values of the a_s coefficient.[9,24,25,42,43,45–47] The only exception being the aromatic carbon for which no shielding term was required. Alternatively the π electrons may be considered as responsible for the shielding effects. As these electrons have a node in the $y - z$ plane (Figure 4.9) the shielding term would include an orientation term (Equation (4.5)) where R and θ_1 are as shown in Figure 4.10. Both these alternatives need to be considered.

$$\text{Shielding} = \cos^2 \theta_1 / R^6 \qquad (4.5)$$

Table 4.6 *Compounds studied with the numbering*

Number	Compound	Number	Compound
1	Ethylene	18	4-Methyl cyclohex-1-ene
2	Propene	19	1,4-Dimethyl cyclohex-1-ene
3	E-Pent-2-ene	20	Methylene cyclohexane
4	Z-Pent-2-ene	21	Methylene cyclopentane
5	Isobutene	22	Cycloheptene
6	Butadiene	23	Endo norbornyl-5n,6n-norbornene
7	t-Butyl ethylene	24	Styrene
8	Pent-1-ene	25	9-Vinyl anthracene
9	Z-Hex-3-ene	26	5-Methylene-2-norbornene
10	E-Hex-3-ene	27	Camphene
11	Cyclopentene	28	Bicyclopentadiene
12	Cyclohexene	29	α-Pinene
13	Cyclohexa-1,3-diene	30	β-Pinene
14	Cyclohexa-1,4-diene	31	7,7-Dimethyl norbornene
15	Pent-1,4-diene	32	Norbornene
16	Tetrahydroindene	33	Norbornadiene
17	Isotetralin	34	Bicycle-2,2,2-oct-2-ene

4.3.3 Observed versus Calculated Shifts

The molecules studied are identified in Table 4.6 and shown with the atom numbering in Scheme 4.1. Compounds **1, 2, 11, 13, 14, 18, 19, 20, 22, 24, 25, 26, 27, 28, 29** and CDCl$_3$ solvent were obtained commercially and the spectra obtained at 400 MHz. The data for compounds **3, 4, 6, 8, 9, 10, 12, 15, 16, 17, 21, 33** and **34** was obtained from the 300 MHz spectra in the *Aldrich Library of FTNMR Spectra*.[48] The assignments for all these spectra are straightforward and the chemical shifts are accurate to ±0.01 ppm. The data for the remaining compounds **5, 7, 23, 30, 31** and **32** are from the literature and the appropriate references are given in the tables.

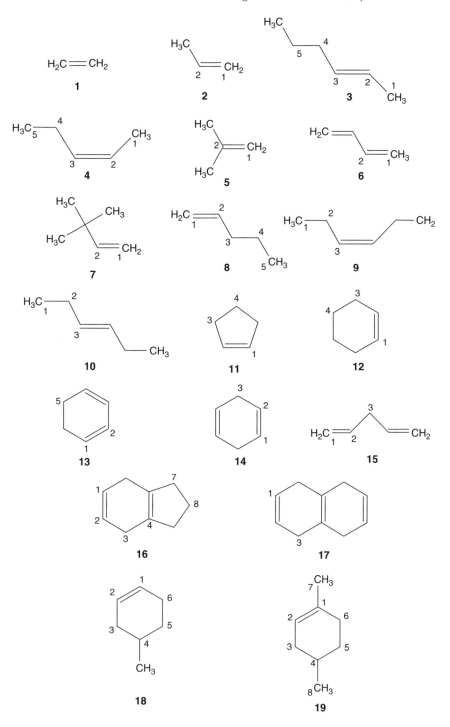

Scheme 4.1 *Molecules studied and their numbering.*

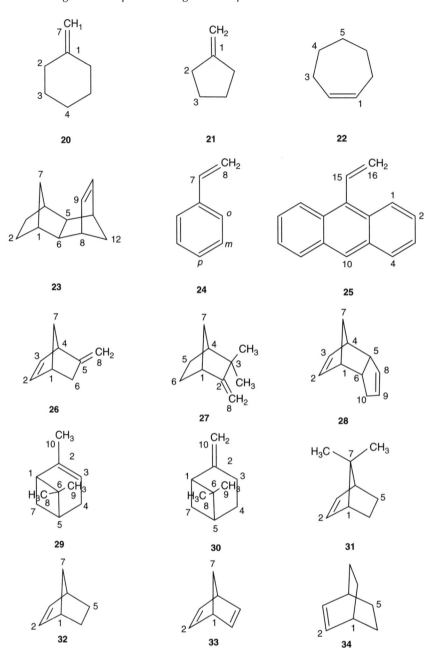

Scheme 4.1   (*Continued*).

In order to quantify the olefin shielding, the compounds must be of a known fixed geometry. The geometries of **1, 2, 5, 6, 7, 20, 21, 22** and **33** were obtained by optimizations using *ab initio* calculations at the HF/6-31G* level.[44] The rest of the geometries were obtained by optimizations using the PCMODEL programme.[26]

The acyclic olefines **3, 4, 8, 9** and **10** can exist in a number of rotational forms. The predominant form in these compounds is the *trans* (anti) conformation of the carbon chain and this conformer is the one considered in these molecules. Similarly in butadiene only the stable s-*trans* conformer[49] was considered. In the cyclic series **18** and **19** can exist in a number of possible conformations. MM calculations showed that the preferred conformer in both cases is the half-chair with the 4-methyl group equatorial. The calculated ax–eq energy difference is 1.6 and 2.4 kcal mol^{-1} for **18** and **19,** respectively and thus the equatorial conformer is > 90 % populated in both cases. In styrene the dihedral angle of the olefine group was given as 30° by PCMODEL and 0° by *ab initio* calculations and both geometries were considered. However in **25** both approaches gave similar geometries with the vinyl group orthogonal to the anthracene ring. For **20** the room temperature spectrum gave only three signals as the C3, C4 and C5 protons overlap and thus a variable temperature experiment was performed. At −120°C in a 1:1 mixture of CD$_2$Cl$_2$ and CFCl$_3$ the ring inversion slowed sufficiently (T_c = −80°C) to observe all the different protons. Lessard *et al.*[50] had previously observed this for some 2-substituted methylenecyclohexanes using ^{13}C NMR. In order to check for any solvent effects the ^{1}H spectrum at room temperature in the solvent mixture was compared against the spectrum in CDCl$_3$. No appreciable differences were observed so it was assumed that the low temperature shifts could be used in the calculations.

The ^{1}H assignments of all the compounds are given in Tables 4.8–4.12. Full details of all the assignment experiments and spectra are given elsewhere.[41,55] The assignment of bicyclopentadiene **28** agrees with a previous assignment[51] and those of α- and β-pinene (**29** and **30**) were assigned previously[52,53] and confirmed recently by a complete analysis of the ^{1}H spectrum.[54]

Tables 4.8–4.12 comprise a large dataset of alkene proton chemical shifts and this dataset was used to test the various theories for alkene shielding detailed earlier in the context of the CHARGE model. In this model the parameters *A* and *B* (Equation (3.5)) for each γ effect have to be determined as well as the long-range shielding, i.e. the anisotropy and van der Waals effects. This was achieved by separating the γ effects into two groups. Those involving the olefinic protons were obtained first, and subsequently the remaining γ effects together with the anisotropy and the shielding were considered. This is because the alkane protons are affected by both the alkene γ effects and the C=C anisotropy and van der Waals shielding, but the alkene protons are not affected by the latter, except for compounds with more than one olefin group.

The values of the parameters were obtained by use of a non-linear least mean square program CHAP8[56] which compares the observed vs. calculated chemical shifts. The values obtained for the *A* and *B* parameters of Equation (3.5) are given in Table 4.7. Note that the cos θ term averages to zero for a methyl group, thus only the constant *A* is obtained.

Both the anisotropy and van der Waals effects are considered as long-range effects in CHARGE, thus the effect of the C=C bond on protons ≤ three bonds distant is included in the γ effects above. The only protons that experience an anisotropy or shielding effect are three bonds or more from the C=C bond in this model.

Table 4.7   *A and B values (Equation (3.5)) for each γ effect*

H..C Fragment		A	B
H—C=C—C—		−0.155	0.017
H—C=C—C=		−0.428	−0.089
H—C—C=C (‖)		−0.006	−0.044
H—C—C—C— (‖)		0.175	−0.343
H—C—C—C= (‖)		0.131	−0.066
H—C—C=C	−CH	0.183	0.021
	−CH-2	0.093	0.178
	−CH-3	0.190	—
H—C—C—C=	−CH	0.024	−0.362
	−CH-2	−0.039	−0.294
	−CH-3	0.026	—

To determine the appropriate anisotropy and shielding functions a number of approaches were used. The first step was to decide whether the anisotropy was due to both parallel and perpendicular anisotropies or only of one of them. The calculations were performed with both the parallel and perpendicular contributions. The result showed that the parallel anisotropy was almost zero. Indeed the observed–calculated rms was the same whether two anisotropies were used or only the perpendicular one. Therefore the anisotropy of a C=C bond is due to the perpendicular effect only, and the parallel effect can be neglected.

The next step was to determine whether the anisotropy and the shielding have to be calculated from the middle of the C=C bond as suggested by Conroy[33] or at the carbon atoms as suggested by Pople.[34] In addition the shielding term could either be the simple r^{-6} term of Equation (3.11) or the more complex function of Equation (4.5). Thus a number of different approaches were attempted. The results were as follows. The complex shielding function of Equation (4.5) gave poorer results than the simple r^{-6} term and was eliminated. The remaining options gave very similar agreement with the observed data. It was more convenient in the context of the CHARGE model to take the shielding at each carbon atom and the anisotropy at the middle of the C=C bond and this was the option employed. In this case the shielding of a γ proton (e.g. H.C.C.C=C) is given by the γ effect of Table 4.7 from the olefinic carbon plus the anisotropy and steric effects from the C=C bond. Thus protons three bonds or more from the C=C bond experience anisotropy effects from the bond and shielding effects from both the sp² carbons. This option on iterating the parameters gave values of −20.09 Å³ for the anisotropy (i.e. −12.1 × 10⁻⁶ cm³ mol⁻¹) and 82.5 Å⁶ for the shielding together with the γ effects of Table 4.7. For the dataset considered of 172 chemical shifts in Tables 4.8–4.12 spanning a range of ca. 0.5 to 8.4δ the rms error of calculated vs. observed shifts was 0.11 ppm.

These results show that there is *deshielding* above the C=C bond at small distances due to the van der Waals term and *shielding* for large distances due to the bond anisotropy. On

Table 4.8   *Observed versus calculated chemical shifts(δ) for acyclic alkenes[a]*

Compound	Proton	Observed	Calculated
1	—	5.405[c]	5.407
2	1_{cis}[b]	4.941[c]	4.903
	1_{trans}[b]	5.031	4.929
	2	5.834	5.841
	Me	1.725	1.667
3	2	5.42[d]	5.345
	4	1.98	2.057
	Me_5	0.96	0.937
	Me_1	1.63	1.682
4	2	5.40[d]	5.341
	4	2.05	2.006
	Me_5	0.96	0.919
	Me_1	1.60	1.622
5	1	4.65[e]	4.712
	Me	1.72	1.702
6	1_{cis}[b]	5.08[d]	5.096
	1_{trans}[b]	5.19	5.191
	2	6.31	6.310
7	1_{cis}[b]	4.82[f]	4.920
	1_{trans}[b]	4.91	4.977
	2	5.85	6.172
	Me	1.00	1.092
8	1_{cis}[b]	4.93[d]	4.928
	1_{trans}[b]	4.98	4.946
	2	5.80	5.809
	3	2.02	1.846
	4	1.41	1.228
	Me	0.90	0.897
9	Me	0.96[d]	0.918
	2	2.02	2.051
	3	5.34	5.358
10	Me	0.97[d]	0.938
	2	2.00	2.066
	3	5.43	5.335
15	1_{cis}[b]	5.03[d]	4.981
	1_{trans}[b]	5.05	5.060
	2	5.84	6.009
	3	2.80	2.688

[a] See numbering in Scheme 4.1.
[b] See text.
[c] This work.
[d] Puchert and Behnke[48].
[e] Partenheimer.[57].
[f] Streitweiser *et al.*[58].

the other hand, there is always a *deshielding* effect in the plane of the C=C bond. The figures obtained here for the anisotropy and shielding show that along the *x*-axis (Figure 4.9) the shielding is positive for distances < 2.0 Å. At this distance the shielding changes sign to become negative. The maximum negative value of the shielding occurs at ca. 2.5 Å. This is in good agreement with both the observed data and with the results from the *ab initio* calculations mentioned earlier which found a change in the sign of the shielding at ca. 2.8 Å.

Acyclic alkenes. The observed and the calculated chemical shifts for the acyclic alkenes (**1–10** and **15**, Scheme 4.1.) are given in Table 4.8. The nomenclature *cis–trans* refers to the hydrogen, not to the alkane substituent. The calculated shifts are in good agreement with the observed, the majority of shifts being within 0.05 ppm. The CH proton in **7** is 0.3 ppm out (calc. 6.16 vs. obs. 5.85δ).

This chemical shift has the influence of the π density and a γ effect (H.Csp2.C.C) from three methyl groups. In t-butyl alkanes a similar enhanced γ effect was explicitly included but it was not felt necessary to include this here for only one chemical shift. The only other error larger than 0.2 ppm is for **8** and this could be due to conformational isomerism in this compound.

Monocyclic alkenes. The observed and calculated chemical shifts for the monocyclic alkenes are given in Table 4.9. The calculated chemical shifts are also in good agreement with the observed shifts though here both the spread of chemical shifts and the differences are greater than for the acyclic alkenes. Some of these differences may well be due to uncertainties in the calculated geometries of these molecules. This could be the case for H-5 in cycloheptene **22** in which the calculated shift is very different from the observed (1.45 vs. 1.72δ). It is generally considered that cycloheptene is largely in the chair form (conformer **1**, Figure 4.11)[49,59] which was the conformer used in the calculations, but the literature is ambiguous on this question.[60] The molecule can adopt up to five different conformations (Figure 4.11) which are rapidly equilibrating by pseudorotation even at very low temperatures. However, the olefinic protons and the other methylene protons are in agreement with the observed data.

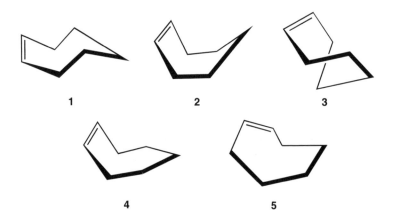

Figure 4.11   *Possible cycloheptene conformers.*

Table 4.9 *Calculated versus observed chemical shifts(δ) for monocyclic alkenes*[a]

Compound	Proton	Observed	Calculated
11	1	5.74[b]	5.765
	3	2.31	2.093
	4	1.82	1.721
12	1	5.68[c]	5.747
	3	1.99	2.057
	4	1.61	1.555
13	1	5.894[b]	5.879
	2	5.798	5.859
	5	2.151	2.245
14	1	5.70[b]	5.650
	3	2.67	2.642
16	1	5.73[c]	5.702
	3	2.63	2.602
	7	2.25	2.132
	8	1.82	1.892
17	1	5.71[c]	5.729
	3	2.53	2.591
18	1	5.650[b]	5.761
	2	5.650	5.758
	3_{eq}	2.080	2.091
	3_{ax}	1.640	1.503
	4	1.680	1.627
	5_{eq}	1.710	1.838
	5_{ax}	1.240	1.036
	6_{eq}	2.060	2.044
	6_{ax}	2.060	2.142
	Me	0.950	0.948
19	2	5.350[b]	5.554
	3_{eq}	2.040	2.123
	3_{ax}	1.610	1.542
	4	1.610	1.666
	5_{eq}	1.700	1.856
	5_{ax}	1.200	1.084
	6_{eq}	1.900	1.921
	6_{ax}	1.980	1.937
	Me_7	1.650	1.692
	Me_8	0.950	0.981
20	2_{eq}	2.271[b]	2.371
	2_{ax}	1.964	1.874
	3_{eq}	1.820	1.789
	3_{ax}	1.255	1.232
	4_{eq}	1.740	1.714
	4_{ax}	1.328	1.247
	7	4.571	4.725

Table 4.9 (*Continued*)

Compound	Proton	Observed	Calculated
21	2	2.250[c]	2.264
	3	1.650	1.619
	6	4.820	4.733
22	1	5.794[b]	5.622
	3	2.120	2.063
	4	1.504	1.429
	5	1.723	1.447

[a] See numbering in Scheme 4.1.
[b] This work.
[c] Puchert and Behnke[48].

Aromatic alkenes. The observed and calculated chemical shifts for the aromatic alkenes styrene **24** and 9-vinyl anthracene **25** are presented in Table 4.10 and again the general agreement is very good. The calculated shifts for styrene are given for the non-planar molecular mechanics geometry. These are in better agreement with the observed shifts than the planar geometry predicted by *ab initio* calculations. In the latter the *ortho* protons and the near alkene protons experience additional downfield shifts due to H..H repulsion between the alkene and aromatic ring, but the *meta* and *para* proton shifts are the same as in Table 4.10. The available geometric evidence[49] does not preclude a slightly non-planar structure for styrene but our results support this structure.

In 9-vinyl anthracene both programs give the same structure with orthogonal vinyl and aromatic groups. It is very encouraging that the model reproduces these shifts also to a very good degree of accuracy.

Table 4.10 *Observed versus calculated chemical shifts (δ) for **24** and **25**[a]*

Compound	Proton	Observed[b]	Calculated
24	*ortho*	7.414	7.620
	meta	7.328	7.432
	para	7.253	7.402
	7	6.722	6.727
	8$_{trans}$	5.758	5.723
	8$_{cis}$	5.246	5.251
25	1,8	8.320	8.079
	2,7	7.465	7.510
	3,6	7.465	7.537
	4,5	7.996	7.994
	10	8.386	8.517
	15	7.476	7.357
	16$_{cis}$	6.010	5.932
	16$_{trans}$	5.629	5.519

[a] See numbering in Scheme 4.1.
[b] Abraham *et al.*[41].

Norbornenes and bicyclooctenes. The observed and calculated chemical shifts for the norbornenes and bicyclooctene compounds (**23, 26–28, 31–34**) are given in Table 4.11. In **26** H-8$_b$ refers to the proton facing C-4 and H-8$_a$ is facing C-6. In **27** H- 8$_a$ is facing C-1 and H-8$_b$ is facing C-3. Finally in **28** H-10$_a$ is facing C-1 and H-10$_b$ C-9 (Scheme 1). The calculated chemical shifts are generally in reasonable agreement with the observed data, but there are a number of exceptions. This is not surprising as the ¹H chemical shifts of the parent hydrocarbons have proved difficult to quantify in the CHARGE routine.[9,24,25]

Table 4.11   *Observed versus calculated chemical shifts (δ) for norbornenes and bicyclooctene[a]*

Compound	Proton	Observed	Calculated
23	7$_{syn}$	1.970[b]	1.627
	7$_{anti}$	0.480	0.458
26	1	2.968[c]	3.247
	2	6.128	5.946
	3	6.073	5.861
	4	3.156	3.419
	6$_{exo}$	2.252	2.438
	6$_{endo}$	1.756	2.173
	7$_{syn}$	1.595	1.677
	7$_{anti}$	1.421	1.497
	8$_a$	4.717	4.717
	8$_b$	4.988	4.786
27	1	2.670[c]	2.714
	4	1.900	2.106
	5$_{exo}$	1.383	1.305
	5$_{endo}$	1.701	1.795
	6$_{exo}$	1.638	1.605
	6$_{endo}$	1.236	1.501
	7$_{syn}$	1.694	1.504
	7$_{anti}$	1.204	0.999
	8$_a$	4.717	4.710
	8$_b$	4.493	4.736
	Me$_{exo}$	1.020	1.015
	Me$_{endo}$	1.050	0.979
28	1	2.878[c]	2.887
	2	5.984	5.786
	3	5.935	5.727
	4	2.785	2.945
	5	3.214	3.023
	6	2.729	2.693
	7$_{syn}$	1.478	1.425
	7$_{anti}$	1.301	1.389
	8	5.507	5.695
	9	5.476	5.547
	10$_a$	2.184	2.180
	10$_b$	1.622	2.037
31	2	5.900[d]	5.862
	Me$_{syn}$	0.900	0.901
	Me$_{anti}$	0.950	0.905

Table 4.11   (*Continued*)

Compound	Proton	Observed	Calculated
32	1	2.841[b]	2.788
	2	5.985	5.871
	5_{exo}	1.603	1.652
	5_{endo}	0.951	1.379
	7_{syn}	1.313	1.627
	7_{anti}	1.073	1.306
33	1	3.580[e]	3.553
	2	6.750	5.873
	7	2.000	1.749
34	1	2.480[e]	2.702
	2	6.230	5.773
	5_{exo}	1.230	1.445
	5_{endo}	1.500	1.596

[a] See numbering in Scheme 4.1 and text.
[b] Marchand *et al.*[61].
[c] This work.
[d] Bessiere-Chretien and Grison[63].
[e] Tori *et al.*[62].

However there are some interesting points to note. Compound **23** is of particular interest as the 7syn proton (syn to the olefine group) is only ca. 2 Å from and almost vertically above the olefine group, thus it provides a crucial test of any shielding theory. Marchand and Rose[61] obtained the ¹H spectrum of this compound and identified the AB pattern of the H-7 protons from decoupling experiments. However they assigned the 7syn proton to the more shielded resonance at 0.48δ based on the Jackman shielding cone for the C=C bond anisotropy (Figure 4.9). We have reversed this assignment. The more shielded proton is the 7anti and the 7syn is the deshielded proton nearer to the C=C bond. This is strikingly confirmed by the calculated shifts in Table 4.11. Inspection of the CHARGE output shows that the 7syn proton is strongly deshielded by the van der Waals deshielding due to the olefin carbons whilst the anisotropy term is larger for the 7anti proton. This beautifully confirms the shielding pattern obtained here for the C=C group which alters sign along the *x*-axis (see later).

However there are also additional shielding mechanisms in these molecules which are not included in the model. For example, the calculated shifts for the olefinic protons for **33** at 5.87δ are almost 1 ppm less than the observed shifts (6.75δ). Tori *et al.*[62] noted the unusual deshielding effects upon bridge methylenes of norbornadienes. They showed by UV spectroscopy a considerable transannular interaction between the two double bonds. This transannular interaction could affect the ¹H chemical shifts of the olefin protons involved in this interaction as well as the bridge methylenes, which are also not well calculated. However the calculated methine proton chemical shifts are in agreement with the observed data.

In **34** the olefinic ¹H shifts are again not as well calculated as expected (obs. 6.23, calc. 5.81δ). There will also be considerable transannular interactions in this compound between the olefinic group and the endo protons (Scheme 4.1) and this may be a reason for this deviation. However the rest of the ¹H chemical shifts are calculated in good agreement with the observed data.

Pinenes. The calculated and observed chemical shifts for α- and β-pinene (**29** and **30**) are given in Table 4.12. The calculated shifts are generally in fair agreement with the observed data. There are some deviations which mainly concern the protons near the four-membered rings. The cyclobutane ring has not yet been included in the CHARGE model and there may be shielding effects from this fragment which are not covered in these calculations. However the general picture is reasonably well reproduced. In particular Me-9 is calculated as more shielded than Me-8 which is the observed assignment in both molecules.

Table 4.12   *Observed versus calculated proton chemical shifts (δ) for α-pinene and β-pinene*

α-pinene (**29**)   β-pinene (**30**)

Proton	Compound **29** Observed	Calculated	Proton	Compound **30** Observed	Calculated
1	1.931[a]	2.049	1	2.430[b]	2.631
7_a	2.333	2.414	7_a	2.310	1.886
7_b	1.151	1.101	7_b	1.420	1.531
5	2.067	2.522	5	1.970	2.072
4_{eq}	2.231	2.475	4_{eq}	1.820	1.675
4_{ax}	2.152	2.076	4_{ax}	1.850	1.893
3	5.185	5.564	3_{eq}	2.230	2.358
Me$_8$	1.264	1.042	3_{ax}	2.510	2.304
Me$_9$	0.835	0.984	10_a[c]	4.500	4.736
Me$_{10}$	1.658	1.777	10_b	4.570	4.737
			Me$_8$	1.240	1.032
			Me$_9$	0.730	0.993

[a] This work.
[b] See Abraham *et al.*[52] and Kolehmainen *et al.*[54].
[c] See text.

Conclusions. The agreement between the observed and calculated ^{1}H chemical shifts is encouraging. The incorporation of the olefin γ effects together with the calculation of the C=C anisotropy and shielding allows the prediction of the ^{1}H chemical shifts for alkenes, thus extending the CHARGE model to these important compounds. The results demonstrate clearly that the parallel contribution to the anisotropy can be neglected and that the only anisotropic contribution is due to the perpendicular anisotropy.

4.4 Alkyne Chemical Shifts

4.4.1 Introduction

The magnetic anisotropy of the C≡C bond was first proposed by Pople to explain the high-field shift of the acetylene proton compared to that of ethylene. He subsequently obtained an estimate of $\Delta\chi^{C\equiv C}$ of -19.4×10^{-6} cm^3 mol^{-1} from approximate MO theory. In a review of the early investigations Bothner-By and Pople[23] noted other values of $\Delta\chi^{C\equiv C}$. Reddy and Goldstein[64] obtained a value of -16.5×10^{-6} using the linear relationship they found between ^{1}H shifts and the corresponding ^{13}C—^{1}H couplings in compounds where the anisotropic effects were negligible. The anisotropic effects of other groups including the C≡C group were then extrapolated from these linear plots. In a similar manner Zeil and Buchert[65] examined the ^{1}H chemical shifts of a variety of acetylenes and nitriles. Assuming that ^{1}H chemical shifts were linearly dependent on the substituent electronegativity plus a constant shift arising from the diamagnetic anisotropy gave a value of -36×10^{-6}. Subsequently Shoemaker and Flygare[66] obtained a value of the anisotropy of the acetylene group as -7.7×10^{-6} from the second-order Zeeman effect in the microwave spectra of propyne and its isotopic species.

Mallory and Baker[67] showed that regions of deshielding existed alongside C≡C bonds by the observation of low-field ^{1}H chemical shifts in the aromatic compounds 4-ethynylphenanthrene, 5-ethynyl-1, 4-dimethylnaphthalene and 5-ethynyl-1, 4-diethylnaphthalene. They concluded that the deshielding effect of the C≡C bond fell off approximately as $1/r^3$.

Abraham and Reid[68] presented a systematic investigation of the ^{1}H chemical shifts of acetylenes and assigned the ^{1}H spectra of a variety of aliphatic and aromatic acetylenes. These data provided the basis for a quantitative analysis of acetylene SCSs using the CHARGE model for the calculation of ^{1}H chemical shifts. Using this model they showed that ^{1}H chemical shifts are influenced by both the magnetic anisotropy and steric effects of the acetylene group.

4.4.2 C≡C Bond Anisotropy and Shielding

The acetylene group has in principle steric, electric field and anisotropic effects on protons more than three bonds away plus for aromatics an effect on the π-electron densities. All these have to be incorporated into the model. The magnetic anisotropy of a bond with cylindrical symmetry such as C≡C is given from the simple McConnell equation (Equation (3.14)). Here the angle 'θ' is the angle between a vector **R** and the symmetry axis and $\Delta\chi^{C\equiv C}$ the molar anisotropy of the C≡C bond; $(\Delta\chi^{C\equiv C} = \chi^{C\equiv C}_{parl} - \chi^{C\equiv C}_{perp})$ where $\chi^{C\equiv C}_{parl}$ and $\chi^{C\equiv C}_{perp}$ are the susceptibilities parallel and perpendicular to the symmetry axis, respectively. The shielding effect of the triple bond is illustrated in Figure 4.12.

The major electric field of the acetylene group is due to the ≡C.H bond as the C≡C bond is non-polar. The electric field calculation for any C.H bond is automatically included in the model. There are possible steric effects of the acetylene group on the neighbouring protons and of the near aliphatic protons on the acetylene proton. These are both given by Equation (3.11) with different steric coefficients a_s which may be of either sign. Thus the

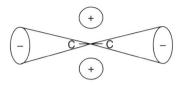

Figure 4.12   *The anisotropic shielding ($\Delta\delta$) in an axially symmetric molecule such as acetylene.*

unknowns to be obtained are $\Delta\chi$, the molar anisotropy of the C≡C bond and the steric coefficients a_s.

For protons three bonds or less from the C≡C group it is necessary to determine the orientational dependence of the γ ¹H chemical shift with respect to the α acetylene carbon. This is simulated by a γ substituent effect from the acetylene carbon (H.C.C.C≡) following Equation (3.5), in which the coefficients A and B may differ for the C≡C group in aromatic vs. saturated compounds. Also in CHARGE the β effect is given by Equation (3.3) which was sufficient for the calculation of charge densities but not always sufficiently accurate to reproduce the ¹H chemical shifts. Thus the β effect from the acetylene carbon atom (H.C.C≡) needs to be obtained. As there is no orientation dependence in this case only one coefficient is required.

Aromatic compounds. For the aromatic acetylenes it is necessary to include the shifts due to the aromatic ring current and the π-electron densities in the aromatic ring. The treatment given in Chapter 3 is incorporated unchanged here. For the aromatic acetylenes it is necessary to obtain the appropriate values of the factors h_r and k_{rs}, the Huckel integrals for the C≡C group (Equation (3.7)). The π-electron densities and dipole moments from *ab initio* calculations are very dependent on the basis set used. The 3-21G basis set gave the best agreement with the observed dipole moments (Table 4.13) and the π densities from this basis set were used to parameterize the Huckel calculations. The CN group contains an sp hybridized carbon atom and the parameters for this group had already been derived.[1] Thus the values of h_r(Csp) and k_{rs}(Csp²–Csp) used for nitriles were used for the acetylene calculations as the Huckel integrals for Csp operates for both of these functional groups. A value of k_{rs} of 1.60 (Csp–Csp) gave π-electron densities for the aromatic acetylenes in reasonable agreement with those from the *ab initio* calculations.

The accuracy of the π densities calculated in the CHARGE program can be examined by calculating the dipole moments of some acetylenes. The calculated vs. observed (in parenthesis) dipole moments[69] (D) of propyne, but-1-yne, tert-butylacetylene, phenylacetylene and para-ethynyltoluene are 0.50 (0.75), 0.50 (0.81), 0.52 (0.66), 0.36 (0.72) and 1.26 (1.02) and the general agreement is support for the π density calculations. The electron densities (total and π) and dipole moments calculated for propyne and phenylacetylene by CHARGE and the *ab initio* method are given in Table 4.13.

Values of h_r and k_{rs} for X–Csp have been determined for a number of different substituents C≡C X. Values of h_r for F, Cl and O for olefins (C=C.X) were obtained previously from π-electron densities calculated using *ab initio* methods at the 3-21G level for a range of olefinic compounds.[55] These were left unchanged for the acetylenes and the values of k_{rs} for the ≡C.X bond varied for the best agreement with the *ab initio* π-electron densities. Values of 0.74 (Csp–F), 0.57 (Csp–Cl) and 1.00 (Csp–O) gave reasonable agreement with *ab initio* calculations. Again, the accuracy of the calculated charges can be examined by calculating the dipole moments of these molecules. The calculated vs. observed (in parenthesis) dipole moments (D) of fluoroacetylene, chloroacetylene, propynal and methoxyacetylene are 0.79(0.75), 0.74(0.44), 2.56(2.46), and 1.62(1.93). The value of k_{rs} for the Csp.Csp² bond is already known from the phenyl acetylene data in Table 4.13. Also, the calculated vs. observed (in parenthesis) chemical shifts of the acetylene proton in fluoroacetylene, chloroacetylene and propynal are 1.33(1.63), 1.95(1.80) and 3.61(3.47). The good agreement of the calculated vs. observed chemical shifts for these molecules is strong support for the above treatment.

Table 4.13 *Total and π (in parenthesis) charges (me) and dipole moments (D) for propyne and phenylacetylene*

Atom			Method		
	STO-3G	3-21G	6-31G	CHARGE	Observed
(a) Propyne					
C_β	−136(−21.7)	−419(−22.0)	−488(−24.7)	−106(−22.4)	
C_α	−37 (11.3)	−47 (12.2)	−29 (13.9)	−62 (22.4)	
μ(D)	0.50	0.69	0.68	0.50	0.75
(b) Phenylacetylene					
C_β	−125(−5.1)	−363(−14.2)	−531(−16.5)	−83(−10.6)	
C_α	−40(−0.9)	−60(−0.1)	−155(2.4)	−46(−0.7)	
C_1	2 (−21.0)	−44(−32.6)	−156(−26.7)	−24(−0.6)	
C_2	−54 (8.6)	−215 (18.5)	−148 (14.9)	−57 (4.5)	
C_3	−63 (0.3)	−230(−1.3)	−209 (0.1)	−72(−0.3)	
C_4	−59 (9.1)	−237 (12.6)	−188 (10.8)	−73 (3.6)	
μ(D)	0.50	0.65	0.64	0.36	0.72

4.4.3   Observed versus Calculated Shifts

The molecules studied here with the atom numbering are shown in Figure 4.13. Full details of the compound preparation and the experimental details are given elsewhere.[68] The assignments of all the compounds investigated are given in Tables 4.14–4.18 together with the calculated ¹H chemical shifts. The ¹H NMR data for but-1-yne (**2**), but-2-yne (**3**), pent-1-yne (**4**), hex-3-yne (**5**), t-butylacetylene (**6**), para-ethynyltoluene (**14**), and 2-ethynylpropene (**18**) were from Puchert and Behnke[48] and that for 1-ethynylnaphthalene (**15**) from Hanekamp and Klusener.[70] The spectrum of cyclohexylacetylene (**10**) was obtained at −60 °C, at which temperature the rate of interconversion of the conformers was slow on the NMR time scale. The integral ratio for H1a to H1e was 1:6.2 to give ΔG (eq–ax) 0.70 kcal mol⁻¹, in fair agreement with previous measurements. Eliel[49] quotes 0.41–0.52 kcal mol⁻¹. The equatorial conformer was fully assigned at low temperature but for the axial conformer only H1a, 2e and 2a were assigned as the remaining protons were obscured by the resonances of the protons of the dominant equatorial conformer. The conformations of 2-ethynyl-endo-norbornan-2-ol (**8**) and 2,2′-ethynyl-bis-bornan-2-ol (**9**) were confirmed by X-ray crystal structures with the configuration at C-2 as shown in Figure 4.13.[68]

The *ab initio* geometries obtained were of some interest. The GAUSSIAN software at the MP2/6-31G* level gave values of the H.C≡and C≡C bond lengths in acetylene of 1.061 and 1.203 Å res. in complete agreement with the experimental values (1.061 and 1.203 Å).[56] The same basis set gave corresponding values for phenylacetylene of 1.057 and 1.188 Å, but for para tolylacetylene the values were 1.067 and 1.223 Å. This large change on the introduction of a para methyl group seemed odd and these geometries did not give good results when used in CHARGE. In particular the acetylene proton shift is identical in these aromatic compounds (Table 4.18) but was not calculated to be so with these geometries. Using the recommended DFT/B3LYP[72] method with the 6-31G** basis

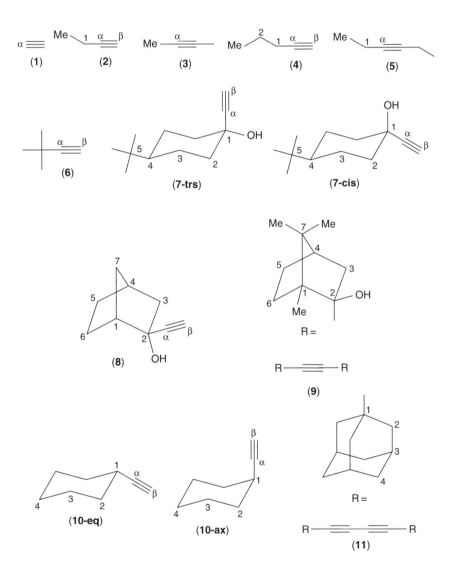

Figure 4.13 *Molecules studied and their numbering.*

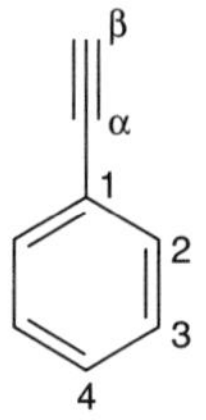

Figure 4.13   *(Continued).*

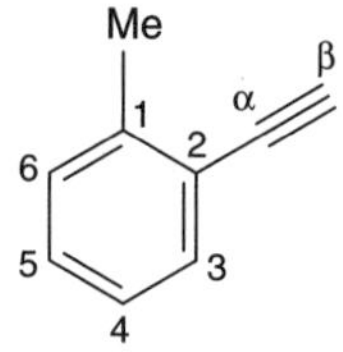

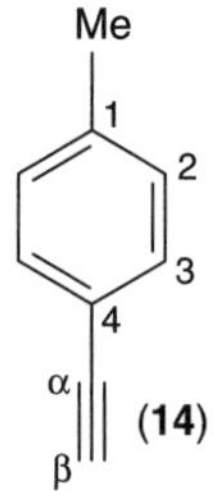

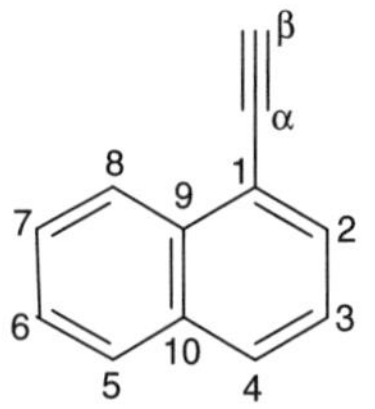

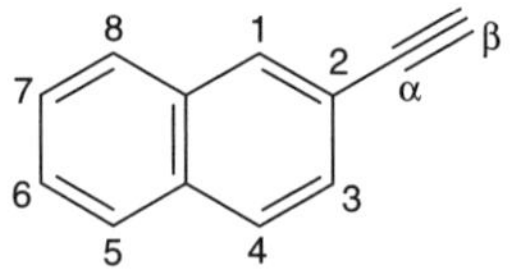

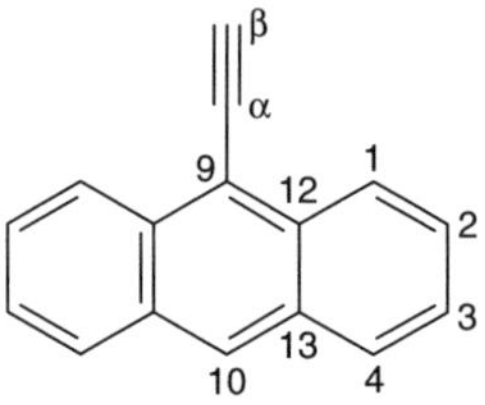

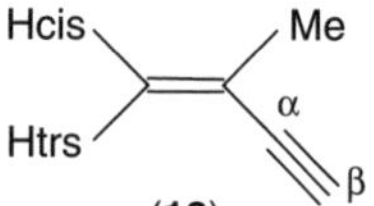

set in GAUSSIAN gave bond lengths of 1.065 and 1.210 Å for both compounds and these values were used as standard for all the aromatic acetylenes. It is well known[72] that the DFT technique treats electron correlation much better than the MP2 method and this could be the explanation of the above result.

The data in Tables 4.14–4.18 provide a rigorous test of the application of both the CHARGE model and also of the theories of C≡C SCSs. The geometries are well determined, thus the only empirical parameters to be obtained are those required for the model.

It is first necessary to consider how the acetylene (H.C≡) protons will be calculated. These could be reproduced in CHARGE by the appropriate values of the integral for the H.C(sp) bond. The near effects of anisotropic (or polar) bonds have been reproduced in this manner in previous parts of this series as attempting to calculate anisotropic (or polar) effects at such short distances by means of simple geometric functions (Equations (3.13–3.14)) is not a feasible option. However if this procedure was adopted here the charge on the acetylene proton would be approximately equal to that in ethane reflecting the near equality of their chemical shifts. This is obviously not the case as the acetylene proton is more acidic and the C.H bond more polar than even the olefinic proton. Thus the anisotropic contribution has been included in the chemical shift calculation for these protons. The procedure adopted was that the values of $\Delta\chi^{C\equiv C}$ and the steric coefficient together with the coefficients of the γ effects were obtained from the shifts of all the protons except the acetylene protons. The appropriate parameters for these protons were then included. This gave the correct chemical shift for the acetylene protons and an acceptable value of the proton charge (see later).

The parameters required for the calculations are the anisotropy of the C≡C bond, the sp carbon steric coefficient $a_S^{C \equiv C}$, the γ effect of the sp carbon atom, i.e. H.C.C.C≡(coefficients A and B, Equation 3.5) and the β effect of the β acetylene carbon, i.e. H.C.C≡. The γ effects may differ for aliphatic and aromatic acetylenes. This gives a total of five parameters for the aliphatic series plus a possible three more for the aromatic compounds. The acetylene ^{1}H chemical shifts were then fitted by the appropriate values of the ≡C.H exchange integral and the γ effect H.C≡C.X plus a steric parameter a_S for the steric effect Hsp^3 → Hsp.

The iterations were carried out on all the observed ^{1}H chemical shift data by use of the non-linear mean squares program (CHAP8[56]). The anisotropy of the C≡C bond was taken from both the centre of the C≡C bond and from each carbon atom, but the steric effect of the sp carbon atoms was taken as usual from the carbon atoms. The iterations gave better results when the anisotropy was taken from each carbon of the C≡C bond. Also both the values of the anisotropy, steric coefficent and the coefficients A and B (Equation (3.5)) for the γ effects were identical when the iterations were performed with either the aliphatic compounds alone or the aromatic compounds, thus the final iteration was performed including all the compounds and using only five parameters. The values of these parameters were as follows. The anisotropy = −9.18 ppm Å^3 at each carbon atom, i.e. $\Delta \chi^{C \equiv C}$ = −18.36 ppm Å^3 molecule^{-1}, i.e. −11.1 × 10^{-6} cm^3 mol^{-1}. The steric coefficient $a_S^{C \equiv C}$ = 56.6 Å^6. The coefficients for the γ effects (H.C.C.C≡) (Equation (3.5)) were A 0.423 and B −0.177 ppm and the enhanced β effect (H.C.C≡) was 1.37. The acetylene protons were then considered. For these protons the iteration gave values of the C—H exchange integral of 42.8 (cf. 41.4), the γ effect (H.C≡C.C) coefficients were 0.22 and 1.20 for sp^3 and sp^2 carbons, respectively, and the steric coefficient (H.Csp^3 to H.C≡) was 46.5.

The iteration was over 124 chemical shift values of the compounds discussed previously except the acetylene alcohols as the chemical shift of the OH group is concentration dependent (see Chapter 6, Section 6.4). The rms error of the observed–calculated shifts was 0.074 ppm over a chemical shift range from ca. 1–8.5 ppm., a very satisfactory result.

4.4.4   Acetylene SCSs

The data presented here for the acetylenes may be combined with the ^{1}H chemical shifts of the parent compounds given previously[42,55] to give the acetylene SCSs in these compounds. These are shown in Figure 4.14 for eq-cyclohexyl acetylene (**10eq**), 1-ethenyl trans-4-t-butyl cyclohexanol (**7-trs**), 1 and 2 ethenylnaphthalene (**15, 16**) and the norbornane (**8**) and bornane (**9**). The SCSs for **7-trs**, **8** and **9** are obtained as the chemical shifts for **7, 8** and **9** minus the ^{1}H shifts of trans-4-t-butylcyclohexanol, endo norborneol and isoborneol.[71] These SCSs are of some interest, they are both shielding and deshielding but the larger SCSs are always deshielding. The γ effect of the C≡C group (i.e. H.C.C.C≡C) is also deshielding with for the saturated compounds considerable orientational dependence without any obvious pattern, except that the γ SCS of the norbornane and bornane derivatives **8** and **9** is greater for the 120° orientation than for the eclipsed orientation for both the exo and endo compounds. This intriguing observation is valid for all norbornane substituents so far studied.[46,47]

The long-range (> 3 bonds) effects of the C≡C group are large but decrease rapidly with distance. For **10-eq** the C≡C SCS is almost zero for all long-range protons. There is a large 1,3-diaxial interaction of the acetylene and H-3ax in **7-trs**. Similar large effects are observed at the 7syn protons in **8** and the 6-endo protons in **9**. All these protons are in a similar environment to the triple bond, i.e. essentially orthogonal to the C≡C bond. As there is no electric field effect of the C≡C bond these SCSs can be due to either the C≡C anisotropy or a steric effect or both. Significantly the C≡C SCSs at protons situated along the C≡C bond (e.g. the 3ax and 3eq protons in **10eq**, the 7syn proton in **9**, etc.) are small but always deshielding. This would not be so if the SCSs were solely due to the

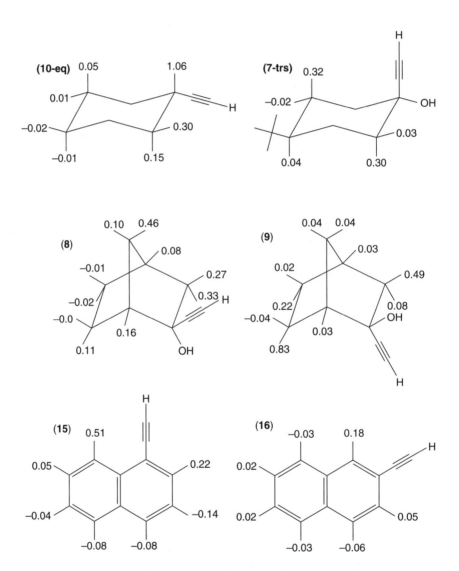

Figure 4.14   *Ethynyl SCSs in aliphatic and aromatic molecules.*

C≡C anisotropy. Similar C≡C SCSs are observed for the aromatic acetylenes **15** and **16** though in these compounds π-electron effects will be present. Again the SCSs are of either sign but the large effects are always deshielding, the largest being again due to the peri interaction in **15**.

The data of Tables 4.14–4.18 provide an examination of both the application of the CHARGE model to alkynes and of the influence of the acetylene group on ¹H chemical shifts. There is generally very good agreement between the observed and calculated ¹H chemical shifts. In the aliphatic compounds the model reproduces very well the sizeable low field shifts of protons situated at the side of the acetylene group, e.g. H-3a in axial cyclohexanes SCS (**7tr**), obs. 0.32, calc. 0.43 ppm, H-7syn in exo-norbornanes (**8**), obs. 0.46, calc. 0.49 ppm and H-6n in endo-bornanes (**9**) obs. 0.83, calc. 1.15 ppm (Figure 4.13). The calculated values are due to both anisotropy and steric effects (see later). The smaller γ effects are again mostly to low field and are also well reproduced by the combination of the anisotropy and the γ effect of Equation (3.5).

Table 4.14 *Observed versus calculated ¹H chemical shifts(δ) for acetyleneᵃ (**1**), but-1-yne (**2**), but-2-yne (**3**), pent-1-yne (**4**), n- hex-3-yne (**5**) and tert-butylacetylene (**6**)*

¹H Number	2		3		4		5		6	
	Obs.	Calc.	Obs.	Calc.	Obs.	Calc.	Obs.	Calc.	Obs.	Calc.
H-1	2.25	2.18	—	—	2.18	2.02	2.15	2.23	—	—
H-2	—	—	—	—	1.57	1.55	—	—	—	—
Me	1.18	1.10	1.75	1.82	1.00	0.77	1.11	1.13	1.24ᵇ	1.24
C≡C—H	1.97	2.04	—	—	1.95	2.05	—	—	2.07	2.10

ᵃ Acetylene: obs. 1.91, calc. 1.91.
ᵇ t-Butyl.

Table 4.15 *Observed versus calculated ¹H chemical shifts (δ) for equatorial and axial-ethynylcyclohexane (**10-eq**, **10-ax**) and 1-ethynyladamantane (**11**)*

Molecule	¹H Number	Obs.	Calc.
10-eq	1a	2.246	2.094
	2e	1.977	1.877
	2a	1.355	1.408
	3e	1.734	1.617
	3a	1.200	1.084
	4e	1.666	1.600
	4a	1.170	1.138
	C≡C—H	2.182	2.100
10-ax	1e	2.871	2.667
	2e	1.775	1.877
	2a	1.481	1.519
	C≡C—H	2.278	2.137
11	γ	1.861	1.810
	δ	1.941	1.943
	E	1.681	1.646
	A	1.681	1.639

Table 4.16 *Observed ¹H chemical shift (δ) for 1-ethynyl-trans/cis-4-t-butyl cyclohexan-1-ol (**7-trs**, **7-cis**) and observed versus calculated C≡C-SCSs*

Compound	Proton	δ (obs.)	SCS (obs.)	SCS (calc.)
7-trs	2e	2.040	0.033	0.293
	2a	1.514	0.297	0.328
	3e	1.762	−0.020	0.051
	3a	1.367	0.321	0.543
	4a	1.000	0.035	0.0
7-cis	2e	2.037	0.204	0.222
	2a	1.705	0.216	0.263
	3e	1.596	0.048	0.047
	3a	1.379	0.020	−0.038
	4a	1.010	0.017	0.026

Table 4.17   *¹H chemical shifts (δ) for (**8**) and (**9**) and observed versus calculated C≡C SCSs*

Compound	Proton	δ(obs.)	C≡C–SCS	
			Obs.	Calc.
8	1	2.407	0.155	0.229
	3x	2.140	0.269	0.255
	3n	1.360	0.334	0.406
	4	2.250	0.079	−0.006
	5x	1.561	−0.009	−0.003
	5n	1.318	−0.017	−0.091
	6x	1.380	−0.002	0.003
	6n	1.979	0.109	−0.142
	7s	1.802	0.462	0.493
	7a	1.389	0.099	−0.055
9	3x	2.228	0.489	0.336
	3n	1.822	0.083	0.195
	4	1.750	0.029	−0.059
	5x	1.695	0.020	0.056
	5n	1.180	0.222	0.159
	6x	1.468	−0.037	0.134
	6n	1.835	0.827	1.153
	Me (1)	0.940	0.034	0.110
	Me (7s)	1.057	0.039	−0.057
	Me (7a)	0.870	0.042	−0.016

There are some discrepancies in the calculated values of chemical shifts. The shifts of both H-1e in **10-ax** and H-1a in **10-eq** are ca. 0.2 ppm larger than the observed values (Table 4.15). These are the only methine (HC.C≡) protons in the data set and this may be a general result. Further data would be necessary to test this.

The observed and calculated shifts for H-2e in **10-ax** are in reasonable agreement (Table 4.15) as are the values for H-2e in **7-cis** (Table 4.16). In the analogous compound **7-trs** the corresponding SCSs are obs. 0.03, calc. 0.29 ppm. It may be that

Table 4.18 *Observed versus calculated ¹H chemical shifts (δ) and observed versus calculated C≡C SCSs for aromatic acetylenes (12–18)*

Molecule	Proton	δ(obs.)	δ(calc.)	SCS(obs.)	SCS(calc.)
Phenylacetylene (**12**)	2,6	7.492	7.544	0.151	0.207
	3,5	7.311	7.337	−0.030	0.000
	4	7.341	7.343	0.0	0.006
	C≡CH	3.069	3.191		
o-Ethynyltoluene (**13**)	3	7.460	7.484	0.200	0.178
	4	7.138	7.155	−0.027	−0.041
	5	7.245	7.289	−0.015	−0.017
	6	7.202	7.005	0.022	−0.022
	Me	2.454	2.494	0.111	0.252
	C≡CH	3.271	3.156		
p-Ethynyltoluene (**14**)	2, 6	7.100	7.016	−0.080	−0.011
	3, 5	7.400	7.496	0.140	0.190
	Me	2.340	2.251	−0.003	−0.033
	C≡CH	3.020	3.124		
1-Et naphthalene (**15**)	2	7.700	7.692	0.223	0.216
	3	7.340	7.478	−0.137	0.002
	4	7.760	7.856	−0.084	0.031
	5	7.760	7.814	−0.084	−0.011
	6	7.440	7.478	−0.037	0.002
	7	7.530	7.515	0.053	0.039
	8	8.350	8.340	0.506	0.515
	C≡CH	3.430	3.298		
2-Et naphthalene (**16**)	1	8.028	8.067	0.184	0.242
	3	7.524	7.652	0.047	0.176
	4	7.788	7.810	−0.056	−0.015
	5	7.810	7.803	−0.034	−0.022
	6	7.500	7.467	0.023	−0.009
	7	7.500	7.462	0.023	−0.014
	8	7.810	7.832	−0.034	0.007
	C≡CH	3.142	3.225		
9-Et anthracene (**17**)	1	8.522	8.478	0.513	0.475
	2	7.602	7.598	0.135	0.043
	3	7.504	7.546	0.037	−0.009
	4	8.001	8.003	−0.008	0.0
	10	8.447	8.410	0.016	−0.022
	C≡CH	3.990	3.594		
2-Ethynylpropene (**18**)	H$_{trans}$	5.300	5.233	0.359	0.337
	H$_{cis}$	5.390	5.479	0.359	0.553
	Methyl	1.900	1.788	0.175	0.149
	C≡CH	2.870	3.164		

in **7-trs** there is an interaction between the geminal hydroxy and acetylene groups. In this case the SCSs for each group cannot be obtained simply by subtracting the shifts in this compound from those of the parent alcohol (or acetylene). There is a similar anomaly in the obs. vs. calc. SCSs for H-3x/3n in **9** but not for **8**. It is of interest that the anomalous results occur for compounds in which the acetylene group is sterically hindered (cf. similar effects in the bromo and iodo SCSs, Chapter 6,

Section 6.3). This intriguing possibility would need to be further tested with more definitive examples.

In the aromatic acetylenes again the large SCSs of the acetylene groups due to the analogous peri-planar interactions are also well reproduced. For example, H-8 in 1-ethenyl napthalene (**16**), obs. 0.51, calc. 0.51 ppm, H-1,5 in 9-ethenylanthracene (**17**), obs. 0.51, calc. 0.48 ppm. The other major SCSs in the aromatic compounds are at the *ortho* protons and again these effects are due to the anisotropy plus γ effects. The SCSs at the other ring protons due mainly to π effects are much smaller reflecting the small interaction between the acetylene and the aromatic π systems.

An interesting anomaly occurs with H-3 in 1-ethynylnaphthalene (**15**). The observed SCS (-0.137 ppm) contrasts with the calculated value (0.00 ppm). The calculated SCS at this proton is as expected the same as the SCS for the *meta* proton in phenylacetylene and this agrees exactly with the observed value for this proton. An exactly similar effect was found for the cyano group. It would appear that both the C≡C and CN SCSs operate differently in naphthalene and benzene.

There is generally very good agreement between observed and calculated shifts for the acetylene protons but the model does not fully account for the value in 9-ethynylanthracene (**17**), cf. obs. 3.99, calc. 3.59 ppm. This may be due to enhanced π effects at this position or to H (aromatic)–H (acetylene) steric effects which would be expected to give a low-field shift. As no other molecule in the data set experiences these interactions it was not felt necessary to include them.

4.4.5   Contributions to the Acetylene SCSs

It is of interest to consider the actual magnitudes of the contributions to the acetylene SCSs. The acetylene proton has a partial atomic charge of $+0.088$ electrons corresponding to a ≡C—H dipole moment of 0.45D. This charge gives rise from Equation (3.10) to a chemical shift of 7.47δ. Thus as expected the acetylene proton is more 'acidic' than olefinic or aromatic protons. The difference between this value and the calculated shift (1.90δ) is due entirely to the C≡C anisotropic contribution (-5.65 ppm). In the other compounds other effects are present and Tables 4.19 and 4.20 give the observed vs. calculated C≡C SCSs for the aliphatic and aromatic acetylenes, respectively, together with the calculated anisotropic, steric and electric field contributions.

For the alkyl acetylenes (Table 4.19) the major contribution for the α and β protons is the C≡C anisotropy. All the other contributions (C—H electric field, C≡C steric, C steric and H-steric) are very small for the compounds given with the exception of the acetylene protons in which there is a significant π shift. (This does not appear in acetylene itself as there is no π excess in acetylene.)

In the SCSs of the H-2e/H-2a protons of all the compounds in Table 4.19 the components do not add up to give the calculated values of the SCSs. This is due to the electronic γ effects which are calculated separately and which affect protons that are three bonds or less from the C≡C group.

The large SCS for H-3a in axial-ethynylcyclohexane has been estimated from compound **7tr** as 0.32 ppm (obs.) and 0.43 ppm (calc.). The calculated SCS is made up of a ≡C steric contribution of 0.185 ppm plus an anisotropic contribution of 0.125 ppm plus some other very small contributions. For the other protons with large SCSs a similar pattern is found,

Table 4.19 *Observed versus calculated C≡C SCSs with the electric field, steric and anisotropic contributions for eq/ax-ethynylcyclohexane (**10-eq** and **10-ax**) and 1-ethynyladamantane (**11**)*

Compound	Proton number	Obs.	Calc.	C—H electric field	C≡C– anis.	C≡C– steric	C– steric	H– steric	Pi shift
10-eq	1a	1.056	0.906	−0.053	−0.590	0.0	0.016	−0.046	
	2e	0.297	0.245	−0.019	−0.074	0.027	0.0	0.0	
	2a	0.145	0.225	−0.025	−0.072	0.028	0.0	−0.023	
	3e	0.054	−0.024	0.028	−0.059	0.0	0.0	0.0	
	3a	0.011	−0.110	0.011	−0.011	0.014	0.0	−0.019	
	4e	−0.014	−0.041	0.016	−0.056	0.0	0.0	0.0	
	4a	−0.020	−0.057	0.016	−0.062	0.0	0.0	−0.01	
	C≡C—H	—	—	−0.027	−5.556	0.0	0.05	0.031	−0.169
10-ax	1e	1.231	1.029	−0.045	−0.560	0.0	0.0	0.0	
	2e	0.095	0.244	−0.019	−0.072	0.028	0.0	0.0	
	2a	0.291	0.332	−0.033	−0.174	0.0	0.0	−0.034	
	C≡C—H	—	—	−0.064	−5.550	0.0	0.098	0.052	−0.170
11	γ	0.111	0.137	−0.024	−0.071	0.028	0.0	0.0	
	δ	0.071	−0.012	0.028	−0.059	0.0	0.0	0.0	
	e	0.069	−0.036	0.017	−0.062	0.0	0.0	0.0	
	a	0.069	−0.042	0.015	−0.056	0.0	0.0	0.0	

Table 4.20 *Calculated versus observed C≡C SCSs, with the steric, anisotropy, electric field, ring current and π-shift contributions for phenylacetylene (**12**) and 1/2-ethynylnaphthalenes (**15** and **16**)*

Compound	¹H number	SCS (obs.)	SCS (calc.)	C≡C– steric	C≡C– anis.	C—H electric field	Ring current	π-shift
12	2,6	0.151	0.207	0.029	−0.072	−0.020	0.004	0.043
	3,5	−0.030	0.0	0.008	−0.068	0.045	0.005	0.013
	4	0.0	0.006	0.002	−0.063	0.033	−0.001	0.035
	C≡C—H	—	—	0.0	−5.582	−0.004	0.196	−0.108
15	2	0.223	0.216	0.029	−0.080	−0.020	0.0	0.065
	3	−0.137	0.002	0.008	−0.070	0.046	0.0	0.023
	4	−0.084	0.031	0.002	−0.062	0.034	0.0	0.058
	5	−0.084	−0.011	0.0	−0.036	0.014	0.0	0.010
	6	−0.037	0.002	0.0	−0.010	0.010	0.0	0.003
	7	0.053	0.039	0.007	0.023	0.0	0.0	0.012
	8	0.506	0.515	0.326	0.210	0.084	0.0	0.0
	C≡C—H	—	—	0.0	−5.581	0.009	0.318	−0.136
16	1	0.184	0.242	0.032	−0.061	−0.021	0.0	0.068
	3	0.047	0.176	0.027	−0.078	−0.020	0.0	0.024
	4	−0.056	−0.015	0.007	−0.069	0.045	0.0	0.006
	5	−0.034	−0.022	0.0	−0.038	0.012	0.0	0.004
	6	0.023	−0.009	0.0	−0.022	−0.001	0.0	0.013
	7	0.023	−0.014	0.0	−0.019	−0.001	0.0	0.006
	8	−0.034	0.007	0.005	−0.029	0.018	0.0	0.014
	C≡C—H	—	—	0.0	−5.581	−0.004	0.246	−0.123

e.g. for H-7s in **8** the calculated SCS of 0.49 ppm is made up of 0.37 ppm (steric) and −0.11 ppm (anisotropy) and for H-6n in **9** the corresponding values are 1.153, 0.57 and 0.27 ppm. The results show categorically that the largest contribution to these SCSs is due to the C≡C steric term and not the C≡C anisotropy. Amazingly the C≡C steric term has not been considered in any previous investigation.

The aromatic acetylenes have other mechanisms which may affect the ¹H chemical shifts, in particular, the ring current and π-electron effects and Table 4.20 gives the observed versus calculated SCSs for selected molecules with the electric field, ring current and π shift contributions.

We have assumed in this investigation that the introduction of the acetylene group has no effect on the parent hydrocarbon ring current and thus there are no ring current effects on the C≡C SCSs. The agreement obtained here is strong support for this assumption. In contrast the C≡C group does affect the π-electron densities and this has a significant effect on the SCSs. The data of Table 4.20 show the similarities between the aromatic and aliphatic acetylenes. In particular the large peri planar interaction between the 1-acetylene and H-8 in **15** giving a calculated SCS of 0.49 ppm is predominantly due to the steric contributuion (0.415 ppm) with only a small anisotropic term (0.10 ppm). The remaining SCSs for the ring protons are quite small with the π shifts and electric field effects roughly comparable. The ring current contribution to the SCSs of the aromatic protons is as stated above zero but Table 4.20 includes the actual ring current shift at the acetylene protons and the π shifts which are both significant

As stated previously, various values of the C≡C diamagnetic anisotropy have been given ranging from −7.7 to −36($\times 10^{-6}$ cm³ mol⁻¹). The value found here of −11.1 × 10^{-6} cm³ mol⁻¹ is a middle value which is in reasonable agreement with both Pople's original estimate of −19.4 and the value of −7.7 of Shoemaker and Flygare.

4.4.6   Naphthyl and Phenanthryl Acetylenes

It is of some interest to see whether the large low-field shifts observed by Mallory and Baker in the proton NMR of 4-ethynylphenanthrene (**19**), 5-ethynyl-1, 4-dimethylnaphthalene (**20**) and 5-ethynyl-1,4-diethylnaphthalene (**21**) are predicted by our model.

(19)   (20)   (21)

They observed large low-field shifts for H-5 in **19** (1.63 ppm from H-5 in phenanthrene), the 4-methyl protons in **20** (0.49 ppm) and the methylene protons of the C-4 ethyl group of **21** (0.55 ppm) due to the deshielding effect of the C≡C group.

The calculated ¹H shifts vs. the observed δ values (in parenthesis) for H-5 in **19**, the methyl protons in **20** and for the CH_2 protons in **21** are 9.38 (10.34), 2.90 (3.01) and 3.39 (3.62).

There is excellent agreement between the observed and calculated shifts for the methyl and methylene protons in **20** and **21**, but the calculated value for H-5 in **19** is too small by almost 1 ppm. This proton is in very close proximity to the triple bond. The distance between the centre of the triple bond and H-5 is calculated as 2.208 Å using *ab initio* methods. This compares with the values of 1.55 Å from Dreiding models and 2.408 Å from PCModel.[26] The *ab initio* geometry calculated at the B3LYP/6-31G** level may not be absolutely correct and small changes in bond lengths and angles at this close distance will have a very significant influence on the calculated ¹H chemical shifts. It would be of interest to obtain the crystal geometry and input this into CHARGE. However the simple Equations (3.11) and (3.14) for the shielding and anisotropy of the C≡C bond are also likely to be less accurate for the close distances observed in this case. The major contribution to the low-field shift of this proton is again the steric term (0.71 ppm vs. 0.34 ppm for the anisotropy) and a simple r^{-6} term would not be expected to be very accurate at these short internuclear distances.

Mallory and Baker[67] concluded that the C≡C shielding was proportional to r^{-3} and that the shielding was from the centre of the triple bond. In the CHARGE scheme the steric term is proportional to r^{-6} but the anisotropy is proportional to r^{-3} and both terms are calculated at each carbon atom. Placing the anisotropy in the middle of the acetylene bond and using an r^{-3} steric term both gave poorer agreement for the data set considered here.

4.5 Summary

It is convenient to summarize the results for hydrocarbons given in this chapter. The CHARGE model gives a quantitative description of ¹H chemical shifts for a diverse range of hydrocarbons and this provides the basis for the extension to substituted compounds.

The ¹H chemical shifts of more than 30 alkanes including acyclic, monocyclic (cyclohexanes), bi and tri cyclic compounds are recorded and analysed using the CHARGE model. The results show that the major contributions are the C.C.CH γ effect and the H..H and C..H shielding. The electric field is a minor contribution. There is no reason to invoke C—C anisotropy in describing ¹H chemical shifts. There may be a possible contribution from eclipsed C.C fragments but this could also be due to electronic effects. There are however still some significant anomalies in certain molecules. Possible improvements that could be made are a more sophisticated carbon γ orientation dependence as there is no theoretical reason to use only a cos θ type dependence. Also a proper theoretical interpretation of the 'push–pull' effect would be a first step in a more comprehensive treatment.

The ¹H chemical shifts of 34 alkenes including acyclic, cyclic and aromatic olefins are recorded. The results for methylene cyclohexane are at 153 K to freeze the ring inversion. The incorporation of the olefin γ effects together with the calculation of the C=C anisotropy and shielding allows the prediction of the ¹H chemical shifts, thus extending the CHARGE model to these important compounds. The results demonstrate clearly that the parallel contribution to the anisotropy can be neglected and that the only anisotropic contribution is due to the perpendicular anisotropy. The results also show that there is *deshielding* above the C=C bond at small distances (< 2.0 Å) due to the van der Waals term and *shielding* for larger distances due to the bond anisotropy. On the other hand, there is always a *deshielding* effect in the plane of the C=C bond.

The ¹H chemical shifts of 18 alkynes including cyclohexyl acetylene at 213 K to distinguish the axial and equatorial conformers are recorded. The results show that the C≡C SCSs over more than three bonds is determined largely by the C≡C bond anisotropy and steric effect for both aliphatic and aromatic compounds. In all the compounds considered the steric term is the major contribution with the anisotropy a significant, but smaller contribution. Protons < 3 bonds from the triple bond require in addition the inclusion of electronic β and γ effects from the acetylene carbons in both aliphatic and aromatic acetylenes. The γ effect of the acetylene carbon atom has an orientational dependence. For the data set of 88 ¹H chemical shifts spanning a range of ca. 8 ppm the rms error (obs.–calc. shifts) was 0.074 ppm.

References

1. Abraham, R. J.; Edgar, M.; Glover, R. P.; Warne, M. A.; Griffiths, L., *J. Chem. Soc. Perkin Trans. 2* 1996, 333.
2. *HyperNMR.* Hypercube Inc.: Waterloo, ON, Canada, 1995.
3. Pople, J. A.; Beveridge, D. V., *Approximate Molecular Orbital Theory.* McGraw-Hill: New York, NY, 1970.
4. Blizzard, A. C.; Santry, D. P. „*J. Chem. Phys.* 1971, **55**, 950.
5. Li, S.; Allinger, N. L. ,*Tetrahedron* 1988, **44**, 1339.
6. Abraham, R. J.; Edgar, M.; Griffiths, L.; Powell, R. L., *J. Chem. Soc. Chem. Commun.* 1993, 1544.
7. Winstein, S. P.; Carter, R.; Anet, F. A. L.; Bourn, A. J. R., *J. Am. Chem. Soc.* 1965, **87**, 5247.
8. Anderson, J. E., *J. Chem. Soc. Chem. Comunm.* 1996, 93.
9. Abraham, R. J.; Griffiths, L.; Warne, M. A., *J. Chem. Soc. Perkin Trans. 2* 1997, 31.
10. Daneels, D.; Anteunis, M., *Org. Magn. Reson.* 1974, **6**, 617.
11. Pretsch, E.; Simon, W., *Helv. Chim. Acta* 1969, **52**, 2133.
12. Moritz, A. G; Sheppard, N., *Mol. Phys.* 1962, **5**, 361.
13. Bothner-By, A. A.; Naar-Colin, C., *J. Am. Chem. Soc.* 1958, **80**, 1728.
14. Bothner-By, A. A.; Naar-Colin, C., *Ann. NY Acad. Sci.* **1958**, **70**, 833.
15. Hall, L. D., *Tetrahedron Lett.* 1964, **23**, 1457.
16. Musher, J. I., *J. Chem. Phys.* 1961, **35**, 4.
17. Musher, J. I. , *Mol. Phys.* 1963, **6**, 93.
18. Guy, J.; Tillieu, J., *J. Chem. Phys.* 1956, **24**, 1117.
19. Cavanaugh, J. R.; Dailey, B. P. , *J. Chem. Phys.* 1961, **34**, 4.
20. Yamaguchi, I.; Brownstein, S., *J. Chem. Phys.* 1964, **68**, 1572.
21. Zürcher, R. F., *Prog. Nucl. Magn. Reson. Spectrosc.* 1967, **2**, 205.
22. Bothner-By, A. A., *Disc. Faraday Soc.* 1962, **34**, 66.
23. Bothner-By, A. A.; Pople, J. A. , *Ann. Rev. Phys. Chem.* 1965, **16**, 43.
24. Abraham, R. J.; Griffiths, L.; Warne, M. A., *Magn. Reson. Chem.* 1998, **36**, S179.
25. Abraham, R. J., *Prog. Nucl. Magn. Reson. Spectrosc.* 1999, **35**, 85.
26. Gilbert, K., *Serena Software.* 9.0 ed.;: Bloomington, IN, 2005.

27. Abraham, R. J.; Fisher, J.; Loftus, P., *Introduction to NMR Spectroscopy.* 2nd ed.; John Wiley & Sons, Ltd: Chichester, 1988.

28. Dodziuk, H.; Jaszunski, M.; Schilf, W., *Magn. Reson. Chem.* 2005, **43**, 639.

29. Marshall, J. L.; Walters, S. R.; Barfield, M.; Marchand, A. P.; Marchand, N. W.; Serge, A. L., *Tetrahedron* 1976, **32**, 537.

30. Wiberg, K. B.; Lowry, B. R.; Nist, B. J., *J. Am. Chem. Soc.* 1962, **84**, 1594.

31. Schneider, H. J.; Buchheit, U.; Becker, N.; Schmidt, G.; Siehl, U., *J. Am. Chem. Soc.* 1985, **107**, 7027.

32. Jackman, L. M. , *Applications of N.M.R. in Organic Chemistry.* Pergamon Press: London, 1959.

33. Conroy, H., *Advances in Organic Chemistry: Methods and Results.* Interscience: New York, NY, 1960.

34. Pople, J. A., *J. Chem. Phys.* 1962, **37**, 53.

35. Franczus, B.; Baird, W. C.; Chamberlain, N. F.; Hines, T.; Snyder, E. I., *J. Am. Chem. Soc.* 1968, **90**, 3732.

36. ApSimon, J. W.; Craig, W. G.; Demarco, P. V.; Mathieson, D. W.;.Saunders, L.; Whalley, W. B., *Tetrahedron* 1967, **23**, 2399.

37. Alkorta, I.; Elguero, J., *New J. Chem.* 1998, **22**, 381.

38. Martin, N. H.; Allen, N. W.; Minga, E. K.; Ingrissa, S. T.; Brown, J. D., *J. Am. Chem. Soc.* 1998, **120**, 11510.

39. Martin, N. H.; Allen, N. W.; Minga, E. K.; Ingrissa, S. T.; Brown, J. D., *Struct. Chem.* 1999, **10**, 375.

40. Martin, N. H.; Allen, N. W.; Minga, E. K.; Ingrissa, S. T.; Brown, J. D., *Modelling NMR Chemical Shifts.* American Chemical Society: Washington, DC, 1999.

41. Abraham, R. J.; Canton, M.; Griffiths, L. , *Magn. Reson. Chem.* 2001, **39**, 421.

42. Abraham, R. J.; Canton, M.; Reid, M.; Griffiths, L., *J. Chem. Soc. Perkin Trans. 2* 2000, 803.

43. Abraham, R. J.; Smith, P. E., *J. Comput. Chem.* 1987, **9**, 288.

44. Frisch, M. J.; Trucks, G. W.; Schlegel, H. B.; Scuseria, G. E.; Robb, M. A.; Cheeseman, J. R.; Montgomery, J. A.; Jr., T. V.; Kudin, K. N.; Burant, J. C.; Millam, J. M.; Iyengar, S. S.; Tomasi, J.; Barone, V.; Mennucci, B.; Cossi, M.; Scalmani, G.; Rega, N.; Petersson, G. A.; Nakatsuji, H.; Hada, M.; Ehara, M.; Toyota, K.; Fukuda, R.; Hasegawa, J.; Ishida, M.; Nakajima, T.; Honda, Y.; Kitao, O.; Nakai, H.; Klene, M.; Li, X.; Knox, J. E.; Hratchian, H. P.; Cross, J. B.; Bakken, V.; Adamo, C.; Jaramillo, J.; Gomperts, R.; Stratmann, R. E.; Yazyev, O.; Austin, A. J.; Cammi, R.; Pomelli, C.; Ochtersk, J. W.; Ayala, P. Y.; Morokuma, K.; Voth, G. A.; Salvador, P.; Dannenberg, J. J.; Zakrzewski, V. G.; Dapprich, S.; Daniels, A. D.; Strain, M. C.; Farkas, O.; Malick, D. K.; Rabuck, A. D.; Raghavachari, K.; Foresman, J. B.; Ortiz, J. V.; Cui, Q.; Baboul, A. G.; Clifford, S.; Cioslowski, J.; Stefanov, B. B.; Liu, G.; Liashenko, A.; Piskorz, P.; Komaromi, I.; Martin, R. L.; Fox, D. J.; Keith, T.; Al-Laham, M. A.; Peng, C. Y.; Nanayakkara, A.; Challacombe, M.; Gill, P. M. W.; Johnson, B.; Chen, W.; Wong, M. W.; Gonzalez, C.; Pople, J. A., *Gaussian 03*, D. Gaussian Inc.: Wallingford, CT, 2006.

45. Abraham, R. J.; Ainger, N. J., *J. Chem. Soc. Perkin Trans. 2* 1999, 441.

46. Abraham, R. J.; Griffiths, L.; Warne, M. A., *J. Chem. Soc. Perkin Trans. 2* 1997, 881.

47. Abraham, R. J.; Griffiths, L.; Warne, M. A., *J. Chem. Soc. Perkin Trans. 2* 1998, 1751.

48. Puchert, C. J.; Behnke, J., *Aldrich Library of 13C and 1H FT NMR Spectra.* Aldrich Chemical Company: Milwaukee, WI, 1993.

49. Eliel, E. L.; Wilen, S. H., *Stereochemistry of Carbon Compounds.* J. Wiley & Sons, Inc.: New York, NY, 1994.

50. Lessard, J.; Tan, P. V. M.; Martino, R.; Saunders, J. K., *Can. J. Chem.* 1977, **55**, 1015.

51. Ramey, K. C.; Lini, D. C., *J. Magn. Reson.* 1970, **3**, 94.

52. Abraham, R. J.; Cooper, M. A.; Salmon, J. R.; Whittaker, D., *Org. Magn. Reson.* 1972, **4**, 489.

53. Coxon, J. M.; Hydes, G. J.; Steel, P. J., *J. Chem. Soc. Perkin Trans .2* 1984, 1351.

54. Kolehmainen, E.; Lihia, K.; Laatikainen, R.; Vepsäläinen, J.; Niemitz, M.; Suontamo, R., *Magn. Reson. Chem.* 1997, **35**, 463.

55. Canton, M., *Ph.D. Thesis.* University of Liverpool, 2000.

56. Kuo, S. S., Chapter 8, in *Computer Applications of Numerical Methods.* Addison Wesley: London, 1972.

57. Partenheimer, W., *J. Am. Chem. Soc.* 1976, **98**, 2779.

58. Streitweiser, A.; Xie, P.; Speers, P.; G., W. P., *Magn. Reson. Chem.* 1998, **36**, S209.

59. Binsch, G., *Top. Stereochem.* 1968, **3**, 164.

60. St-Jacques, M.; Varizi, C., *Can. J. Chem.* 1971, **49**, 1256.

61. Marchand, A. P.; Rose, J. E., *J. Am. Chem. Soc.* 1968, **90**, 3724.

62. Tori, K.; Hata, Y.; Muneyuki, R.; Takano, Y.; Tsuji, T.; Tanida, H., *Can. J. Chem.* 1964, **42**, 926.

63. Bessiere-Chretien, Y.; Grison, C., *Bull. Soc. Chim. France* 1971, 1454.

64. Reddy, G. S.; Goldstein, J. H., *J. Chem. Phys.* 1963, **39**, 3509.

65. Zeil, w.; Buchert, H., *Z. Physik. Chem.* 1963, **38**, 47.

66. Shoemaker, R. L.; Flygare, W. H., *J. Am. Chem. Soc.* 1969, **91**, 5417.

67. Mallory, F. B.; Baker, M. B., *J. Org. Chem.* 1984, **59**, 1323.

68. Abraham, R. J.; Reid, M., *J. Chem. Soc. Perkin Trans.* 2 2001, 1195.

69. McClellan, A. L., *Table of Experimental Dipole Moments*, Volume 3. Rahara Enterprises: El Cerrito, CA, 1989.

70. Hanekamp, J. C.; Klusener, P. A. A., *Synth. Commun.* 1989, **19**, 2677.

71. Abraham, R. J.; Barlow, A. P.; Rowan, A. E., *Magn. Reson. Chem.* 1989, **27**, 1024.

72. Foresman, J. B.; Frisch, Æ., *Exploring Chemistry with Electronic Structure Methods.* 2nd ed.; Gaussian, Inc.: Pittsburgh, PA, 1996.

5

Modelling ^{1}H Chemical Shifts, Aromatics

5.1   Aromatic Hydrocarbons

5.1.1   Introduction

The influence of the π-electron densities and ring currents of aromatic compounds on their ^{1}H chemical shifts have been investigated since the beginning of ^{1}H NMR.[1] Despite this wealth of investigation there was for many years no general calculation of the ^{1}H chemical shifts of aromatic compounds and the structural chemist had to rely on ^{1}H data banks for the identification of aromatic compounds by NMR.

Pauling[2,3] introduced the concept of an aromatic ring current to explain the diamagnetic anisotropy of crystalline benzene. Pople[4] later extended this to explain the difference in the ^{1}H chemical shifts of benzene and ethylene and he further showed that the equivalent dipole model of this ring current gave a surprisingly good account of this difference. More sophisticated ring current models for benzene were then developed. The classical double-loop[5,6] and double dipole models[7,8] mimic the π-electron circulation by placing the current loops (and equivalent dipoles) above and below the benzene ring plane. A value of ± 0.64 Å was found to be appropriate. The equations of Haigh and Mallion[9,10] give the shielding ratios directly from quantum mechanical theory. Schneider et al.[11] presented a detailed experimental examination of the double-loop and Haigh and Mallion ring current models, though not the simple equivalent dipole model (see later). The calculations gave good agreement with the experimental data, thus the effect of the benzene ring current on the chemical shifts of neighbouring protons is reasonably well understood.

However, the ^{1}H chemical shifts in condensed aromatic compounds and substituted benzenes were not well calculated. Bernstein et al.[12] in their initial calculations of the chemical shifts of condensed aromatic compounds assumed the same ring current for each

---

benzenoid ring but this was subsequently considered to be an over simplification. Thus it is first necessary to calculate the π-electron current density for each benzenoid ring and then to calculate the effects of these currents on the chemical shifts of the ring hydrogens. The quantum mechanical method for calculating the π-electron current densities was first given by Pople[13] and McWeeny[14] subsequently extended the London–Pople theory. McWeeny's work gives not only the circulating current density but also the effect of this circulating current at the proton in question. It should be noted that all these theories were based on simple Huckel theory.

Early experimental investigations to test these theories were not helped by the complex ¹H spectra of many condensed aromatic hydrocarbons at the low applied magnetic fields then in use and also by the quite large concentration effects on the ¹H chemical shifts due to the propensity of these large planar rings to stack in solution. However, three systematic investigations attempted to overcome these difficulties. Jonathan et al.[15] analysed the ¹H spectra of several condensed aromatics at infinite dilution in CCl_4 or CS_2. They then used the London–Pople theory to calculate the current intensity in the benzenoid rings and the Johnson–Bovey Tables[5] to obtain the ring current shifts. They also estimated C—C and C—H anisotropic effects and found that these could be ignored. They obtained 'only fair agreement' with the observed shifts. Varying the separation of the π-electron loops gave a poorer fit with the observed shifts. They noted that other interactions were affecting the ¹H shifts and in particular noted a high frequency shift for close hydrogens which was suggested to be due to van der Waals contact. They did not attempt to quantify this.

Cobb and Memory[16] and Haigh and Mallion[9,10] performed two similar but more extensive investigations. The ¹H spectra of several condensed aromatic compounds in dilute solution were analysed and the McWeeny equation used to obtain the ring current densities and shielding ratios. They both ignored σ bond anisotropies in this calculation. Both investigations obtained reasonable correlations for 'non-overcrowded protons' between the observed ¹H shifts and the ratio of the π-electron shielding for any hydrogen compared to benzene (H'/H'_b in the nomenclature of Haigh and Mallion.[9,10]). The more comprehensive data of Haigh and Mallion[9,10] when converted to the δ scale may be written as $\delta_{obs} = 1.56 \, (H'/\, H'_b) + 5.66$ with an rms error of 0.06 ppm over a range of ca. 1.6 ppm. However, the differences between the calculated and observed data for the 'crowded' protons were ca. 0.5–0.7 ppm with one of 1.2 ppm., all to high frequency of the calculated value. Again they attributed these shifts to steric effects but did not quantify or define these effects.

Subsequently Westermayer et al.[17] used a double dipole model to test the observed shifts. They correlated the resulting geometric factors with the observed shifts to obtain a value for the benzene diamagnetic susceptibility anisotropy. They stated that superior results for the sterically crowded protons were obtained but it is not clear why this should be the case as no steric term was introduced.

Although it is obvious which protons are crowded (for example, $H_{4,5}$ in phenanthrene) it is not obvious whether this interaction is also present in the other 'less crowded' protons. Thus the simple question of whether the difference between the α and β ¹H chemical shifts in naphthalene is due to ring currents, π-electron densities or steric effects has not been satisfactorily answered. Pople in his original studies[13] calculated the ring current intensities in the five- and seven-membered rings of azulene but did not calculate the ¹H chemical shifts in non-alternant hydrocarbons.

Abraham *et al.*[18] recorded the ¹H chemical shifts of a selection of condensed aromatic compounds in $CDCl_3$ and showed that these differ by a small but significant amount from the earlier data in CCl_4 solution. These provided sufficient data for an analysis of the chemical shifts based on the CHARGE model[1] and they showed that this model can be extended to provide a quantitative calculation of the ¹H shifts in condensed aromatic compounds, including two non-alternant hydrocarbons and in monosubstituted benzenes. Two alternative calculations of the ring current intensity in the benzenoid rings were presented together with a dipole model of the benzene ring current. In model A the ring current intensity in the individual benzenoid rings is a function of the number of adjoining rings whereas in model B the molecular ring current is given by the classical Pauling treatment as proportional to the molecular area divided by the molecular perimeter. All the hydrogens in the condensed aromatic compounds are considered and the 'crowded' ¹H chemical shifts reproduced by a simple steric effect. The effects of substituents in monosubstituted benzenes are well reproduced for the *ortho*, *meta* and *para* protons on the basis of calculated π-electron densities plus the steric, anisotropic and electric field effects of the substituents. The model also reproduces the high field shifts of hydrogens situated over the benzene ring thus providing a general calculation of ¹H chemical shifts of condensed aromatic compounds and will now be discussed in detail.

5.1.2   Ring Currents, π-Electron Densities and Steric Effects

To allow for the calculation of the ¹H chemical shifts of aromatic compounds a number of modifications had to be made. The program must automatically recognize five- and six-membered aromatic rings including the heterocyclic rings of pyrrole, furan and thiophene. Once these were recognized the ring current at any proton can be calculated using Equation (3.17). Next a method must be established to determine the empirical parameter fc of that equation.

Two alternative methods were considered to calculate the value of fc for any given compound. The first method (model A) was based on inspection of the calculated ring current intensities.[9,10,15] Haigh and Mallion[9,10] did not publish the calculated ring current intensities for the common aromatic compounds, but a selection of their calculated values for some less common condensed aromatic compounds is given in Table 5.1.

Inspection of these data shows that the changes in the ring current intensity are a function of the number and orientation of the rings attached to the benzenoid ring. In model A the ring current intensity in any given benzenoid ring is assumed to be only a function of the number and orientation of the rings attached to the ring considered. This may be quantified by the number and orientation of the substituent sp^2 carbon atoms attached to the ring in question (R_o). The authors defined (a) the number of attached sp^2 carbons on each ring carbon atom and (b) the relative position of these attached atoms in the benzene ring. Thus for benzene each carbon atom has two carbon neighbours, thus $R_o = 12$. For either ring in napthalene two of the carbon atoms have three carbon neighbours, thus $R_o = 14$. The middle rings of anthracene and phenanthrene both have $R_o = 16$ but the relative positions of the substituent carbons differ in the two cases. These are defined as R_o equals 16a and 16b. This analysis gives seven different ring systems (Table 5.1) of which six are present in the molecules indicated in Figure 5.1. Only the molecules with the rings itemized A,B in Figure 5.1 are included in Table 5.1 as these are the only molecules for which the ring

Table 5.1   *Calculated ring current intensities in condensed aromatic hydrocarbons*

Molecule	Ring type (R_o)[a]		Ring current intensity (*fc*)		
		b	*c*	Model A	Model B[d]
Benzene (**1**)	12	1.00	1.00	1.00	1.00
Naphthalene (**2**)	14	1.093	1.048[e], 1.094[f], 1.121[g]	0.950	0.925
Anthracene (**3**)	Ring A 14	1.085	1.119[h], 1.197[i], 1.104[j]		0.943
	Ring B 16a	1.280	1.291[e], 1.311[f], 1.299[g], 1.298[h], 1.170[j]	0.818	
Phenanthrene (**4**)	Ring A 14	1.133			0.943
	Ring B 16b	0.975	0.877[g], 0.876[h]	0.745	
Triphenylene (**5**)	Ring A 14	1.111			0.876
	Ring B 18	0.747			
Pyrene (**6**)	Ring A 15	1.329	1.337[k], 1.292[l]	0.786	0.878
	Ring B 16b	0.964			
Perylene (**7**)	Ring A 15	0.979			0.681
	Ring B 18	0.247	0.603[f], 0.606[m]	0.173	
Coronene (**8**)	Ring A 16b	1.460		1.06[a]	1.008
	Ring B 18	1.038	0.745[n], 0.684[l]		
	17	—	1.297[k], 1.226[m], 1.310[j]		

[a] See text.
[b] See Jonathan *et al.*[15].
[c] See Haigh and Mallion[9,10].
[d] See Abraham *et al.*[18].
[e] Hexacene.
[f] 1,2,3,4-Dibenzotetracene.
[g] 1,2,7,8-Dibenzotetracene.
[h] 1,2,9,10-Dibenzotetracene.
[i] 3,4,8,9-Dibenzopyrene.
[j] 6,7-Benzopentaphene.
[k] Anthanthrene.
[l] 1,12-Benzoperylene.
[m] 1,2,4,5-Dibenzopyrene.
[n] 1,2,6,7-Dibenzopyrene.

current intensities were given in the literature.[9,10] However, all the molecules measured were included in the iteration (see later).

Inspection of Table 5.1 shows that with few exceptions the separation of the ring current densities into the different ring types gives a reasonably constant value for each ring type. The only serious exception is the calculated values for ring type 18[15] (i.e. all substituted carbons) which are very different for perylene and coronene. The values reported[9,10] for the similar molecules 1,12-benzoperylene and 1,2,4,5-dibenzopyrene are much more consistent. It would be possible to average the calculated values reported[9,10] for each ring type and use these averages in the calculation. In view of the approximations inherent in these calculations it was decided to parameterize the current density for each ring type separately to obtain the best agreement with the observed shifts. These optimized values are given in Table 5.1 (column 5) and will be considered later. An alternative method of calculating the molecular ring current (model B) is to use the Pauling model[2] in which the carbon skeleton is considered as a conducting electrical network in which for any current loop the e.m.f. is

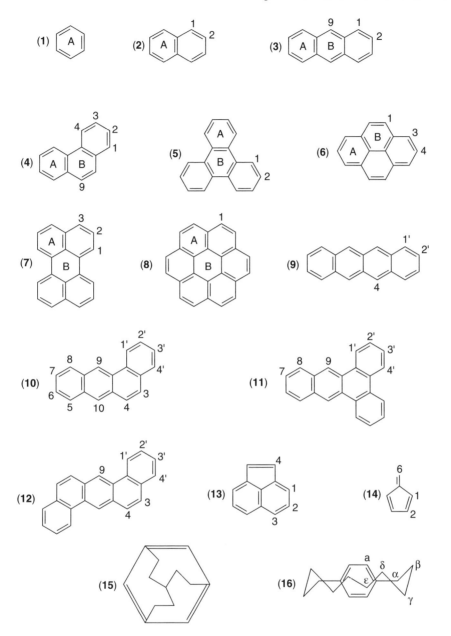

Figure 5.1 *Molecules studied and their nomenclature.*

proportional to the area enclosed and the resistance proportional to the number of bonds. On this basis if the condensed aromatics are considered to consist of a number of regular hexagons the ring current for any molecule is simply proportional to the number of hexagons in the molecule divided by the number of bonds in the perimeter of the molecule. Thus for benzene, naphthalene and anthracene the ring current ratio is 1:6/5:9/7. The Pauling model gives too large a value for the diamagnetic anisotropy of condensed aromatics[7] so that as in method A the Pauling model was used to separate the various molecular types and the ring current for each molecular type was parameterized against the experimental data. Although the same experimental data are used in both models the two procedures give different answers. For example, in model B anthracene and phenanthrene have identical ring currents which is not the case in model A. Conversely in model A the fully substituted benzenoid rings in perylene (**7**) and coronene (**8**) have identical ring currents whereas in model B they differ as the molecular area/perimeter ratio differs for the two compounds.

π-Electron densities. The π-electron densities are calculated in the CHARGE program from Huckel theory.[19,20] The standard coulomb and resonance integrals for the Huckel routine are given by Equation (3.7). The Huckel routine was modified by the ω technique to model the very polar π systems of the nucleic acid bases.[21,22] The ω technique involves varying the coulomb integral for each atom according to the charge on that atom. This is shown in Equation (5.1) where α_r is the coulomb integral, α_r^0 the initial coulomb integral, q_r the excess π charge on atom r and ω a constant. Equation (5.1) 'cuts in' at a given value of the excess π charge on atom r. For the nucleic acid bases the appropriate value of ω was 1.40 and and the cut-in threshold 0.2 electrons.[21,22]

$$\alpha_r = \alpha_r^0 - q_r\omega \qquad (5.1)$$

For alternant aromatic hydrocarbons this calculation gives π-electron densities at every carbon equal to 1.0 as in benzene. Thus the excess π-electron density is zero. This is in reasonable agreement with the results of more sophisticated calculations. For example, the excess π densities at the α and β carbons of naphthalene is calculated as −0.8 and −4.1 me (millielectrons) from *ab initio* calculations with the 6-31G* basis set.

For the non-alternant hydrocarbons fulvene and acenaphthylene the Huckel routine gives large excess π densities at certain carbon atoms, much larger than those calculated by *ab initio* methods in which iteration procedures restrict the tendency in the Huckel routine to separate the π charges. The ω technique was modified to correct this by decreasing the 'cut in' point of Equation (5.1) from 0.2 electrons to 0.01 electrons and increasing the value of ω to 6.0. This simple modification gave reasonable results for these two compounds, though the dipole moments are still on the high side. For example, fulvene 0.92D (calc.) vs. 0.44D (obs.)[23] and acenaphthylene 0.93D (calc.) vs. 0.3D (obs.).[23] As these hydrocarbons have quite different π densities and geometries from the alternant hydrocarbons both the ring current of the five-membered ring and the ring current density of the attached six-membered ring were parameterized separately.

Having obtained the π-electron density in the benzene ring it is then necessary to determine the effect of the π-electron density at a given carbon atom on the ¹H chemical shifts. This is given in CHARGE by Equation (3.18).

These modifications were the only ones needed to apply the CHARGE routine to aromatic compounds. It is still necessary to calculate the charge densities at the aromatic protons in CHARGE and thus to quantify the appropriate α, β and γ effects. Also the long-range interactions present in the aliphatic molecules (i.e. steric, electric and anisotropic) must also be included and where necessary evaluated. These will be considered subsequently.

The steric effects of both the aromatic carbon and hydrogen atoms are not known and must be determined. It will be shown (see later for a full discussion) that an aromatic carbon atom has no steric effect on a close aromatic hydrogen but two aromatic hydrogens in close proximity exert a *de*shielding effect on each other. We assume that this can be represented by the simple r^{-6} term (Equation (3.11)) thus only the appropriate value of a_s in Equation (3.11) for the aromatic H—H steric shift needs to be obtained. The electric field and anisotropies of the polar and anisotropic groups involved are calculated in an identical manner to that for any aliphatic C—H bond and thus no further parameterization is necessary.

5.1.3   Observed versus Calculated Shifts

Condensed aromatics. All assignments are given in Table 5.2. The molecular geometries were calculated using *ab initio* methods at the RHF/6-31G* level.[34] For molecules too large to be handled conveniently by this basis set, for example, perylene, a smaller basis set was used (3-21G). For the largest molecules, for example, coronene and the two cyclophanes (**15**) and (**16**) molecular mechanics calculations were used.

The ^{1}H spectra of the compounds all consisted of well-separated peaks at 400 MHz (except for toluene), thus the assignments of the compounds were straightforward. For toluene the ^{1}H spectrum of toluene-d_8 was first obtained. The dilute ^{1}H spins only couple to the ^{2}D nuclei and the spectrum consists of three broad singlets at 7.165, 7.170 and 7.254δ. The assignment of the ^{1}H spectrum follows and hence the more accurate ^{1}H chemical shifts in Table 5.2. The data obtained in CDCl$_3$ solution are given and compared with previous investigations in CCl$_4$ solution in Table 5.2. Jonathan *et al.*[15] only reported the shift differences from benzene and 7.27 ppm (the benzene value in CCl$_4$) has been added to them. There is generally good agreement between the data sets in Table 5.2 but it is noteworthy that there is a small but almost constant difference in the ^{1}H chemical shifts in CDCl$_3$ solution compared to CCl$_4$. Averaging over all the aromatic hydrogens in Table 5.2 gives a value of 0.086 ppm (±0.01) to high frequency in CDCl$_3$ solution. This is also the case for ethylene but here the difference is slightly less. The aliphatic protons of the methyl groups in toluene and t-butylbenzene do not show this effect but have the same shifts in the two solvents. The constant value of this difference means that data in CCl$_4$ solution can be converted directly to CDCl$_3$ solution by merely relating the shifts to benzene. Furthermore the accurate SCS values reported earlier for the monosubstituted benzenes in CCl$_4$ solution may be used with confidence to investigate the application of the CHARGE model to these compounds and these data are reproduced in Table 5.4. Also given in Table 5.4 are the SCS values obtained in this investigation for selected compounds in dilute CDCl$_3$ solution. The excellent agreement between the sets of SCS values confirms this assumption.

The data collected in Tables 5.2 and 5.3 provide a rigorous test of the application of both the CHARGE model and also of present ring current theories to these compounds. The compounds listed in these tables are all of fixed conformation. The GAUSSIAN94 (6-31G*/3-21G) and the PCMODEL calculations gave molecular geometries for the aromatic hydrocarbons in excellent agreement with the experimental geometries, where known. For example, benzene C.C 1.397, C.H 1.087 Å (MP2/6-31G*), vs. 1.395 and 1.087 Å (PCMODEL) and 1.396 and 1.083 Å (experimental).[30]

In the CHARGE model the α, β and γ effects of the substituents are considered to be due to electronic effects and therefore they are modelled on a simple empirical basis. The α effect of an sp^2 carbon is given from the difference in the electronegativities of the carbon and hydrogen atoms divided by the appropriate exchange integral. The value of

Table 5.2   *Observed and calculated ¹H chemical shifts (δ) for aromatic compounds*

Compound	Proton	Observed			Calculated	
		$CDCl_3^a$	$CCl_4^{b,c}$		Model A	Model B
Ethylene		5.405	5.352^d	—	5.407	
Benzene (**1**)		7.341	7.27^b	(7.27)^c	7.331	7.342
Naphthalene (**2**)	1	7.844	7.73	7.81	7.931	7.829
	2	7.477	7.38	7.46	7.524	7.493
Anthracene (**3**)	1	8.009	7.93	8.01	7.948	7.946
	2	7.467	7.39	7.39	7.524	7.533
	9,10	8.431	8.36	8.31	8.495	8.407
Phenanthrene (**4**)	1	7.901	7.80	un	7.930	7.968
	2	7.606	7.51	un	7.509	7.544
	3	7.666	7.57	un	7.566	7.600
	4,5	8.702	8.62	8.51	8.455	8.433
	9,10	7.751	7.65	7.71	7.839	8.085
Triphenylene (**5**)	1	8.669	8.61	8.56	8.587	8.707
	2	7.669	7.58	7.61	7.613	7.654
Pyrene (**6**)	1	8.084	8.00	8.06	7.976	8.253
	3	8.190	8.10	8.16	7.930	8.156
	4	8.010	7.93	7.99	7.546	7.785
Perylene (**7**)	1	8.196	8.11	8.09	8.361	8.250
	2	7.466	7.38	7.41	7.515	7.404
	3	7.656	7.57	7.60	7.845	7.630
Coronene (**8**)	1	8.90^e	8.82	8.84		8.900
1,2-Benzanthracene (**9**)	1′	8.840	8.77		8.698	8.553
	2′	7.685	7.59		7.708	7.627
	3′	7.651	7.525		7.638	7.557
	4′	7.849	7.755		8.102	8.004
	3	7.616	7.55		7.987	8.117
	4	7.800	7.72		8.027	8.200
	5	8.048	8.03		8.101	7.977
	6	7.540	7.465		7.637	7.544
	7	7.564	7.47		7.647	7.553
	8	8.133	8.03		8.169	8.038
	9	9.174	9.08		9.125	9.052
	10	8.370	8.275		8.561	8.572
2,3-Benzanthracene (**10**)	1′	8.00^f			8.082	7.947
	2′	7.39			7.619	7.522
	4	8.67			8.581	8.546
1,2,3,4-Dibenz anthracene (**11**)	1′	8.791	8.675		8.685	8.758
	2′	7.670	7.54		7.649	7.634
	3′	7.651	7.53		7.636	7.618
	4′	8.592	8.475		8.637	8.674
	7	7.568	7.455		7.641	7.521
	8	8.097	7.965		8.134	8.008
	9	9.097	9.075		9.103	9.238
1,2,5,6-dibenz anthracene (**12**)	1′	8.874	8.805		8.708	8.502
	2′	7.719	7.625		7.721	7.583
	3′	7.646	7.55		7.649	7.511
	4′	7.914	7.82		8.113	7.944
	3	7.760	7.67		8.016	8.077
	4	7.963	7.88		8.121	8.230
	10	9.155	9.075		9.170	9.107

Acenaphthylene	1	7.812		7.829	7.826
(**13**)	2	7.548		7.474	7.519
	3	7.692		7.708	7.701
	5,6	7.083		7.070	7.024
Fulvene (**14**)	1,4	6.228*ᵍ*		6.384	6.317
	2,3	6.531		6.421	6.404
	6	5.892		6.015	5.960
Toluene	*ortho*	7.180	7.061*ʰ*	7.080	
	meta	7.260	7.140	7.284	
	para	7.165	7.042	7.172	
	Me	2.343	2.337*ⁱ*	2.343	
t-Butylbenzene	*ortho*	7.390	7.281*ʰ*	7.279	
	meta	7.297	7.180	7.358	
	para	7.165	7.052	7.218	
	Me	1.325	1.319*ⁱ*	1.332	

[a] See Abraham et al.[18].
[b] See Haigh and Mallion[9,10].
[c] See Jonathan et al.[15].
[d] See Matter et al.[24].
[e] See Puchert and Behnke[25].
[f] SeeNetka et al.[26].
[g] See Nuchter et al.[27].
[h] See Hayamizu and Yamamoto[28,29].
[i] See Emsley et al., Vol. 2, Appendix B[1].

this integral was chosen to reproduce the observed chemical shift of ethylene (Table 5.2). This gives a partial atomic charge for the ethylene protons of +0.075e which corresponds to a C—H bond dipole of 0.4D in reasonable agreement with the usual quoted range (ca. 0.6–0.7D).[31,32] The β effect is calculated directly from the carbon electronegativity and proton polarizability,[33] thus the only other electronic effect to be considered is the γ effect (H.C.C.C) of the unsaturated carbon atoms in the aromatic compounds. For the condensed aromatic compounds considered here the only values of the CCCH dihedral angle θ are 0° and 180° (Table 5.2) and thus Equation (3.5) may be simplified to $A + B\cos\theta$ with the coefficients A and B to be obtained from the observed data.

Long-range effects. The interactions considered to be responsible for the long-range effects of the aromatic ring have been documented earlier as steric plus magnetic anisotropy (i.e. ring current) effects. There is also a small electric field effect due to the C—H dipoles which is calculated by CHARGE directly from the partial atomic charges. The theoretical treatment given earlier can now be tested against the observed data in the tables.

In substituted alkanes the steric effect of all non-hydrogen atoms was deshielding on the near protons, but H—H interactions gave a shielding effect. This was confirmed both experimentally and theoretically. In contrast it is immediately obvious from both the results of previous investigations[9,10,16] and the data presented here that H—H steric interactions in the aromatic systems considered here give rise to *de*shielding effects on the ¹H chemical shifts. A further unambiguous demonstration that steric effects on ¹H chemical shifts in aromatic systems are totally different from those in saturated systems came from the observation of the chemical shift of the unique CH hydrogen in the cyclophane (**15**). This hydrogen occupies a position along the symmetry axis of the benzene ring and occurs at −4.03δ. Because of it's proximity to the benzene ring plane (it is ca. 1.9 Å above the ring plane) it is an excellent test of any ring current theory and was used by

Schneider *et al.* in their investigation of the different ring current models.[11] It is also in close proximity to the benzene ring carbon atoms, the average C..H distance being ca. 2.20 Å. Any deshielding effect from the aromatic carbon atoms comparable to that found for saturated carbon atoms would have a pronounced deshielding effect on this proton. For example, using the steric coefficient found previously for saturated carbon atoms (a_s in Equation (3.11) = 220.0 ppm Å^6) would give a value for the CH chemical shift of +6.0δ! Clearly there is no significant deshielding steric effect from the aromatic carbon atoms at this hydrogen.

Schneider *et al.*[11] termed this a 'soft' steric effect in contrast to the 'hard' steric effect of H—H interactions. This is supported by the results for 10-paracyclophane (Table 5.3) in which there is good agreement between the observed and calculated shifts again with no sp^2 carbon steric effect. This result was adopted in the CHARGE routine so that there is no steric effect on the ^{1}H chemical shifts from any aromatic carbon atom. Note that this is not the case for olefinic carbon atoms (Chapter 4, Section 4.3).

Thus the parameters to be determined from the observed results in Table 5.2 were the coefficients A and B for the carbon γ effect, the appropriate H.H steric coefficient (Equation (3.11)), the ring current equivalent dipole μ and the factors *fc* (Equation (3.17)) for the condensed rings. There are six factors for both model A and model B (Table 5.1) making a total of 10 unknown parameters. The values of the unknown parameters were achieved using a non-linear least mean squares program (CHAP8)[34] to give the best fit with the observed data. The data set used comprises all the condensed aromatics of Table 5.2, a total of 57 proton shifts and thus the iteration is over-determined. The initial iteration for model A clearly showed that coronene was an exception and this was removed from the subsequent iteration. With this amendment the program iterated satisfactorily with reason-able rms error and definition. For model B coronene is a separate case and the iteration performed satisfactorily. The iteration gave A = −0.107, B = 0.143, the H..H steric coeffi-cient a_s = +24.55 ppm Å^6, μ = 26.2 ppm Å^3 and the *fc* values in Table 5.1. In fulvene and acenapthalene the ring current of the five-membered ring (μ$_P$) and also the factors (*fc*) for the benzenoid rings in acenapthalene were parameterized separately. This gave μ$_P$ = 11.6 ppm Å^3 and *fc* = 0.81. These iterations are for two unknowns and seven observed shifts, thus the iterations are still over determined.

Cyclophanes. It was also of interest to determine whether the equivalent dipole ring cur-rent calculation given here could be used to determine the benzene ring current effect for protons at the side and over the benzene ring. These data were used by Schneider *et al.*[11] to determine the accuracy of the various ring current models. The authors considered two illustrative examples: the unique CH proton in the tribridged cyclophane (**15**)[11] and the protons in [10]-paracyclophane (**16**).[35] The geometries of both compounds were modelled by PCMODEL and GAUSSIAN. (**15**) is a rigid strained molecule but in (**16**) the methylene chain exists in two equivalent rapidly interconverting staggered conformations. Thus the two hydrogens on each methylene group in the alkyl chain have the same observed shift and the calculated shifts have to be averaged. The calculations used Equation (3.17) to determine the ring current shifts with the value of the equivalent dipole obtained above. The CH proton of (**15**) is observed at −4.03δ (calc. −4.03δ) and the data for (**16**) is given in Table 5.3.

The general agreement of the observed vs. calculated shifts in Tables 5.2 and 5.3 is very good. Although the calculated values for models A and models B for the individual protons vary appreciably (Table 5.2), the overall agreement for both models is similar. For the 57 data points of Table 5.2 the rms error (obs. vs. calc. shifts) is 0.13 ppm (model A) and 0.12 ppm (model B) over a range of 3.3 ppm.

Table 5.3   *Observed versus calculated ¹H chemical shifts (δ) in [10]-paracyclophane (16)*

Carbon atom (CH₂)	Observed (CH₂)	Calculated (average)	
α	2.62	2.453	2.606
		2.759	
β	1.54	1.806	1.699
		1.592	
γ	1.08	1.631	1.270
		0.909	
δ	0.73	1.133	0.894
		0.655	
ε	0.51	0.626	0.525
		0.424	
Aromatic	7.04	7.102	7.088
		7.074	

The analogous calculation using only the benzene ring current (i.e. all fc values $= 1.0$) gives much poorer agreement (rms $= 0.28$ ppm) showing that it is necessary to take account of the variation in the ring current density for a proper description of the ¹H chemical shifts.

The calculation also provides new insight into the interpretation of these chemical shifts as the different interactions responsible for the calculated values are separately identified in the CHARGE program. The ring current calculations provide further evidence for the accuracy of the simple equivalent dipole model of the benzene ring current. The value of μ of 26.2 ppm Å³ is very similar to that obtained from the classical circulating current model (27.6).[36,37] The calculations also confirm previous studies[37] in demonstrating that the ring current effect is not the only factor responsible for the difference between the ethylene and benzene ¹H shifts. The experimental difference of 1.93 ppm (Table 5.2) is made up of 1.77 ppm from the ring current and 0.17 ppm from the electronic effects of the β and γ carbon atoms of benzene. This was allowed for in some previous ring current calculations by using cyclohexadiene rather than ethylene as the appropriate olefinic model[37] and the above calculations support this approach. It is also pertinent to note the excellent agreement obtained with the simple equivalent dipole model. On this basis the use of the more complex double dipole and double loop models does not appear to be justified. Interestingly Mallion[38] came to the same conclusion many years ago.

The values of the separate ring current factors (fc) in Table 5.1 may be compared with the values obtained previously.[9,10,15] The trends are similar supporting the original compartmentalization of these factors though the values obtained here are mostly much nearer to the benzene value ($fc = 1$) than the previous calculations. This is exactly to be expected as Huckel theory usually over estimates any electron separation. The only exception is the value for coronene. In model A the outer rings are of type 16β (i.e. analogous to the middle ring of phenanthrene) but this value of the ring current density (0.745, Table 5.1) gives a much too low value for the ¹H chemical shift. A value of fc of 1.06 reproduces the experimental chemical shift. In model B this problem does not arise as coronene is a separate case, and the iteration gives a value of $fc = 1.008$, very close to the benzene value and the Huckel calculated value.

It is encouraging that the calculated shifts for the non-alternant hydrocarbons of fulvene and acenaphthylene are in good agreement with the observed shifts (Table 5.2) as this suggests that the approach adopted here can be extended to these systems. The value of the ring current of the five-membered ring obtained here (11.6 ppm Å³) may be used to obtain the current density in the five-membered ring as the equivalent dipole $\mu = iA$ where A is the area of the current loop. After allowing for the area of the five-membered ring compared to benzene this gives a current density of $0.63i_b$, much less than benzene. More data on similar systems would be necessary to confirm this result.

Substituted benzenes. The influence of the substituents on the ¹H chemical shifts in the benzene ring has also been investigated for many years and again there is still no quantitative calculation of these effects. Following the classic work of Castellano *et al.*[39-42] and Hayamizu and Yamamoto[28,29] who completely analysed the complex ¹H spectra of a wide range of monosubstituted benzenes in dilute solution in CCl₄ the SCSs are known accurately and tables of these SCSs are an integral part of any text on NMR spectroscopy.[36,43,44] The theoretical interpretations of these effects have concentrated on the correlation between the SCSs and the calculated π (and also σ) electron densities on the adjacent carbon atoms. This follows the excellent correlation found between the ¹³C SCSs and the π-electron densities at the *para* carbon atom in monosubstituted benzenes. Correlations with π-electron densities calculated by various methods have been reported such as the *ab initio* calculations of Hehre *et al.*[45]. They used the STO-3G basis set and showed that the ¹³C SCSs could be well interpreted on the basis of calculated electron densities but this was not the case for the ¹H SCSs. The *para* hydrogen SCS could be correlated with the total charge density at the *para* carbon atom but the *meta* hydrogen SCS did not correlate well with the calculated *meta* carbon charge densities but with the sum of the charges at the hydrogen and attached carbon atoms. They noted that 'this lack of consistency indicates either that the calculations are unrealistic or that the ¹H SCS depend to a very significant extent on factors other than electron densities at the H and attached C atoms'. They omitted the *ortho*-H SCS presumably on the grounds that these other effects are even more important at these hydrogens. They also noted that strongly electronegative substituents caused polarization of the π system without charge transfer, leading to changes in the π densities around the ring and called this the π inductive effect. They also found various correlations between the calculated charge densities and the Taft σ_I and σ_R values. This reflects the results of other investigations who have attempted to correlate substituent parameters with the ¹H SCSs.[28,29,46-48] Despite all these endeavours there is still no calculation of ¹H SCSs in substituted benzenes reliable enough to be of use to the structural chemist.

For the calculation of the ¹H shifts in the substituted benzenes the appropriate values of the coefficients h_r and k_{rs} in Equation (3.7) for the orbitals involving hetero atoms have to be found. Values of the coefficients were determined so that the π densities calculated by the Huckel routine reproduced the π densities obtained by *ab initio* calculations. The only other modification necessary to the Huckel routine concerns the effect of saturated substituents (for example, CX$_3$) on the π-electron densities in the benzene ring. This is usually termed hyperconjugation. This effect was reproduced in the Huckel calculation by regarding it as an example of the π inductive effect mentioned earlier. In this case an equation corresponding to Equation (5.1) was used to vary the coulomb integral of the aromatic carbon atom connected to an sp^3 carbon. In this way changes to the π-electron density of the benzene ring due to both electron donating substituents such as CH$_3$ and electron withdrawing substituents such as CF$_3$ were handled by the same procedure. The values of the coefficients in Equation (5.1) needed to model the effect of the alkyl substituents on the π densities were $\alpha_r^0 = \alpha_r + 0.15$, $\omega = -0.50$ without any cut-off.

The electric field and anisotropic effects of the substituents have already been determined. There is also a steric effect of the side-chain protons on the *ortho* protons of the benzene ring. The steric effect of alkane protons on olefinic protons was determined (Chapter 4, Section 4.3)[49] to be deshielding and this result was used here. The steric effect of the OH and NH protons in alcohols and amines has been found to be zero[68] and again this result was incorporated into the present calculations.

The determination of the π density contributions (Equation (3.18)) for substituted benzenes allows the calculation of the ^{1}H SCSs of the monosubstituted benzenes and these results are given with the observed shifts in Table 5.4. There is generally excellent

Table 5.4 *Observed[a] versus calculated ^{1}H SCSs ($\Delta\delta_H$) of substituted benzenes*

Substituent	^{1}H Substituent chemical shifts ($\Delta\delta_H$)					
	ortho		*meta*		*para*	
	Obs.	Calc.	Obs.	Calc.	Obs.	Calc.
H	0.00	0.00	0.00	0.00	0.00	0.00
CH$_3$	−0.20	−0.27	−0.12	−0.06	−0.22	−0.17
	−0.16		−0.08		−0.18[b]	
t-Bu	0.02	−0.06	−0.08	0.02	−0.21	−0.12
	0.05		−0.04		−0.19[b]	
F	−0.29	−0.23	−0.02	0.02	−0.23	−0.21
Cl[b]	0.00	−0.03	−0.04	0.03	−0.10	−0.10
Br[b]	0.16	0.15	−0.10	0.03	−0.04	−0.06
I[b]	0.33	0.37	−0.27	0.00	−0.04	−0.06
OH	−0.56	−0.53	−0.12	−0.13	−0.45	−0.42
OCH$_3$	−0.48	−0.44	−0.09	−0.12	−0.44	−0.41
NH$_2$	−0.75	−0.62	−0.25	−0.24	−0.65	−0.65
CF$_3$	0.32	0.28	0.14	0.18	0.20	0.20
	0.29		0.14		0.21[c]	
CHO	0.56	0.54[d]	0.22	0.20[d]	0.29	0.26
CO.CH$_3$	0.62	0.61[d]	0.14	0.21[d]	0.21	0.28
CO.OCH$_3$	0.71	0.91[d]	0.11	0.21[d]	0.21	0.26
CN	0.36	0.30	0.18	0.23	0.28	0.27
	0.32		0.14		0.27[e]	
NO$_2$	0.95	0.81	0.26	0.23	0.38	0.25

[a] From Hayamizu et al.,[28,29] unless stated otherwise.
[b] From Abraham et al.[18].
[c] From Canton[49].
[d] Averaged, see Table 5.5 and text.
[e] From Abraham et al.[50].

agreement between the observed and calculated shifts. Although for convenience the SCSs are given in Table 5.4, as the ^{1}H chemical shift of benzene is calculated accurately (Table 5.2) the chemical shifts of all the substituted benzenes are calculated to the same accuracy as the SCS values in Table 5.4. It can be seen that the great majority of the observed shifts are reproduced to < 0.1 ppm. This was the first quantitative calculation of this data and showed that the CHARGE program can be applied to the prediction of the ^{1}H chemical shifts of substituted benzenes. Also this agreement together with the separation of the different interactions in CHARGE allows the SCSs in the benzene ring to be analysed further in terms of the constituent interactions and Table 5.5 gives the contributions to the ^{1}H SCSs for selected substituents.

In this table the contributions for the anisotropic substituents (for example, C=O) are given for each separate proton (for example, H_2 and H_6) although these are averaged in Table 5.4 to compare with the observed (averaged) data. The large effect of the carbonyl anisotropy is clearly apparent in these tables. The orientation of the carbonyl is such that the oxygen atom is syn to H_6 .The calculations are supported by and also show very clearly the origin of the large *ortho* proton deshielding in o-methoxy benzaldehyde (H_6, 7.82δ)[25] compared to o-hydroxybenzaldehyde (H_6,7.50δ) where the carbonyl group is now hydrogen bonded to the hydroxyl group.

Table 5.5 also shows that the carbonyl anisotropy is also the major factor in the *meta* proton SCS of benzaldehyde (vs.H_3 and H_5). This demonstrates the importance of these 'other' effects which are of course not included in any of the correlations of electron densities etc. with ^{1}H SCS. Indeed it is important to stress the difference between the present calculations and the correlations with Hammett σ[46], the Swain–Lupton F and R values,[51] etc. The CHARGE calculations are ground state calculations whilst the other parameters are derived from pH and rate constants and therefore reflect energy differences between the anion or the transition state and the ground state of the molecule, a totally different quantity. However in view of the numerous correlations of these quantities with the ^{1}H SCSs it is useful to consider these correlations with the present calculations. The correlation between the ^{1}H SCSs and Hammett σ_I and σ_R^0 values for a similar set of substituents to those in Table 5.5 was given as:[45]

$$\text{SCS } (para) = 0.27\sigma_I + 1.25\sigma_R{}^0$$

$$\text{SCS } (meta) = 0.24\sigma_I + 0.446\sigma_R{}^0$$

and a similar analysis of the SCSs in terms of the Swain–Lupton$F$ and R values gives:

$$\text{SCS } (para) = 0.142F + 0.926R$$

$$\text{SCS } (meta) = 0.098F + 0.376R$$

These equations are reasonably consistent implying in general a much greater resonance effect on the *para* H SCSs than on the *meta* H SCSs.

Inspection of the data in Table 5.5 shows a much more diverse pattern. Indeed the major disadvantage of such correlations is that they obscure the large differences in the SCS components of the various groups which all need to be considered individually. For example, the OH group has no anisotropic or steric effect and both the *meta* and *para* SCSs are dominated by the π-electron shift. This is much greater in the *para* position but the meta SCS is still dominated by the π effect. In contrast in benzaldehyde the electric field

Table 5.5 *Calculated contributions to ¹H SCSs (Δδ$_H$) in substituted benzenes*

Substituent			Calculated contribution				
			γ Effect	Steric	Anisotropic	Electric field	π Shift
CH$_3$	ortho		−0.144	—	—	—	−0.064
	meta		—	—	—	—	−0.132
	para		—	—	—	—	−0.183
F	ortho		0.128	—	—	—	−0.360
	meta		—	—	—	0.115	−0.137
	para		—	—	—	0.088	−0.332
OH	ortho		−0.128	—	—	—	−0.494
	meta		—	0.011	—	—	−0.188
	para		—	0.005	—	—	−0.456
CHO	ortho	H-2	0.144	—	−0.125	0.360	0.195
		H-6	0.144	—	0.767	0.153	0.195
	meta	H-3	—	—	−0.043	0.062	0.073
		H-5	—	—	0.107	0.069	0.073
	para		—	—	0.010	0.049	0.181
CN	ortho		−0.230	—	—	0.372	0.151
	meta		—	—	—	0.127	0.056
	para		—	—	—	0.097	0.138
NO$_2$	ortho		0.096	—	—	0.606	0.105
	meta		—	—	—	0.143	0.043
	para		—	—	—	0.105	0.115

and anisotropy contributions equal the π shift for the *meta* proton and are a significant but minor contribution for the *para* proton. The nitro and cyano groups differ from both of these in that they appear to have no anisotropic effect but the electric field effect is predominant at the *meta* proton and equal to the π shift at the *para* proton. Further investigations have confirmed this result for the cyano group[50] and nitro group (Chapter 6). Clearly each substituent group must be considered separately in order to evaluate the separate steric, electric and anisotropic contributions at the various protons. Finally it is of interest to consider the discrepancies in the observed versus calculated data of Table 5.4. The most interesting deviation is that due to Br and I. The calculated values for the halobenzenes are taken from recent studies[52] which include a refined model of the C–X (X = Cl, Br, I) anisotropy (Chapter 6, Section 6.3). The calculated values for the *ortho* and *para* SCSs for all the halogens are in excellent agreement with the observed data. However the *meta* SCS for Br and I are much more deshielding than calculated. Similar exceptional behaviour was observed for H-3eq in eq-halocyclohexanes.[52,53] Again there is a large deviation from the calculated value for the Br and I substituents. The equatorial proton is in a similar W orientation to the halogen atom as the *meta* proton in the substituted benzenes and it may be that there is an additional long-range (four bond) mechanism for the halogen atoms in this specific orientation.

In this chapter we have been concerned so far with monosubstituted benzenes. It is obvious that determining the proportions of the interactions responsible for the SCSs is much easier to do in monosubstituted benzenes than in di- or tri-substituted benzenes.

However the question remains as to whether the CHARGE program can equally well calculate the ^{1}H chemical shifts in multi-substituted benzenes as other effects such as *ortho–ortho* repulsion, π–π interactions (for example, in *para*-nitroaniline) etc. may distort these calculations.

To check this, the ^{1}H spectra of 113 substituted benzenes in the *Aldrich Spectral Catalogue* were assigned and the chemical shifts obtained and compared with the shifts calculated by CHARGE. In these calculations the molecular geometries were obtained using the MMFF94 force field in PCMODEL. These results[69] are given in Appendix A2 and agreement in Figure A2.1. The overall rms error of the 407 protons in the database was 0.087 ppm, an impressive result when it is considered that only one force field and one program with no further parameterization was used.

5.2   Heteroaromatics

5.2.1   Introduction

Heteroaromatic compounds comprise an important group of compounds in organic chemistry. They are of considerable commercial importance and the extent to which the properties of these compounds are determined by their aromatic character has interested chemists for many years.[54]

The discovery of the aromatic ring current of benzene[4] allowed in principle the determination of the aromaticity of any molecule by measuring its ring current. In a pioneering study Abraham and Thomas[37,55] compared the chemical shifts of the H-2 and 2-methyl protons in furan, thiophene, thiazole, imidazole and benzene and their methyl derivatives with those of similarly constituted protons in the 4,5-dihydro compounds where there is no ring current. They proposed that the observed differences in the ^{1}H chemical shifts were a measure of the ring currents in these compounds and found that the ring currents in furan and thiophene 'did not differ significantly' from the benzene ring current. Elvidge[56] using polyenes as models obtained values of the ring currents in furan, pyrrole and thiophene of 46, 59 and 75% that of benzene.

De Jongh and Wynberg[57] used the same method as Pople[4] but averaged the shifts of H-2 and H-3 in the dihydro compounds. They obtained values of the ring currents in furan and thiophene between those of previously reported values.[37,55,56]

Since this early work no calculation of the ^{1}H shifts in these compounds has been given. In particular the calculation of these chemical shifts using the *ab initio* GIAO method has not been reported to date. Lampert *et al.*[58] compared the observed versus calculated NMR chemical shifts for phenol and benzaldehyde and for 13 substituted derivatives, using a variety of *ab initio* calculations (theories/basis sets). The calculated shielding of the aromatic protons with respect to methane varied by ca. 0.5–1.0 ppm. depending on the theory and basis set used and this may well represent the limit of accuracy of such calculations.

A calculation of the ^{1}H shifts in condensed aromatic hydrocarbons has been given above based on ring current and π-electron effects and the same procedure was applied to heteroaromatics. The ^{1}H chemical shifts of a number of heteroaromatic and related compounds in CDCl$_3$ solution were obtained and analysed by Abraham and Reid.[59] It was convenient for the purposes of parameterization to include some related compounds. For example, vinyl methyl ether and thio ether were useful additions to the oxygen and sulfur compounds and aniline for the nitrogen heterocycles, etc. The molecules considered here are shown with the atom numbering in Figures 5.2(a–f) and are as follows:

- **Figure 5.2(a)**: vinylmethylether (**1**), phenol (**2**), anisole (**3**), furan (**4**), 4,5-dihydrofuran (**5**), 2-methylfuran (**6**), 2-methyl-4,5-dihydrofuran (**7**), 2,5-dimethylfuran (**8**), 3-methylfuran (**9**) and benzofuran (**10**).
- **Figure 5.2(b)**: vinylmethylsulfide (**11**), thiophenol (**12**), thiophene (**13**), 4,5-dihydrothiophene (**14**), 2-methylthiophene (**15**), 2-methyl-4,5-dihydrothiophene (**16**), 2,5-dimethylthiophene (**17**), 3-methylthiophene (**18**) and thionaphthene (**19**).
- **Figure 5.2(c)**: pyrrole (**20**), *N*-methylpyrrole (**21**), 2-methylpyrrole (**22**), 2,5-dimethylpyrrole (**23**), 1,2,5-trimethylpyrrole (**24**), 3-methylpyrrole (**25**), indole (**26**) and *N*-methyl (**27**), 2-methyl (**28**), 3-methyl (**29**) and 7-methylindoles (**30**).
- **Figure 5.2(d)**: aniline (**31**), pyridine (**32**) and 2-, 3- and 4-picolines (**33, 34, 35**).
- **Figure 5.2(e)**: quinoline (**36**), 2-methylquinoline (**37**), 2-methyl-3,4-dihydroquinoline (**38**), 3-methyl (**39**), 4-methyl (**40**) and 6-methylquinolines (**41**), isoquinoline (**42**), 1-methylisoquinoline (**43**), 1-methyl-3,4-dihydroisoquinoline (**44**), 3-methylisoquinoline (**45**), pyrimidine (**46**), pyrazine (**47**) and pyridazine (**48**).
- **Figure 5.2(f)**: imidazole (**49**), 2-methylimidazole (**50**), 2-methyl-4,5-dihydroimidazole (**51**), thiazole (**52**), 2-methylthiazole (**53**), 2-methyl-4,5-dihydrothiazole (**54**) and oxazole (**55**).

This large set of rigid molecules with fully assigned ¹H NMR spectra provides sufficient data for an analysis of the ¹H chemical shifts in heteroaromatics based on the CHARGE model.

5.2.2   Theory and Application to Heteroaromatics

As before it is necessary to identify and separate the various mechanisms responsible for the ¹H chemical shifts in these molecules. These are the ring current shifts, the π-electron densities, the direct α, β and γ effects of the hetero atoms and the long-range steric, electrostatic and anisotropic effects at the protons. The theory follows the treatment given in Chapter 3 with the following specific amendments.

The major contributions to the ¹H chemical shifts in heteroaromatic compounds are ring current and π-electron effects, with smaller contributions due to the α, β and γ effects of the hetero atom and the long-range contributions. Subroutines were added to the CHARGE program in order to identify the heteroaromatic systems. It was then necessary to determine the π-electron densities at each atom and the ring currents in the compounds investigated.

Ring currents. To determine the ring current density *fc* for the different heteroaromatic ring systems under investigation two methods were used. For those systems in which a 2-methyl substituent was present in both the aromatic and dihydroaromatic compound the method of Abraham and Thomas[37,55] was used to determine the ring current using the C-2 methyl shifts. If the appropriate dihydro compound was not available the ring current density *fc* was obtained by including this factor in the parameterization, using all the ¹H chemical shifts in the ring systems.

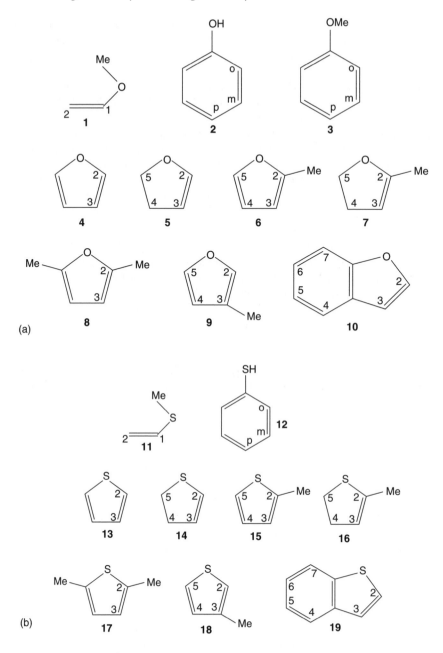

Figure 5.2   (a) Oxygen heteroaromatics and related molecules. (b) Sulfur heteroaromatics and related molecules. (c) Pyrroles and indoles. (d) Monocyclic amines. (e) Bicyclic amines. (f) Difunctional bases.

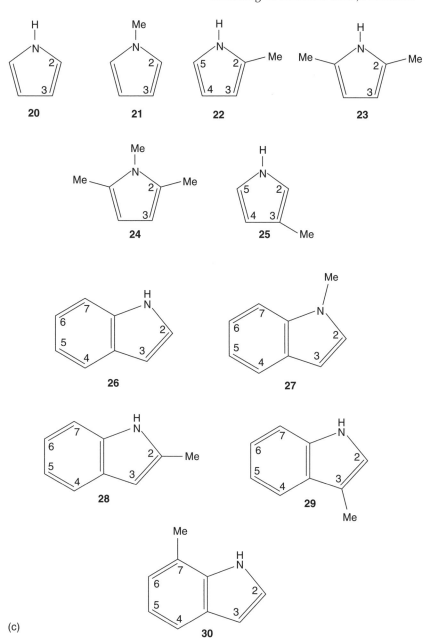

Figure 5.2 (*Continued*).

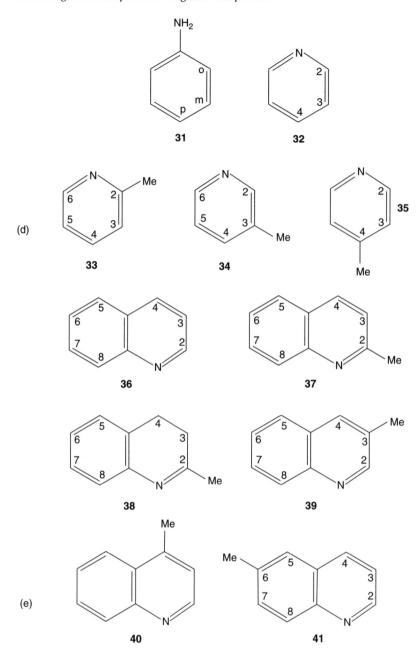

Figure 5.2 (*Continued*).

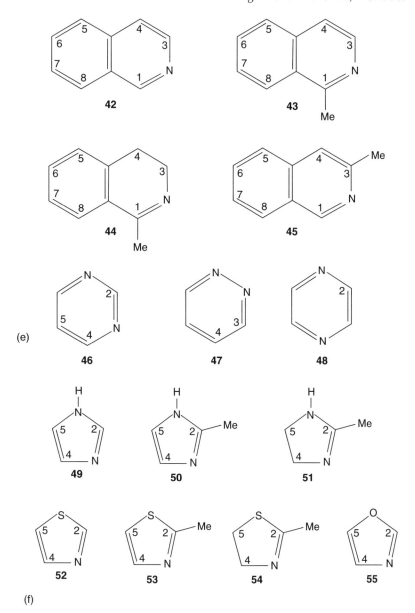

(e)

(f)

Figure 5.2 (*Continued*).

The equivalent dipole μ of a current loop of radius A and current i is given by Equation (5.2) and the ratio of the ring current in a heterocyclic ring to that in benzene by Equation (5.3):

$$\mu = iA \qquad (5.2)$$

$$i/i_B = \mu/\mu_B. A/A_B \qquad (5.3)$$

where μ_B and A_B are the benzene ring current and area, respectively. Using this equation the ratio of the ring currents in the heteroaromatic molecules to that in benzene will be determined (see later).

π-Electron densities. The π-electron densities were reproduced from those calculated from *ab initio* calculations. The results from *ab initio* calculations were as usual dependent on the basis set used.[8,12] The 3-21G basis set at the B3LYP level gave the best values of the dipole moments for the compounds investigated and therefore the π-electron densities from this basis set were used to parameterize the Huckel calculations.

The π systems in the heteroaromatic compounds investigated are quite diverse. For example, the π systems of vinylmethylether, furan and phenol are very different. It was therefore necessary to treat these π systems separately in CHARGE. Similarly the nitrogen atom in aniline is non-planar and therefore in a different hybridization to that of the planar nitrogen atom in pyrrole and pyridine. These were differentiated by determining the appropriate values of the atomic orbital coefficients h_r and k_{rs} (Equation (3.17)) the Huckel integrals for Csp$_2$–X, where X=O, S, N for the various π systems.

The accuracy of the π-electron densities calculated in CHARGE may be examined by comparing the calculated π-electron densities and dipole moments of some heteroaromatics with those obtained by *ab initio* theory using various basis sets (Table 5.6).

The good agreement of the calculated versus observed dipoles in Table 5.6 is strong support for the calculations. The values of k_{rs} and h_r used for the various Csp$_2$–X bonds in these molecules are given in Table 5.7. The π-electron densities for phenol, thiophenol and aniline were calculated previously.[19,20] These modifications were the only ones needed to apply the CHARGE routine to these heteroaromatic compounds. It was still necessary to calculate the charge densities at the various protons in the molecules and thus to quantify the appropriate α, β, and γ effects. Also the long-range effects must be included. These are the steric, electric field and anisotropic effects of the atoms in the molecules. These have all been calculated previously and no further parameterization is required.

The compound geometries were optimized using the GAUSSIAN program at the B3LYP/6-31G** level. The optimized geometries for the heteroaromatics were in excellent agreement with the experimental geometries. For example, the observed versus calculated bond lengths for furan, thiophene, pyrrole and pyridine are given in Table 5.8 and there is complete agreement of the two data sets.

^{1}H NMR spectra at 400 MHz were obtained and assigned for phenol (**2**), anisole (**3**), benzofuran (**10**), thionaphthene (**19**), indole (**26**) and *N*-methyl (**27**), 2-methyl (**28**), 3-methyl (**29**) and 7-methylindoles (**30**), aniline (**31**), pyridine (**32**), 2-picoline (**33**), 3-picoline (**34**), quinoline (**36**), 2-methyl (**37**), 3-methyl (**39**), 4-methyl (**40**) and 6-methylquinolines (**41**), and isoquinoline (**42**) by Abraham *et al.*[59] The ^{1}H chemical shifts for compounds (**8**), (**12**), (**17**), (**18**), (**20**), (**21–24**), (**35**), (**42**) and (**45–48**) were obtained from the *Aldrich Spectra Catalogue*,[25] those for the furans (**4–7**), thiophenes (**13–16**), imidazoles (**50, 51**) and thiazoles (**52–54**) from Abraham and Thomas[37,55] and those for (**1**), (**9**), (**11**), (**22**), (**38**),

Table 5.6 π Charges (milli-electrons) and dipole moments μ (D) for methylvinylether, furan, thiophene, pyrrole, pyridine and indole[a]

Compound	Atom	Method				
		STO-3G	3-21G	6-31G	CHARGE	Observed[c]
Vinylmethyl	C_1	−58	−18	−11	−9	
ether	C_2	−132	−156	−137	−70	
	O	216	236	193	74	
	μ	1.46	1.09	1.319	0.96	1.11
Furan	C_2	−89[b]	−107	−94	−48	
	C_3	−71	−75	−68	−33	
	O	320	364	323	162	
	μ	0.40	0.71	0.97	0.88	0.72
Thiophene	C_2	−113	−130	−133	−61	
	C_3	−58	−35	−32	−18	
	S	342	330	331	157	
	μ	0.57	0.72	0.82	0.70	0.53
Pyrrole	C_2	−100[b]	−125	−92	−75	
	C_3	−91	−93	−87	−57	
	N	383	436	394	264	
	μ	1.90	2.03	1.93	1.59	1.74
Pyridine	C_2	11[b]	22	36	47	
	C_3	−2	−3	−2	3	
	C_4	33	39	41	30	
	N	−51	−78	−110	−119	
	μ	2.07	2.25	2.49	2.02	2.15
Indole	C_2	−83	−76	−66	−48	
	C_3	−97	−106	−102	−70	
	C_4	−13	−11	−13	−9	
	C_5	−29	−37	−36	−21	
	C_6	−18	−22	−21	−13	
	C_7	−52	−57	−55	−21	
	N	392	394	347	234	
	μ	2.15	2.26	2.16	1.78	2.09

[a] μ, phenol 1.56 calc.(1.50 obs.); quinoline 2.20 calc. (1.94 obs.).
[b] From Abraham and Smith[20,37].
[c] From McLellan[23].

(49) and (55) from references [60–66]. Pretsch *et al.*[44] collected many of these chemical shifts in either CCl_4 or $CDCl_3$ solvent (see later). Full details of all the assignments plus spectra are given by Reid.[67] The ^{1}H chemical shifts of all the compounds investigated are given in Tables 5.9–5.17 with the calculated chemical shifts from the CHARGE model.

5.2.3   Observed versus Calculated Shifts

The chemical shifts obtained compare well with those of previous investigations. There is however as in the benzenoid aromatics an almost constant difference of ca. 0.1 ppm in the shifts given in $CDCl_3$ with those measured previously in CCl_4 solution. For example, comparison of the data for quinoline (**36**) with that of Pretsch *et al.*[44] gives for H_2–$H_8\delta(CDCl_3)$–$\delta(CCl_4)$ 0.12, 0.12, 0.14, 0.12, 0.10, 0.10 and 0.07, respectively, averaging

Table 5.7 k_{rs} *(Csp$_2$–X) and* h_r *(X) (X = O, S, N) in heteroaromatic and related compounds*

Compound	k_{rs}	h_r
Phenol	1.45	0.90
Vinylmethylether	1.05	0.59
Furan	1.69	0.59
Benzofuran	1.22	0.59
Thiophenol	1.27	0.66
Vinylmethylsulfide	0.97	0.40
Thiophene	1.27	0.47
Benzothiophene	0.79	0.47
Pyrrole	1.60	1.28
Indole	1.50	1.28
Pyridine	0.30	1.00
Imidazole (C$_2$.N$_3$)	0.16	1.00
Imidazole (N$_1$.C$_2$)	1.60	1.28

Table 5.8 *Observed[19] and (calculated) bond lengths (Å) for heteroaromatics*

Bond length (Å)	Furan	Pyrrole	Thiophene	Pyridine[a]
X$_1$–C$_2$	1.362 (1.364)	1.370 (1.375)	1.714 (1.736)	1.338 (1.339)
C$_2$–C$_3$	1.361 (1.361)	1.382 (1.378)	1.370 (1.367)	1.394 (1.396)
C$_3$–C$_4$	1.431 (1.436)	1.417 (1.425)	1.423 (1.430)	1.392 (1.394)
C$_2$–H	1.075 (1.079)	1.076 (1.080)	1.078 (1.081)	1.087 (1.089)
C$_3$–H	1.077 (1.080)	1.077 (1.081)	1.081 (1.084)	1.088 (1.086)

[a] C$_4$–H, 1.082 (1.086).

to 0.11 ppm. Identical results hold for isoquinoline and indole. This constant shift to high-frequency appears to be a general effect for both non-polar and polar solutes.

The chemical shifts can now be used to test the application of the CHARGE model and also to investigate the shielding mechanisms in these molecules; in particular the effects of ring currents and π-electron densities on the ^{1}H chemical shifts. The only other unknowns in the CHARGE model are the α, β and γ electronic effects of the atoms. The α and β effects are calculated directly from the atom electronegativity and polarizability, but the γ effects are given by $A + B\cos\theta$, where the parameters A and B are obtained from the observed shifts. The values of all the unknown parameters were obtained by iteration. The orientation dependence averages to zero for a methyl group, thus the coefficient $B = 0.0$. Only γ effects on the methyl protons were determined for the alkyl protons. Also the coefficients A and B for the X.C=CH fragment (X = O,S) differ for olefinic, heteroaromatic and benzenoid systems. In the latter there is only one dihedral angle of 0^0, thus only one parameter can be obtained.

For the nitrogen atoms a different procedure is used. The nitrogen atoms in aniline, pyrroles/indoles and pyridines/quinolines are treated differently reflecting the different hybridization of the N atoms in these molecules. These are termed N$_1$, N$_2$ and N$_3$ henceforth.

In the pyridines the β effects of the N$_3$ atom on the *ortho* protons were given by the basic Equation (3.3). However for pyrazine the two bonded nitrogen atoms have an increased β effect (1.35) and in pyrimidine the β effect on H-2, which has two β N$_3$ atoms required a reduced value of the coefficient of 0.83.

Table 5.9   *Observed versus calculated ¹H chemical shifts (δ) for oxygen compounds*

Compound	¹H Number	Observed	Calculated
Vinylmethylether (**1**)	1-*gem*	6.530	6.606
	2-*cis*	4.160	4.224
	2-*trans*	4.000	4.058
Phenol (**2**)	o	6.824	6.877
	m	7.239	7.212
	p	6.927	6.926
Anisole (**3**)	o	6.897	6.859
	m	7.277	7.232
	p	6.934	6.926
	Me	3.789	3.738
Furan (**4**)	2	7.420	7.415
	3	6.380	6.360
4, 5-Dihydrofuran (**5**)	2	6.310	6.153
	3	4.950	4.939
	4	2.580	2.384
	5	4.310	4.224
2-Methylfuran (**6**)	3	5.940	6.058
	4	6.230	6.289
	5	7.270	7.189
	Me	2.280	2.278
2-Methyl-4,5-dihydrofuran (**7**)	3	4.570	4.496
	4	2.580	2.432
	5	4.310	4.273
	Me	1.790	1.867
2, 5-Dimethylfuran (**8**)	3	5.810	5.983
	Me	2.220	2.295
3-Methylfuran (**9**)	2	7.160	7.052
	4	6.220	6.327
	5	7.290	7.450
	Me	2.030	2.172
Benzofuran (**10**)	2	7.607	7.807
	3	6.758	6.671
	4	7.593	7.514
	5	7.225	7.239
	6	7.285	7.312
	7	7.502	7.400

Also in imidazole, thiazole and oxazole the β effects of the hetero atoms on H-2 need to be obtained. Both adjacent heteroatoms influence the chemical shift, hence three separate effects need to be determined.

For those systems in which the 2-methyl shifts could be determined for both the aromatic and dihydro compounds the ring currents were determined directly from the difference in these shifts, i.e. for furan compound **7** vs. **6**, thiophene **15** vs. **16**, quinoline **37** vs. **38**, isoquinoline **43** vs. **44**, imidazole **50** vs. **51** and thiazole **54** vs. **55**. For those systems in which the dihydro compounds were not available the ring current factor *fc* was included in the iteration procedure. These factors are given in Table 5.21 below.

Table 5.10 *Observed versus calculated ¹H chemical shifts (δ) for sulfur compounds*

Compound	¹H Number	Observed	Calculated
Vinylmethylsulfide (**11**)	*gem*	6.460	6.549
	cis	5.200	5.189
	trans	4.970	4.833
Thiophenol (**12**)	o	7.230	7.316
	m	7.190	7.276
	p	7.110	7.081
Thiophene (**13**)	2	7.310	7.263
	3	7.090	7.044
4,5-Dihydrothiophene (**14**)	2	6.170	6.076
	3	5.630	5.717
	4	2.740	2.592
	5	3.220	3.169
2-Methylthiophene (**15**)	3	6.720	6.733
	4	6.870	6.970
	5	7.040	7.017
	Me	2.480	2.470
2-Methyl-4,5-dihydrothiophene (**16**)	3	5.250	5.248
	4	2.790	2.657
	5	3.260	3.195
	Me	1.940	2.009
2,5-Dimethylthiophene (**17**)	3	6.560	6.655
	Me	2.400	2.481
3-Methylthiophene (**18**)	2	6.870	6.898
	4	6.870	7.020
	5	7.190	7.305.
	Me	2.280	2.214
Benzothiophene (**19**)	2	7.422	7.523
	3	7.325	7.347
	4	7.780	7.642
	5	7.330	7.302
	6	7.310	7.340
	7	7.860	7.996

It is important to note that these iterations were always very over-determined. For example in the furan case a total of 26 chemical shifts (Table 5.9) were included in the iteration spanning a range of ca.1.8 to 7.6 ppm with only four parameters (*A* and *B* values) to be determined. The iteration gave an rms error (observed vs. calculated shifts) of 0.073 ppm. For the pyrrole/indole case the ring current factor *fc* was included in the iteration and this gave a total of 49 chemical shifts (Tables 5.13 and 5.14) from 2.0 to 7.7 ppm with six unknown parameters to give an rms error of 0.107 ppm. Similar results were obtained for the iterations for the other systems. The final parameterization for all the systems considered therefore included π-electron densities, ring current and electronic effects operating on all protons in the molecules.

Furans, thiophenes and pyrroles. There is generally very good agreement of the observed vs. calculated chemical shifts. For the 215 data points in Tables 5.9–5.17 the rms error (obs. vs. calc. shifts) is 0.096 ppm over a range of 1.9 to 9.4 ppm. and there are very few

Table 5.11   *Observed versus calculated ¹H chemical shifts (δ) for compounds **20–25** and **31***

Compound	¹H Number	Observed	Calculated
Aniline (**31**)	o	6.650	6.654
	m	7.136	7.132
	p	6.740	6.676
Pyrrole (**20**)	2	6.710	6.708
	3	6.230	6.187
N-Methylpyrrole (**21**)	2	6.670	6.590
	3	6.110	6.155
	N-Me	3.600	3.513
2-Methylpyrrole (**22**)	3	5.890	5.919
	4	6.110	6.112
	5	6.640	6.507
	Me	2.270	2.285
2,5-Dimethylpyrrole (**23**)	3	5.720	5.839
	Me	2.200	2.300
1,2,5-Trimethylpyrrole (**24**)	3	5.750	5.813
	2,5-Me	2.190	2.246
	N-Me	3.330	3.586
3-Methylpyrrole (**25**)	2	6.530	6.400
	4	6.020	6.122
	5	6.650	6.722
	Me	2.090	2.153

calculated chemical shifts with errors > 0.2 ppm. The *N*-methyl in (**24**) is the only such error in Table 5.11 (0.25 ppm) and this may be due to steric effects.

Indoles, quinolines and isoquinolines. In the indoles (Table 5.12) the only significant error (ca. 0.2 ppm) is for H-4 which is calculated consistently less than the observed shift. The agreement for the quinolines (Table 5.14) and isoquinolines (Table 5.15) is particularly noteworthy with most of the calculated shifts accurate to < 0.1 ppm. There are larger differences in the calculated vs. observed shifts in Table 5.17, for example, H-4 in oxazole (**55**) and 2-methylthiazole (**53**). This latter value is intriguing as H-4 in thiazole (**52**) is calculated accurately.

5.2.4   Ring Currents and π-Electron Shifts

The calculations also provide an insight into the interpretation of these ¹H chemical shifts as the different interactions responsible for the calculated values are separately identified in the CHARGE model. It is of interest to examine the individual contributions to the chemical shifts and Tables 5.18–5.20 give the observed versus calculated chemical shifts for selected molecules, together with the electric field, ring current and π shift contributions. The results in these tables clearly demonstrate the significant ring current and π contributions to the ¹H chemical shifts in these molecules. The ring current shifts of the ring protons in furan are ca. 1.60 ppm and that of the methyl protons ca. 0.50 ppm. Similar effects are observed

Table 5.12 *Observed versus calculated ¹H chemical shifts (δ) for compounds 26–30*

Compound	¹H Number	Observed	Calculated
Indole (**26**)	2	7.207	7.321
	3	6.558	6.643
	4	7.647	7.489
	5	7.115	7.212
	6	7.185	7.263
	7	7.396	7.358
N-Methylindole (**27**)	2	7.001	7.204
	3	6.466	6.611
	4	7.615	7.488
	5	7.092	7.211
	6	7.204	7.258
	7	7.292	7.330
	N-Me	3.742	3.813
2-Methylindole (**28**)	3	6.216	6.325
	4	7.508	7.443
	5	7.059	7.186
	6	7.104	7.202
	7	7.282	7.347
	Me	2.445	2.469
3-Methylindole (**29**)	2	6.964	6.969
	4	7.584	7.496
	5	7.121	7.212
	6	7.189	7.264
	7	7.301	7.370
	Me	2.335	2.427
7-Methylindole (**30**)	2	7.207	7.326
	3	6.563	6.654
	4	7.498	7.276
	5	7.031	7.143
	6	6.994	7.029
	Me	2.502	2.620

Table 5.13 *Observed versus calculated ¹H chemical shifts (δ) for pyridine (32), 2-picoline (33), 3-picoline (34) and 4-picoline (35)*

¹H Number	32		33		34		35	
	Obs.	Calc.	Obs.	Calc.	Obs.	Calc.	Obs.	Calc.
2	8.609	8.577	2.547[a]	2.518	8.440	8.459	8.440	8.584
3	7.266	7.279	7.014	7.162	2.320[a]	2.319	7.080	7.162
4	7.657	7.574	7.571	7.574	7.465	7.454	2.320[a]	2.310
5	7.266	7.279	7.195	7.213	7.159	7.268	7.080	7.162
6	8.609	8.577	8.599	8.597	8.407	8.499	8.440	8.583

[a] Methyl.

Table 5.14   *Observed versus calculated ^{1}H chemical shifts (δ) for quinoline (**36**) and 2-methyl (**37**), 2-methyl-3,4-dihydro (**38**), 3-methyl (**39**) ,4-methyl (**40**) and 6-methylquinolines (**41**)*

^{1}H Number	36		37		38		39		40		41	
	Obs.	Calc.	Obs.	Calc.	Obs.	Calc.	Obs.	Calc.	Obs.	Calc.	Obs.	Calc.
2	8.915	8.865	2.757[a]	2.626	2.10[a]	2.155	8.760	8.790	8.770	8.865	8.825	8.836
3	7.377	7.429	7.295	7.351	2.35	2.59	2.482[a]	2.416	7.212	7.283	7.303	7.419
4	8.139	8.122	8.055	8.141	2.70	2.90	7.876	7.957	2.692[a]	2.531	8.005	8.109
5	7.803	7.841	7.778	7.844	b	7.28	7.714	7.827	7.985	7.822	7.522	7.681
6	7.533	7.509	7.485	7.482	b	7.35	7.489	7.500	7.552	7.499	2.501[a]	2.416
7	7.709	7.571	7.627	7.561	b	7.29	7.627	7.542	7.697	7.569	7.512	7.502
8	8.114	8.060	8.024	8.050	b	7.78	8.066	8.062	8.104	8.067	7.995	8.071

[a] Methyl.
[b] 6.70–7.50.

Table 5.15   *Observed versus calculated ^{1}H chemical shifts (δ) for isoquinoline (**42**) and 1-methyl (**43**), 1-methyl-3,4-dihydro (**44**) and 3-methylisoquinolines (**45**)*

^{1}H Number	42		43		44		45	
	Obs.	Calc.	Obs.	Calc.	Obs.	Calc.	Obs.	Calc.
1	9.251	9.177	2.910[a]	2.759	2.400[a]	2.334	9.150	9.207
3	8.522	8.539	8.370	8.556	3.670	3.831	2.690[a]	2.617
4	7.635	7.621	7.440	7.538	2.710	2.812	7.410	7.464
5	7.808	7.800	7.730	7.817	7.180	7.345	7.680	7.793
6	7.680	7.596	7.600	7.595	7.360	7.477	7.590	7.588
7	7.594	7.533	7.510	7.526	7.300	7.350	7.480	7.507
8	7.955	7.915	8.040	7.897	7.480	7.417	7.880	7.918

[a] Methyl.

Table 5.16   *Observed versus calculated ^{1}H chemical shifts (δ) for pyrimidine (**46**), pyrazine (**47**) and pyridazine (**48**)*

^{1}H Number	46		47		48	
	Obs.	Calc.	Obs.	Calc.	Obs.	Calc.
2	9.250	9.248	—	—	8.600	8.476
3	—	—	9.220	9.316	8.600	8.476
4	8.770	8.856	7.560	7.646	—	—
5	7.270	7.245	7.560	7.646	8.600	8.476
6	8.770	8.856	9.220	9.316	8.600	8.476

for the methyl group in 2-methylquinoline (0.51 ppm), 1-methylisoquinoline (0.65 ppm), 2-methylthiazole (0.54 ppm) and 2-methylimidazole (0.49 ppm).

The introduction of a methyl group has a large effect on the π-electron density in the heterocyclic rings and thus on the chemical shifts. All the protons in the 2-methyl and 3-methyl derivatives of furan and pyrrole are shielded with respect to the parent compound, especially protons that are γ to the methyl group. This is clearly due to the increased

Table 5.17 Observed versus calculated ^{1}H chemical shifts (δ) for imidazole (**49**), 2-methyl (**50**) and 2-methyl (**50**) and 2-methyl-3,4-dihydroimidazoles (**51**), thiazole (**52**), 2-methyl (**53**) and 2-methyl-3,4-dihydrothiazoles (**54**) and oxazole (**55**)

^{1}H Number	49		50		51		52		53		54		55	
	Obs.	Calc.	Obs.	Calc.	Obs.	Calc.	Obs.	Calc.	Obs.	Calc.	Obs.	Calc.	Obs.	Calc.
2	7.74	7.78	2.44[a]	2.40	1.95[a]	1.95	8.88	8.84	2.74[a]	2.81	2.20[a]	2.27	7.90	7.82
4	7.13	6.99	6.97	6.90	3.60	3.42	7.98	8.08	7.64	8.10	4.22	3.99	7.15	7.49
5	7.13	6.99	6.97	6.90	3.60	3.42	7.41	7.40	7.17	7.17	3.32	3.17	7.68	7.60

[a] Methyl.

π-electron density in the heterocyclic ring of the methyl compounds. A similar but smaller effect is observed in thiophene. Large π shifts are also observed in the benzo derivatives but the differences in the chemical shifts of the ring protons in the benzo derivatives compared to the parent heterocycles are due mainly to the increased ring current shift.

The ring current calculations again provide evidence for the accuracy of the simple equivalent dipole model of the benzene ring current. The calculations show that the ring current is not the only factor in the difference between the H-2 and H-3 protons in aromatic heterocycles (furan, thiophene, etc.) and their non-aromatic derivatives.

The difference in the experimental chemical shift of H-2 in furan and 4,5-dihydrofuran is 1.11 ppm. This is due to 1.60 ppm from the ring current but the π electrons compensate to some extent as the π shift is −0.55 ppm compared to −0.47 ppm in dihydrofuran. The remainder is due to σ electronic effects from the olefinic carbon atoms.

Examination of Tables 5.18–5.20 shows the significant changes in the chemical shifts of H-2 and H-3 as the heteroatom varies from oxygen, sulfur and nitrogen. The ring current contributions to the shifts of H-2 and H-3 remain fairly constant throughout but there are very different π contributions, due to the different π-electron densities in these molecules. There are also different γ effects in furan, thiophene and pyrrole due to the different hetero atoms in these systems.

In benzofuran, benzothiophene and indole there is a similar pattern to furan, thiophene and pyrrole with a constant but increased ring current contribution to the chemical shifts. This is also the case for quinoline compared to those in pyridine.

The chemical shifts of the difunctional bases, imidazole, thiazole and oxazole (Table 5.17) are of some interest. H-2 in thiazole is deshielded by ca.1.0 ppm compared to that in imidazole and oxazole. The ring current effect on H-2 is similar in these molecules and

Table 5.18 *Calculated versus observed chemical shifts (δ) with C—H electric field, ring current and π-shift contributions for furan (**4**), 2-methyl (**6**) and 3-methylfurans (**9**) and benzofuran (**10**)*

Compound	¹H Number	Observed	Calculated	C—H electric field	Ring current	π-shift
4	2	7.420	7.415	−0.110	1.600	−0.549
	3	6.380	6.360	−0.057	1.507	−0.487
6	3	5.940	5.923	−0.126	1.514	−0.733
	4	6.230	6.289	−0.024	1.503	−0.585
	5	7.270	7.189	−0.077	1.595	−0.799
	Me	2.280	2.278	−0.054	0.500	0.000
9	2	7.160	6.917	−0.180	1.596	−0.855
	4	6.220	6.193	−0.121	1.513	−0.464
	5	7.290	7.450	−0.082	1.607	−0.534
	Me	2.030	2.172	−0.077	0.466	0.000
10	2	7.607	7.807	−0.030	1.905	−0.536
	3	6.758	6.671	−0.079	1.958	−0.664
	4	7.593	7.514	−0.151	1.967	−0.175
	5	7.225	7.239	−0.060	1.771	−0.250
	6	7.285	7.312	−0.046	1.762	−0.184
	7	7.502	7.400	−0.121	1.985	−0.246

Table 5.19   *Calculated versus observed chemical shift (δ) with calculated contributions for thiophene (13), 2-methylthiophene (15), 3-methylthiophene (18) and benzothiophene (19)*

Compound	^{1}H Number	Observed	Calculated	C—H electric field	Ring current	π-shift
13	2	7.310	7.263	−0.095	1.679	−0.641
	3	7.090	7.044	−0.052	1.764	−0.333
15	3	6.720	6.598	−0.130	1.775	−0.588
	4	6.870	6.970	−0.023	1.761	−0.430
	5	7.040	7.017	−0.068	1.667	−0.898
	Me	2.480	2.470	−0.045	0.542	0.000
18	2	6.870	6.886	−0.171	1.692	−0.620
	4	6.870	6.866	−0.122	1.773	−0.303
	5	7.190	7.305.	−0.065	1.676	−0.620
	Me	2.280	2.214	−0.072	0.564	0.000
19	2	7.422	7.523	−0.026	1.934	−0.702
	3	7.325	7.347	−0.089	2.235	−0.545
	4	7.780	7.642	−0.162	2.095	−0.122
	5	7.330	7.302	−0.061	1.786	−0.172
	6	7.310	7.340	−0.046	1.778	−0.140
	7	7.860	7.996	−0.115	2.058	−0.175

Table 5.20   *Calculated versus observed chemical shifts (δ) with calculated contributions for pyrrole (20), 2-methylpyrrole (22), 3-methylpyrrole (25) and indole (26)*

Compound	^{1}H Number	Observed	Calculated	C—H electric field	Ring current	π-shift
20	2	6.710	6.708	−0.105	1.645	−0.865
	3	6.230	6.187	−0.054	1.633	−0.830
22	3	5.890	5.784	−0.125	1.641	−1.043
	4	6.110	6.112	−0.023	1.628	−0.928
	5	6.640	6.508	−0.075	1.640	−1.088
	Me	2.270	2.285	−0.051	0.510	0.000
25	2	6.530	6.264	−0.176	1.652	−1.117
	4	6.020	5.988	−0.120	1.639	−0.839
	5	6.650	6.722	−0.077	1.641	−0.872
	Me	2.090	2.153	−0.073	0.503	0.000
26	2	7.207	7.321	−0.029	1.938	−0.618
	3	6.558	6.643	−0.081	2.083	−0.866
	4	7.647	7.489	−0.153	2.016	−0.196
	5	7.115	7.212	−0.060	1.775	−0.254
	6	7.185	7.263	−0.046	1.767	−0.212
	7	7.396	7.358	−0.118	2.002	−0.272

there is a small π contribution to the shift of H-2 in thiazole and oxazole. The main contribution to the deshielding of H-2 in thiazole is due to a large β effect of the sulphur atom.

Ring currents in heteroaromatics. The ring current intensities *fc* and equivalent dipoles (μ) for the systems considered are given in Table 5.21. The ratio of the ring current in these molecules to that in benzene i/i_B can be obtained from the equivalent dipoles using Equation (5.3) once the area of the current loops is known. The areas for benzene, furan and thiophene were taken from the literature,[55] and the program PCModel (MMFF94 force field) was used to calculate the areas of the remaining compounds. The results of these calculations are given in Table 5.21. It should be noted though that the area of the current loop may not be exactly the same as the area of the molecule. With this caveat it is clear from the results in this table that the ring currents in furan, thiophene and pyrrole are essentially identical to that in benzene. In contrast the insertion of an aza nitrogen atom in the aromatic ring as in pyridine does decrease the ring current by ca. 15% and the effect is seen to be cumulative in the diazabenzenes. An analogous effect is observed in the five-membered rings of oxazole, thiazole and imidazole with a decrease in the ring current with respect to the parent heterocycle of ca. 10%. In the bicyclic compounds the data for napthalene are given for comparison. There is a small decrease in the benzenoid ring current compared to napthalene in benzofuran, thiophene and indole but again a larger decrease in the quinoline and isoquinoline systems.

Table 5.21 *Ring currents and equivalent dipoles for heteroaromatics*

Molecule	Ring current intensity, *fc*	Equivalent dipole, μ	Ring current ratio, i/i_B
Benzene	1.00	26.23	1.00
Furan	0.67	17.6	1.04
Thiophene	0.83	21.8	1.08
Pyrrole	0.72	19.0	1.03
Oxazole	0.67	16.6	0.94
Thiazole	0.76	20.0	0.95
Imidazole	0.61	16.0	0.89
Pyridine	0.85	22.22	0.85
Diazabenzenes	0.72	18.83	0.74
Napthalene	0.93	24.39	0.93
Benzofuran[a]	0.90	23.6/17.6	—
Benzothiophenee[a]	0.90	23.6/21.8	—
Indole[a]	0.90	23.6/19.0	—
Quinolines/isoquinolines	0.75	24.39/19.7	0.75

[a] The equivalent dipoles for these compounds are the benzene/heterocyclic rings.

In this section for both scientific reasons and convenience we have been concerned with the parent heteroaromatic compounds, apart from some methyl derivatives. The same question as in the 'aromatic' section may be asked. Can the CHARGE program equally well calculate the ^{1}H chemical shifts in substituted heteroaromatics? The effect of substituents in these compounds may not be the same as in the benzenes for which they were parameterized.

To check this for the pyridine ring, the ^{1}H spectra of a number of substituted pyridines and 2- and 4-pyridone in the *Aldrich Spectral Catalogue* were assigned and the chemical shifts obtained and compared with the shifts calculated by CHARGE. In these calculations the molecular geometries were obtained using the MMFF94 force field in PCMODEL. These results[69] are given in Appendix A3 and inspection shows the generally good agreement between the observed and calculated shifts. Although the compounds include most of the common substituents, there is however one exception. All derivatives of 2-hydroxy and 4-hydroxypyridines are excluded. The reason for this is that these compounds interconvert between the hydroxypyridine and the corresponding pyridone. This interconversion is rapid on the NMR time scale, thus only one spectrum is obtained in which the chemical shifts (and coupling constants) are weighted averages of those for the interconverting species (see Chapter 8, Section 8.5 for a detailed discussion). Thus the prediction of the ^{1}H chemical shifts of, for example, 2-hydroxypyridine in the CHARGE model requires the calculation of the shifts of both the 2-hydroxypyridine and 2-pyridone *plus* the proportions of the two compounds in the solution. The determination of the proportions of the species present in any solvent is beyond the scope of this book.

5.3 Summary

In this chapter aromatics and heteroaromatic compounds are considered. The ^{1}H shifts of a variety of condensed aromatic compounds including benzene, naphthalene, anthracene, phenanthrene, pyrene, acenaphthylene and triphenylene are recorded. The shifts recorded in CDCl$_3$ are 0.086 ± 0.01 ppm greater than for the same proton in CCl$_4$ solution. To apply the CHARGE model an identification of both five- and six-membered aromatic rings based on atomic connectivities was included plus a dipole calculation of the aromatic ring current and deshielding steric term for the crowded protons in these molecules. The ring current shifts in the molecules were calculated assuming that the ring current was proportional to the molecular area divided by the molecular perimeter. The model was applied successfully to the non-alternant hydrocarbons of fulvene and acenaphthylene and to the aliphatic protons near to and above the benzene ring in tricyclophane and [10]-cyclophane.

The Huckel calculation of the π-electron densities in CHARGE was used to calculate the π-electron densities in substituted benzenes. The π inductive effect was used to simulate the effect of CX$_3$ groups (X = H, Me, F) on the benzene ring. These plus the long-range electric field and anisotropy effects of the substituents allowed a precise calculation of the SCSs of a variety of substituents on the benzene ring protons.

The model gave the first accurate calculation of the ^{1}H chemical shifts of condensed aromatic compounds and of the ^{1}H SCSs in the benzene ring. For the data set of 55 ^{1}H shifts spanning 3 ppm the rms error (obs.–calc.) shifts was ca. 0.1 ppm. The SCSs were interpreted in terms of the separate interactions calculated in the program, i.e. π-electron densities and steric, anisotropic and electric field effects. Previous correlations of the ^{1}H SCSs with π-electron densities and substituent parameters are shown to be over-simplified. The relative proportions of these different interactions are very different for each substituent and for each ring proton.

The ^{1}H shifts of 55 heteroaromatic compounds including the five-membered rings of furans, pyrroles, thiophens, indoles, imidazole, oxazole and thiazole and the six-membered heterocyclics pyridine, quinolines, isoquinolines and pyrimidine, pyrazine and pyridazine are recorded and analysed using the CHARGE model. The agreement of the observed vs. calculated ^{1}H chemical shifts is good and shows that the CHARGE model can be applied to heteroaromatic compounds. The ring current calculations provide further evidence for the accuracy of the simple equivalent dipole model of the benzene ring current and also demonstrate that the ring current effect is not the only factor responsible for the difference between the chemical shifts in the aromatic and non-aromatic heteroaromatic compounds. The use of suitable dihydro compounds as reference compounds is a useful method for determining the ring currents in these systems.

References

1. Emsley, J. W.; Feeney, J.; Sutcliffe, L. H., *High Resolution NMR Spectroscopy.* Pergamon Press: London, 1965.
2. Pauling, L., *J. Chem. Phys.* 1936, **4**, 673.
3. Haddon, R. C.; Haddon, V. R.; Jackman, L. M., *Top. Curr. Chem.* 1971, **16**, 103.
4. Pople, J. A., *J. Chem. Phys.* 1956, **24**, 1111.
5. Johnson, C. E.; Bovey, F. A., *J. Chem. Phys.* 1958, **29**, 1012.
6. Waugh, S.; Fessenden, R. W., *J. Am. Chem. Soc.* 1957, **79**, 846.
7. Abraham, R. J., *Mol. Phys.* 1961, **4**, 145.
8. Abraham, R. J.; Medforth, C. J., *Magn. Reson. Chem.* 1987, **25**, 432.
9. Haigh, C. W.; Mallion, R. B., *Mol. Phys.* 1970, **18**, 737.
10. Haigh, C. W.; Mallion, R. B.; Armour, E. A. G., *Mol. Phys.* 1970, **18**, 751.
11. Schneider, H. J.; Rudiger, V.; Cuber, U., *J. Org. Chem.* 1995, **60**, 996.
12. Bernstein, H. J.; Schneider, W. G.; Pople, J. A., *Proc. R. Soc. London A* 1956, **236**, 515.
13. Pople, J. A.. *Mol. Phys.* 1958, **1**, 175.
14. McWeeny, R., *Mol. Phys.* 1958, **1**, 311.
15. Jonathan, N.; Gordon, S.; Dailey, B. P., *J. Chem. Phys.* 1962, **36**, 2443.
16. Cobb, T. B.; Memory, J. D., *J. Chem. Phys.* 1967, **47**, 2020.
17. Westermayer, M.; Hafelinger, G.; Regelmann, C., *Tetrahedron* 1984, **40**, 1845.
18. Abraham, R. J.; Canton, M.; Reid, M.; Griffiths, L., *J. Chem. Soc. Perkin Trans.* 2 2000, 803.
19. Abraham, R. J.; Smith, P. E., *J. Comput. Chem.* 1987, **9**, 288.
20. Abraham, R. J.; Smith, P. E., *J. Comput. Aided Mol. Design* 1989, **3**, 175.
21. Abraham, R. J.; Smith, P. E., *Nucleic Acid Res.* 1988, **16**, 2639.
22. Streitweiser, A., *Orbital Theory for Organic Chemists.* John Wiley & Sons, Inc.: New York, NY, 1961.
23. McClellan, A. L., *Table of Experimental Dipole Moments*, Vol. 3. Rahara Enterprises: El Cerrito, CA, 1989.
24. Matter, U. E.; Pascual, C.; Pretsch, E.; Pross, A.; Simon, W.; Sternhell, S., *Tetrahedron* 1969, **25**, 691.
25. Puchert, C. J.; Behnke, J., *Aldrich Library of 13C and 1H FT NMR Spectra.* Aldrich Chemical Company: Milwaukee, WI, 1993.
26. Netka, J.; Crump, S. L.; Rickborn, B., *J. Org. Chem.* 1986, **51**, 1189.
27. Nuchter, U.; Zimmermann, G.; Francke, V.; Hopf, H., *Liebigs Ann. Rec.* 1997, 1505.
28. Hayamizu, K.; Yamamoto, O., *J. Mol. Spect.* 1968, **28**, 89.
29. Hayamizu, K.; Yamamoto, O., *J. Mol. Spect.* 1968, **29**, 183.
30. *Landholt–Bornstein Structure Data of Free Polyatomic Molecules.* Springer-Verlag: New York, NY, 1976.
31. Exner, O., *Dipole Moments in Organic Chemistry.* George Thieme Pubishers: Stuttgart, 1975.
32. Minkin, V. I.; Osipov, O. A.; Zhdanov, Y. A., *Dipole Moments in Organic Chemistry.* Plenum Press: New York, NY, 1970.

33. Abraham, R. J.; Griffiths, L.; Warne, M. A., *Magn. Reson. Chem.* 1998, **36**, S179.
34. Kuo, S. S., Chapter 8, in *Computer Applications of Numerical Methods*. Addison Wesley: London, 1972.
35. Agarwal, A.; Barnes, J. A.; Fletche, J. L.; McGlinchey, M. J.; Sayer, B. G., *Can. J. Chem.* 1973, **51**, 87.
36. Abraham, R. J.; Fisher, J.; Loftus, P., *Introduction to NMR Spectroscopy*. 2nd ed.; John Wiley & Sons, Ltd.: Chichester, 1988.
37. Abraham, R. J.; Thomas, W. A., *J. Chem. Soc. B* 1966, 127.
38. Mallion, R. B., *J. Chem. Phys.* 1980, **75**, 793.
39. Castellano, S.; Sun, C., *J. Am. Chem. Soc.* 1966, **88**, 4741.
40. Castellano, S.; Sun, C.; Kostelnik, R., *Tetrahedron Lett.* 1967, **46**, 4635.
41. Castellano, S.; Sun, C.; Kostelnik, R., *Tetrahedron Lett.* 1967, **51**, 5205.
42. Williamson, M. P.; Kostelnik, R.; Castellano, S., *J. Chem. Phys.* 1968, **49**, 2218.
43. Lambert, J. B.; Scurvell, H. B.; Kightner, D. A.; Cooks, R. G., *Organic Structural Spectroscopy*. Prentice-Hall, Upper Saddle River, NJ, 1998.
44. Pretsch, P. D.; Clerc, T.; Siebl, J.; Simon, W., *Spectral Data for Structure Determination of Organic Compounds*. 2nd ed.; Springer-Verlag: New York, NY, 1989.
45. Hehre, W. J.; Taft, R. W.; Topsom, R. D., *Prog. Phys. Org. Chem.* 1976, **12**, 159.
46. Brownlee, R. T. C.; Taft, R. W., *J. Am. Chem. Soc.* 1970, **92**, 7007.
47. Hamer, G. K.; Peat, I. R.; Reynolds, W. F., *Can. J. Chem.* 1973, **51**, 897.
48. Hamer, G. K.; Peat, I. R.; Reynolds, W. F., *Can. J. Chem.* 1973, **51**, 915.
49. Canton, M., *Ph.D. Thesis*. University of Liverpool, 2000.
50. Abraham, R. J.; Reid, M., *Magn. Reson. Chem.* 2000, **38**, 570.
51. Swain, C. G.; Lupton, E. P., *J. Am. Chem. Soc.* 1968, **90**, 4328.
52. Abraham, R. J.; Mobli, M.; Smith, R. J., *Magn. Reson. Chem.* 2004, **42**, 436.
53. Abraham, R. J.; Griffiths, L.; Warne, M. A., *J. Chem. Soc. Perkin Trans. 2* 1997, 881.
54. Katritsky, A. R., *Adv. Hetrocyclic Chem.* 2001, **77**, 1.
55. Thomas, W. A., *Ph.D. Thesis*. University of Liverpool, 1965.
56. Elvidge, J. A., *J. Chem. Soc. Chem. Commun.* 1965, 160.
57. De Jongh, H. A. P.; Wynberg, H., *Tetrahedron* 1965, **21**, 515.
58. Lampert, H.; Mikenda, W.; Karpfen, A.; Kählig, H., *J. Phys. Chem. A* 1997, **101**, 9610.
59. Abraham, R. J.; Reid, M., *J. Chem. Soc. Perkin Trans. 2* 2002, 1081.
60. Taskien, E., *Magn. Reson. Chem.* 1995, **33**, 256.
61. Yoshida, K.; Fueno, T., *Bull. Chem. Soc. Jpn* 1987, **60**, 229.
62. Aitken, R. A.; Armstrong, J. M.; Drysdale, M. J.; Rossi, F. C.; Ryan, B. M. *J. Chem. Soc. Perkin Trans. 1* 1999, 593.
63. Hatton, P. M.; Sternhell, S. *J. Heterocyclic Chem.* 1992, **29**, 933.
64. Padialla-Martinez, I. I.; Ariza-Castolo, A.; Contreras, R. *Magn. Reson. Chem.* 1993, **31**, 189.
65. Adam, G.; Andrieux, J.; Plat, M. M. *Tetrahedron Lett.* 1983, **24**, 3609.
66. Shafer, C. M.; Molinsk, T. F. *Heterocycles* 2000, **53**, 1167.
67. Reid, M. *Ph.D. Thesis*, University of Liverpool, 2002.
68. Abraham,R.J.;Byrne,J.J.;Griffiths,L.;Koniouou,R. *Magn. Reson. Chem.* 2005, **43**, 611
69. Byrne, J.J. *Ph.D. Thesis*, University of Liverpool, 2007.

6

Modelling ¹H Chemical Shifts, Monovalent Substituents

6.1 Introduction

Having dealt with the ¹H chemical shifts of saturated and unsaturated hydrocarbons, including aromatics in Chapters 4 and 5 we can now consider the effect of various substituents on the ¹H chemical shifts. The theory of Chapter 3 breaks down the effects of substituents into short-range effects over ≤ three bonds which are obtained by determining the constants in Equation (3.5) from the observed shifts and long-range effects (steric, polar and anisotropic) which are given by the classical Equations 3.11–3.15. For convenience we will divide the substituents into monovalent (F, Cl, OH, etc.) in this chapter and more complex divalent (esters, amides, sulfones, etc.) in Chapter 7. As alkyl groups were included with the alkane chemical shifts in Chapter 4 all the functional groups to be considered are polar. They will therefore produce an electric field at the hydrogens in the molecule and give rise to electric field shifts. In many cases the substituents are anisotropic and will also produce anisotropic shifts and in most cases they will also have steric effects on near protons.

Only one substituent, fluorine, is not anisotropic and so small that the H. F steric effects are minimal (see below), thus fluorine is an ideal substituent to test the electric field theory. This will be dealt with in detail in the next section. The other functional groups with polar, anisotropic and steric effects will be considered subsequently.

6.2 Fluorine Substituent Chemical Shifts

Due to both the relative scarcity of fluoro compounds and the complex ¹H spectra produced due to extensive H—F coupling the ¹H chemical shift data for these compounds was until recently much less comprehensive than that for the other halogens. In their initial attempts to calculate the chemical shifts of fluoro and chloro substituted alkanes Abraham *et al.* observed that there was a pronounced through space effect of the fluorine substituent on ¹H chemical shifts.[1,2] In contrast to the other substituents investigated there was no push–pull effect for the fluorine substituent (cf. Table 1.7 and Figure 1.3) and also the deshielding due to the fluorine atom was better represented by an r^{-3} function than the r^{-6} function used for the other substituents. Fluorine is of comparable size to the hydrogen atom thus the replacement of a proton by a fluorine atom should not produce any steric perturbations. Thus they suggested that the fluorine SCSs in the compounds examined were primarily due to the electric field produced by the fluorine atom and not to steric effects. This original hypothesis was later confirmed by a calculation of the electric field of the substituent using the partial atomic charges calculated by the CHARGE routine.[3] However before considering these results it is useful to give the electric field theory.

6.2.1   Electric Field Theory

The influence of a uniform external electric field (E) on ¹H shielding was first calculated by Marshall and Pople.[4] for the hydrogen atom in which by symmetry only an E^2 term is present. Subsequently Buckingham[5] extended this method to derive the shielding for a C—H proton. The equation on the δ scale is given by Equation (6.1)

$$\delta_{\text{electric}} = A_Z E_Z + BE^2 \tag{6.1}$$

where A_z is the linear electric field coefficient or shielding polarisability and B the quadratic electric field coefficient or shielding hyperpolarisability. For a dipolar (eg. C—X) substituent the linear electric field is proportional to r^{-3} and the quadratic term proportional to r^{-6} where r is the distance from the substituent or centre of the point charge to the proton considered. The quadratic electric field is different in origin from the steric or van der Waals term but has an similar geometric dependence and therefore it is not practical experimentally to distinguish between these effects (see Chapter 4, Section 4.2.4).[6-8]

Buckingham also noted that the value of the linear coefficient is dependent upon the nature of the atom attached to the proton, thus C—H, N—H and O—H protons will have different values of A_Z. For the Csp^3—H bond he suggested a value of 34 ppm au. Subsequent semi-empirical calculations gave values of A_Z from 44 to 83 ppm au.[9-11] for the Csp^3—H bond and then SCF calculations[12,13] gave values between 62.0 and 80.2 ppm au, with an average of 70 ppm au for methane, ethane, acetonitrile, chloromethane and fluoromethane.

Thus the basic theory of the electric field effect was well established, but the experimental determination of the effect of the electric field on ¹H shielding and in particular the relative proportions of the linear and quadratic terms was still unknown. Early investigations on the density dependence of the ¹H shielding in gaseous trifluoromethane[6,7] and on the effect of solvent on the proton shifts of acetonitrile.[14] gave values of A_Z of ca. 50–60 ppm au. Zürcher.[8] in his analysis of the proton SCS in steroids and bicycloheptenes containing Cl, OH, CN and C=O substituents included the bond anisotropy and both the linear and quadratic electric fields of the substituents. He obtained a value of A_Z of 72 ppm au. He included the quadratic term in evaluating Cl SCSs but concluded that this term was not significant and ignored it subsequently.

Further investigations only partially clarified the situation. The ¹H SCSs of ketones, thioketones[14–16] and ethers[67] were interpreted as arising from anisotropy and electric field effects but for alcohols electric field effects were regarded as the dominant term.[8,16–19]

For chloro and bromo substituents Davis *et al*[17] suggested that apart from 1,3-*syn* diaxial protons the SCSs could be explained by electric field effects alone and similar conclusions were drawn for the proton SCSs in halosteroids[16,18] and for halo substituted *trans*-decalins.[20] Studies on halobicycloheptanes[21,22] and halocamphors[23] suggested that linear electric field effects plus steric contributions could explain the SCSs on the remote protons with a short-range mechanism (anisotropy, van der Waals or inductive) needed for the vicinal protons.

As fluorine is a small, highly polar and almost non-polarizable atom fluorine SCSs are the ideal data to examine electric field effects, but there was little systematic data on fluorine SCSs in rigid molecules. The only complete SCS data were for 3α and 3β fluoroandrostan-17-one[16] and 3-endo and 3-exo fluorocamphor[23] and in neither case were the calculated SCSs given. Abraham *et al.*[1,2] analysed the ¹H spectra of a number of fluorocyclohexanes and norbornanes to obtain the fluorine SCSs and this provided sufficient data to test the electric field calculation.[3]

In the CHARGE scheme the effect of the fluorine substituent on atoms up to three bonds away has already been given in Chapter 3, Section 3.5.1. Note that the β proton SCSs in CH_3F, CH_2F_2 and CHF_3 are non-additive (Chapter 3, Figure 3.3) and correction factors were included for CF_2 and CF_3 groups. The γ effect of fluoro substituents was observed to be non orientational and also non-additive and similar correction factors for CF_2 and CF_3 groups were included.

To calculate the electric field of a substituent in a scheme based on partial atomic charges it is computationally simpler and more accurate to directly calculate the electric field at the hydrogen atom due to the partial atomic charges on the substituents rather than the field due to a C—F dipole. Thus for the C—F bond the charge on the carbon atom (δ+) was taken as the same magnitude as the charge on the fluorine (δ−) but of opposite sign (Figure 6.1). The vector components of the electric field were calculated (a) from the fluorine atom to the proton and (b) from the carbon atom to the proton and summed to give the component of the total field along the C—H bond. In Figure 6.1 $\mathbf{E}$ is the field vector and $|\mathbf{E}|$ the magnitude of the field vector; $\mathbf{i}_E$ the unit vector of $\mathbf{E}$ along $\mathbf{E}$; $\mathbf{i}_{CH}$ the unit vector along the C—H bond;

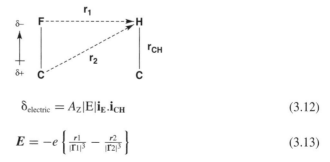

$$\delta_{electric} = A_Z |E| \mathbf{i}_E . \mathbf{i}_{CH} \qquad (3.12)$$

$$E = -e \left\{ \frac{r1}{|\mathbf{r}1|^3} - \frac{r2}{|\mathbf{r}2|^3} \right\} \qquad (3.13)$$

Figure 6.1 *Model used to calculate C—F electric field effects (reproduced from Abraham et al.[3] by permission of the Royal Society of Chemistry).*

e the charge on the substituent atom; $|r_1|$ the magnitude of vector $\mathbf{r_1}$ and $|r_2|$ the magnitude of vector $\mathbf{r_2}$. Operating the vector quantities gives Equations (3.12) and (3.13).

Thus the effect of a C—F bond on a parallel C—H proton is deshielding. Alternately the effect of a C(δ−)—H(δ+) bond on a parallel C—H proton is shielding. This approach differs from the point dipole approximation used by Zurcher[8] in several ways. It is more accurate at close interatomic distances and also the charge on the substituent atom (eg. F) will vary depending on the chemical environment of the substituent as opposed to a fixed C—F dipole. In particular the charge on a fluorine atom decreases in the order $CH_2F > CHF_2 > CF_3$ thus the electric field contribution will decrease in this order also.

6.2.2   Fluoroalkanes

The electric field calculation was included in the CHARGE program with the remainder of the program unchanged[3] and the partial atomic charges on the atoms obtained directly from the CHARGE routine. The observed and calculated fluorine SCSs were then compared. The geometries used were obtained as previously from *ab initio* calculations at the RHF/6-31G* level.[24] The calculated C—F bond lengths for this basis set were in good agreement with the experimental values (cf. fluoroethane 1.372 vs. 1.397 Å[25]).

For the mono fluoro substituted compounds, for which there were 40 data points including ethanes, cyclohexanes, bornanes and steroids it was found that for the long-range fluorine SCSs good agreement was obtained for a value of A_Z of 3.67×10^{-12} esu, i.e. 63 ppm au. This calculation gave SCS effects for the CF_2 groups which were slightly too large and the calculated SCSs for this group was reduced by moving the position of the atomic charge from the fluorine atom towards the bond centre and similarly for the attached carbon atom. This is equivalent to assuming the electron density is more towards the bond centre in CF_2 groups (see later). A displacement of the CF_2 centres by 10% of the bond length gave good results. There were insufficient data available to determine the long-range effects of CF_3 groups in rigid molecules.

It was noted in Chapter 3, Section 3.5.2 that the effect of the two fluorines in a CF_2 group on the SCSs of the vicinal protons is non additive and this effect is clearly evident in the cyclohexanes studied (Figure 6.2). In 4-methyl-1,1-difluorocyclohexane the SCS of H2,6ax is 0.47 ppm which equals the sum of the monofluoro SCSs (0.24 and 0.23 ppm). However the SCS of H2,6eq (0.34 ppm) is *less* than the individual SCSs for the monofluoro compounds (0.35, 0.47 ppm). A similar effect occurs in norbornanes. The SCS for the 1-CH proton in 2,2-difluoronorbornane is +0.24 ppm (Figure 6.2) in contrast to +0.30 ppm and +0.16 ppm for the corresponding proton in 3-endo and 3-exo fluorocamphor.[23] Furthermore the γ SCS varies for CH_3, CH_2 and CH protons whereas the long-range effects on CH and CH_2 protons are the same. Thus the γ (H.C.C.F) proton SCS were calculated as shown in Chapter 3, Section 3.5.2.

The electric field effect is a general effect for all polar C—X bonds, including the C—H bond. Thus the electric field of the C—H bonds was included in the calculations with the additional constraint for these small effects of a cut-off at 6.0 Å (cf. Equation (3.11)) to avoid the calculation of a large number of very small contributions.

The observed and calculated ¹H chemical shifts and fluorine SCSs on the above model for the compounds considered are given in Figure 6.2 and Tables 6.1–6.5. The electric field calculation is unchanged from the original version.[3] In the acyclic compounds (Table 6.1)

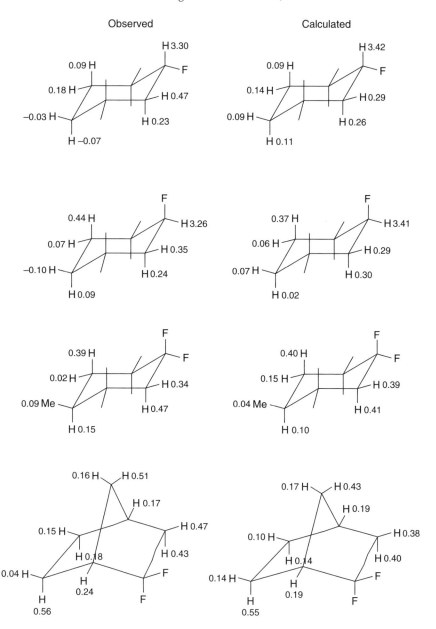

Figure 6.2 *Observed and calculated ^1H SCSs in fluorocyclohexanes and norbornane.[2]*

where more than one possible conformer exists, the data for both forms are given. The *gauche* and *trans* forms of the fluoroethanes have the same calculated shifts as the fluorine γ effect is non-orientational.

From the comparison of the observed and predicted shifts in Tables 6.1 to 6.5 it can be seen that the model replicates the experimental data very well. In particular the agreement between the observed and calculated SCSs for the long-range protons shows that the model of the C—F linear electric field gives a quantitative interpretation of the long-range fluorine SCSs in these systems. The value obtained for A_Z of 63 ppm au is also in excellent agreement with both Zurcher's value and the calculations of Grayson and Raynes further supporting these results.

Table 6.1 *Observed versus calculated ¹H chemical shifts (δ) of acyclic fluoroalkanes*

Molecule		Observed[a]	Calculated
CH_3F	CH_3	4.27	4.26
CH_2F_2	CH_2	5.45	5.46
CHF_3	CH	6.41	6.41
CH_3CH_2F	CH_2	4.55	4.60
	CH_3	1.35	1.26
CH_3CHF_2	CH	5.94	5.76
	CH_3	1.56	1.42
$CH_3CF._3$	CH_3	1.87	1.76
CH_2FCH_2F	CH_2	4.59	4.80
CH_2FCHF_2	CH	5.93	5.85
	CH_2	4.45	4.87
CHF_2CHF_2	CH	5.64	5.88
$CF._3CH_2F$	CH_2	4.55	5.03
$CH_3CH_2CH_2F^b$	CH_2F	4.30	4.65(g), 4.55(t)
	CH_2	1.68	1.59(g), 1.59(t)
	CH_3	0.97	0.95(g), 0.96(t)
$(CH_3)_2CHF$	CH	4.84	4.93
	CH_3	1.34	1.33
$(CF._3CH_2)_2$	CH_2	2.46	2.31(t)

[a] Data from Abraham *et al.*[3].
[b] (g) gauche; (t) *trans* conformer.

Table 6.2 *Observed versus calculated ¹H chemical shifts (δ) of 1-eq and 1-axial fluorocyclohexane[a] and 4-methyl and 4-t butyl-1,1-difluoro cyclohexanes[b]*

Proton	Axial		Equatorial		4-Methyl-		4-ᵗButyl-	
	Obs.	Calc.	Obs.	Calc.	Obs.	Calc.	Obs.	Calc.
1a (CH)	—	—	4.49	4.58				
1e (CH)	4.94	4.99	—	—				
2,6a	1.43	1.49	1.42	1.46	1.67	1.62	1.68	1.62
2,6e	2.03	1.93	2.15	1.93	2.02	2.04	2.09	2.05
3,5a	1.63	1.56	1.28	1.32	1.27	1.26	1.31	1.35
3,5e	1.75	1.71	1.86	1.82	1.70	1.80	1.80	1.88
4a	1.28	1.21	1.12	1.33	1.47	1.53	1.07	1.01
4e	1.58	1.72	1.65	1.75	0.95	0.91[c]	0.89	0.89[d]

[a] Data from Abraham *et al.*[3].
[b] Data from Abraham *et al.*[2].
[c] Methyl.
[d] t-Butyl.

Table 6.3 *Observed[a] versus calculated ¹H Chemical shifts (δ) (ppm) of 3-methyl-1,1-difluoro-cyclohexane and 2,2-difluoronorbornane*

	3-Methyl-			2,2-Difluoro-	
Atom	Obs.	Calc.	Atom	Obs.	Calc.
2a	1.29	1.23	1	2.43	2.35
2e	2.02	1.99	3x	1.94	1.89
3a (CH)	1.72	1.83	3n	1.59	1.52
3e-CH₃	0.96	0.93	4	2.36	2.41
4a	0.91	0.98	5x	1.62	1.65
4e	1.69	1.77	5n	1.34	1.41
5a	1.54	1.65	6x	1.51	1.58
5e	1.76	1.85	6n	1.72	1.80
6a	1.54	1.61	7syn	1.69	1.73
6e	2.05	2.04	7anti	1.34	1.33

[a] Data from Abraham *et al.*[2].

Table 6.4 *Observed versus calculated ¹H chemical shifts (δ) and SCSs (ppm) for fluoro-adamantanes*

	1-Fluoro-			1,3-Difluoro-			1,3,5-Trifluoro-	
Proton	Obs.[a]	Calc.	Proton	Obs.[b]	Calc.		Obs.[b]	Calc.
¹H Chemical shifts								
γ	1.83	1.81	CF.CH₂.CF.	2.13	2.00		2.10*	2.11, 2.14
δ (CH)	2.20	2.16	CF.CH₂.CH	1.87*	1.93		1.78*	2.07
ε-ax	1.62*	1.74	CH	2.45	2.33		2.50	2.50
ε-eq	1.62*	1.77	CH.CH₂.CH	1.55	1.88			
SCSs[c]								
γ	0.08	0.11	CF.CH₂.CF.	0.38	0.39		0.35*	0.49, 0.52
δ (CH)	0.33	0.20	CF.CH₂.CH	0.12*	0.23 (0.19)		0.03	0.45
ε-ax	−0.13*	0.04	CH	0.58	0.37		0.63	0.52
ε-eq	−0.13*	0.07	CH.CH₂.CH	−0.20	0.13			

* Unresolved.
[a] Fort *et al.*[26].
[b] Bhandari *et al.*[27].
[c] Calculated SCSs, cf. adamantane (Chapter 4, Table 4.2).

The 10% reduction in the field required for the difluoro (CF₂) group is also explained on this basis as this non-linear effect is well known in quantum mechanical calculations of fluoro compounds. The geminal fluorine atoms strongly interact with each other, the F.C.F angle is much less than tetrahedral and the CF bond dramatically shortened in the CF₂ and CF₃ groups[2] due to resonance forms such as F^+=CF^-. On this basis the electron distribution in the CF bond would be greater between the atoms in the CF₂ and CF₃ groups than in the CF bond. A similar explanation has been proposed to explain the correction for the β and γ protons.[2] The alternative explanation that the partial atomic charge on the fluorine atom should be reduced by 10% from that calculated in the CHARGE scheme is

Table 6.5 *Observed[a] versus calculated[b] SCSs (ppm) for fluoro-androstanes*

	3α-Fluoro-		3β-Fluoro-	
Atom	Obs.	Calc.	Obs.	Calc.
1α	0.43	0.37	0.09	0.09
1β	−0.14	0.06	0.10	0.14
2α	0.31	0.29	0.47	0.29
2β	0.10	0.29	0.11	0.25
3α	—	—	3.24	3.41
3β	3.13	3.41	—	—
4α	0.34	0.33	0.45	0.31
4β	0.12	0.27	0.22	0.25
5 (CH)	0.50	0.47	0.04	0.05
6α	−0.01*	0.05	0.10*	0.06
6β	−0.09*	−0.01	0.10*	0.06
7α	0.05	0.04	0.01	0.01
7β	0.03	0.00	0.03	0.03
8 (CH)	0.01	−0.01	−0.02	0.02
9 (CH)	0.10	0.07	−0.04	0.00
19-Me	0.01	0.01	0.05	0.05

* Unresolved.
[a] Observed SCS, cf. 3α- and 3β-fluoro-androstan-17-one, Schneider *et al.*[16].
[b] Calculated SCSs, cf. 3α- and 3β-fluoroandrostane vs. 5α-androstane (Chapter 4, Table 4.5).

not supported by the dipole moments calculated by CHARGE which are in good agreement with the observed values.[28]

Detailed inspection of the observed vs. calculated SCS data in the equatorial and axial fluorocyclohexanes (Figure 6.2) shows good agreement except for the C-4 protons, in which the observed SCS is often shielding whereas the calculated SCS is always deshielding. However the H_{4a} SCS in 4-methyl-1,1-difluorocyclohexane is +0.10 (obs.) and +0.15 (calc.) and also the SCS in 3-methyl- and 4-ᵗbutyl-difluoro are deshielding on the 4 protons as expected. The monofluoro spectrum is a complex pattern at low temperatures as the spectra of the two conformers are overlapping. Also solvent and/or temperature effects may be responsible for this anomaly.

In 2,2-difluoronorbornane (Table 6.3) the calculated ¹H shifts are in excellent agreement with the observed data, particularly for the deshielded 6-endo proton (obs. 1.72 ppm vs. calc. 1.80 ppm).

The calculated SCSs for fluoro-adamantanes and 3-fluoro-5α-androstanes (Tables 6.4 and 6.5) are encouraging, in that they generally reflect the observed trends, though with some exceptions. In the fluoro-adamantanes some of the discrepancies may be due to the incomplete resolution of the spectra. However the most distant methylene protons appear to be *shielded* by the introduction of the fluorine. This latter effect is seen consistently for all substituents[26] (including alkyl groups) irrespective of their electron withdrawing/donating abilities, suggesting this is a result of some additional mechanism in the adamantane ring itself.

The observed SCS of the 3α-fluoro substituent in androstane on the 1β proton (−0.14 ppm) is of opposite sign to the same effect in cyclohexanes suggesting that the reported steroid value may be anomalous. The long-range effects on protons in the C and D rings are too small (< 0.03ppm) to be significant and have not been included. These SCSs are for a single substituent in a bifunctional compound and this assumes no interaction between the functional groups. This would appear reasonable for the 3-halo-androstan-17-ones[16] in which the substituent groups are far apart.

The ^1H chemical shifts of the fluoroalkanes considered comprise over 60 data points spanning a range of ca. 0.9 to 6.4δ and were predicted with an rms error of 0.11 ppm, which is not much larger than the experimental error. Thus this investigation showed that fluorine SCSs over more than three bonds are determined primarily by linear electric field effects without the need to invoke other terms.

The determination of the value of A_z for the linear electric field calculation in such good agreement with the theoretical value lends considerable support to the extension of this calculation to other polar groups for which the presence of other effects (e.g. steric, anisotropic) means that this coefficient could not be determined so easily. Thus the approach followed is to use the above theory with the partial atomic charges calculated by CHARGE to obtain the electric field effects of any polar group.

6.2.3   Fluoroalkenes and Aromatics

The additional parameters required for these compounds are those which define the π-electron densities, i.e. the exchange and overlap integrals in the Huckel equations (Equations (3.7)). Also the short range β and γ effects (e.g. FC=CH) will have different values than those for fluoroalkanes. The results for fluorobenzene are given in Chapter 5, Section 5.1.2 and for poly-substituted benzenes (including polyfluorobenzenes) in Appendix B and the calculated shifts are in very good agreement with the observed values. 1-Fluoronaphthalene is conveniently included with the other halo-naphthalenes (Table 6.10 below). Thus we consider here fluoroalkenes and Table 6.6 gives the observed vs. calculated ^1H shifts for a number of these molecules. Ethylene is included for comparison. The geometries of the compounds in Table 6.6 were obtained using the PCMODEL program with the MMFF94 force field. The major problem with these molecules is as mentioned above the calculation of the π-electron densities. The contrasting effects of fluorine as a strong σ acceptor and moderate π donor are very evident in the reactivities of these compounds. Ethylene is susceptible to electrophillic reagents whereas tetrafluoroethylene in contrast is susceptible to nucleaphillic reagents. The ^1H chemical shifts also show large effects. The hydrogens β to the fluorine (CHF) are deshielded with respect to ethylene (6.59 vs. 5.41, Table 6.6) wheras those γ to the fluorine as in 1,1-difluoroethylene are shielded

Table 6.6 *Observed versus calculated ^{1}H chemical shifts (δ) and dipole moments (D) of fluoroalkenes*

Molecule		^{1}H chemical shift		Dipole moment, μ	
		Obs.	Calc.	Obs.	Calc.
CH$_2$=CH$_2$	CH	5.41	5.41	—	—
CHF=CH$_2^a$	CHF	6.59	6.81	1.43	1.71
	CH (*trans*)	4.46	4.43		
	CH (*cis*)	4.80	4.84		
cis CHF=CHF^b	CH	6.30	6.30	2.42	3.02
trans CHF=CHF^b	CH	7.25	7.25	—	—
CH$_2$=CF$_2$^{b}	CH$_2$	3.81	3.90	1.38	1.25
CHF=CF$_2$^{a}	CH	6.59	6.49	1.40	1.62
z-CHF=CH.CH$_3^c$	CH$_3$	1.59	1.56	1.46	1.75
	CH	4.68	4.88		
	CHF	6.38	6.34		
e-CHF=CH.CH$_3$^{c}	CH$_3$	1.50	1.52	1.85	2.09
	CH	5.25	5.29		
	CHF	6.44	6.36		
CH$_2$=CF.CH$_3^c$	CH$_3$	1.85	1.86	1.60	2.01
	CH (F, *cis*)	4.40	4.37		
	CH (F, *trans*)	4.13	3.98		
CF$_2$=CHCH$_2$CH$_2$.CH$_3$^{d}	CH	4.14	4.32		
	CH$_2$ (α)	1.93	2.04		
	CH$_2$ (β)	1.34	1.24		
	CH$_3$	0.92	0.94		
CH$_2$=CH.CH$_2$F^e	CH$_2$	4.69	4.69^f, 4.73^g	1.77^f	1.88
	CH (*trans*)	5.12	5.02 5.13	1.93^g	
	CH (*cis*)	5.24	5.29 5.11		
	CH	5.89	5.85 5.85		
CH$_2$=CH.CHF$_2$^{e}	CHF$_2$	5.98	5.80^f, 5.97^g	2.47^f	2.10
	CH (*trans*)	5.48	5.42 5.34	2.12^g	
	CH (*cis*)	5.57	5.55 5.68		
	CH	5.91	5.84 5.84		
CH$_2$=CH.CF$_3^e$	CH	5.90	5.81	2.45	1.99
	CH (*trans*)	5.56	5.59		
	CH (cis)	5.86	5.92		
Ph.CF$_3$	*ortho*	7.63	7.76	2.61	2.25
	meta	7.48	7.52		
	para	7.63	7.82		

^{a} Data from Smith *et al.*[30].
^{b} Data from Ihrig *et al.*[31].
^{c} CFCl$_3$ solution, data from DeWolf *et al.*[32].
^{d} Data from Wheaton *et al.*[33].
^{e} Data from Bothner-By *et al.*[29].
^{f} Symmetrical rotamer.
^{g} Gauche rotamer.

with respect to ethylene (3.81 vs. 5.41). These large changes are a challenge even for *ab initio* calculations, thus the values of the Huckel parameters h_r, k_{rs} (Equation (3.7)) were obtained by comparison of the observed vs. calculated dipole moments. These are shown in Table 6.6 and are in general agreement. The other parameters required (*A* and *B* values, see Chapter 3, Equation (3.5)) were obtained by iterative methods from the observed data. Note that allyl fluoride and allylidene fluoride both have two populated conformers,[29] thus the observed chemical shifts are averages over the two conformers.

The ¹H chemical shifts obtained (Table 6.6) are in generally good agreement with the observed data demonstrating that the charge program does provide a reasonable account of the ¹H chemical shifts in these highly polar molecules.

6.3   Steric, Anisotropic and Electric Field Effects in Cl, Br and I SCSs

6.3.1   Introduction

In contrast to fluorine, the SCSs of the other halogens have been measured since the early days of NMR. In 1958 Bothner-By *et al.*[34,35] noted the anomalous inductive effects of bromine and iodine substitution on ¹H chemical shifts of alkyl halides compared to chlorine but were unable to explain these observations. In their studies of the effects of halogens on ¹H and ¹³C chemical shifts, Spiesecke and Schneider concluded that the observed shifts could only be related to the electronegativity of the halogens if magnetic anisotropy contributions were allowed.[36] In a subsequent study of halobenzenes, again they found the anisotropy effect to be crucial for the explanation of the observed results. They also found that the effect on the ¹³C shift due to the change in π-density on the carbon atoms by the substituent was proportional to the change in the ¹H shifts. They did, however, find an anomalous effect on the *meta* protons which could not be correlated with chemical shift changes in the attached carbon.[37] Although these studies provided evidence for the existence of the above effects, no attempt was made to determine any values. Subsequently, Hruska *et al.*[38,39] correlated the observed SCSs of the *ortho* protons to the polarizability (*P*) of the C—X bond using the quantity *Q*, according to $Q = P/Ir^3$, where *I* is the first ionization potential of X and *r* the C—X bond length. The good correlation to the polarizability for the *ortho* protons was not observed for the *meta* and *para* protons.[38]

Later studies[8,16–18] used mainly the methyl ¹H shifts in halo steroids, norbornanes, ethyl chloride, etc. They found that the electric field effect of the C—X bond was a major factor in halo SCSs, but could not decide on the presence of other effects such as C—X anisotropy or the steric term. The anomalies were the 1,3-syn diaxial effects from 2β, 4β and 6β-haloandrostanes on the 19-CH₃ protons where errors of ca. 0.3 ppm arose unless the magnetic anisotropy term was included as well. Subsequently more detailed investigations obtained the halo SCSs for all the ¹H nuclei in 3α and 3β-haloandrostane and androstan-17-one[16,18], 2-exo and 2-endo bornanes and norbornanes and 1- and 2- haloadamantane[22] and 3-endo-, 3-exo- and 3,3-dichloro-camphor.[23] The SCSs in 3,3-dichlorocamphor were additive from the monochloro data for all except the nearest protons. In these investigations the electric field term produced results of the correct sign and magnitude in contrast to the anisotropy calculations. Differences for close substituents were considered to derive from inductive through bond effects, with a possible minor role for steric effects. Also the SCSs in 9-halo-trans-decalins[20,40] were consistent with the steroid data and thus predictable by electric field calculations alone, except for the shielded methine γ protons.

Subsequently a systematic study of chloro,[41] bromo and iodo[42] cyclohexanes was reported including the assignments of the ¹H spectra of the individual conformers at −85 °C. The results for the halocyclohexanes at −85 °C and chlorocyclohexane at room temperature in the same solvent are given for comparison in Table 6.7. The room temperature shifts of chlorocyclohexane are as expected the weighted averages of the shifts of the equatorial and axial conformers.

Table 6.7   *¹H chemical shifts (δ) of chloro, bromo and iodo cyclohexanes*[a]

Proton	20 °C Average	Chloro (−85 °C)[b]		Bromo (−85 °C)[b]		Iodo (−85 °C)[b]	
		Eq-Cl	Ax-Cl	Eq-Br	Ax-Br	Eq-I	Ax-I
1ax	3.96	3.88	—	4.09	—	4.18	—
1eq	—	—	4.59	—	4.81	—	4.96
2,6ax	1.66	1.58	1.76	1.75*	1.81*	1.97	1.53
2,6eq	2.06	2.22	2.00	2.33	2.08	2.45	2.06
3,5ax	1.37	1.33	1.77*	1.35	1.79	1.36	1.72*
3,5eq	1.81	1.84	1.55*	1.80*	1.60	1.67	1.62*
4ax	1.32	1.18	1.26*	1.22	1.24*	1.30	1.26
4eq	1.54	1.68	1.75	1.72	1.78*	1.80	1.73*

* Chemical shift from ¹H−¹³C correlations.
[a] Data from Abraham *et al.*[41].
[b] 50:50 CDCl₃.CFCl₃.

Inspection of the halocyclohexane SCSs (Figure 6.3) shows that more than the linear electric field is involved. The Cl, Br and I SCSs are *larger* than the corresponding fluorine SCSs, e.g. H3ax in the axial conformer 0.44 for F vs. 0.58, 0.60 and 0.53 for Cl, Br and I, respectively, even though the partial atomic charge on the fluorine atom is much greater than that on the other halogens. Also the deshielding of H3ax and corresponding shielding of H3e in the Cl, Br, I axial-conformers contrasts with the situation for fluorine and is consistent with the push–pull effect. This suggested that steric interactions are involved. Thus the CHARGE scheme was parameterized using the C−X linear electric field effect on distant protons (δ and beyond) with the A_Z coefficient as determined previously, together with the steric r^{-6} term (Equation (3.11)) and push–pull effect. This gave good results for the haloalkanes studied.[41,42] However a later investigation on 1-halo-napthalenes[43] showed that this scheme needed to be modified to calculate the ¹H chemical shifts of the aromatic protons and in particular the SCS of H-8. This provides a severe test of any model as the Cl.H-8 distance in 1-chloronapthalene (2.76 Å) is much less than the 1,3-diaxial Cl.H distance in axial chlorocyclohexane (2.97 Å). Two alternatives were suggested (Figure 6.4). In model A the steric coefficient for an aromatic halogen differs from that for the corresponding aliphatic halogen. This can be rationalized on the basis that the C−X bond length is much less in aromatic than aliphatic compounds. The second model takes account of the diamagnetic circulation of the halogen electron cloud around the C−X bond. This diamagnetic circulation gives rise to a large *shielding* effect on the C¹³ shift in bromo and iodo compounds and has been termed the 'heavy atom effect'.[44] This anisotropy of the C−X

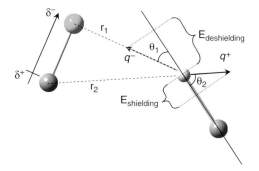

Plate 1 *The geometry dependence of the electric field effect on ^1H chemical shifts (see Figure 3.4).*

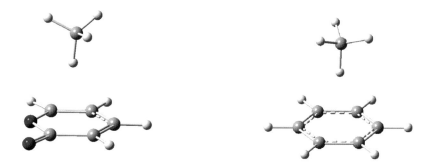

Plate 2 *Comparison of ring current shifts in pyrone and benzene (see Figure 7.8).*

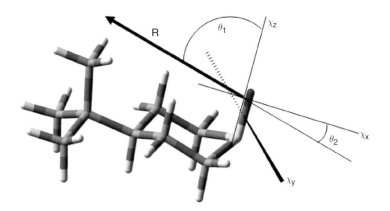

Plate 3 *The anisotropy of the SO bond (see Figure 7.16).*

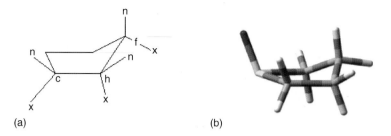

(a)   (b)

Plate 4 *(a) Geometry of tetramethylene sulfoxide (6): c, corner; h, hinge; f, flap; n, endo; x, exo. (b) DFT, B3LYP/6-311+G(3df,2p) optimized geometry (see Figure 7.17).*

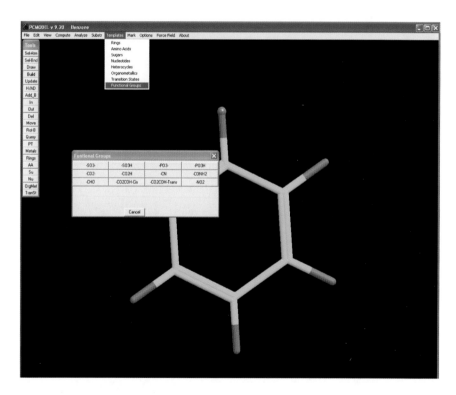

Plate 5 *Substituting a selected atom with a functional group in PCModel (see Figure 9.2).*

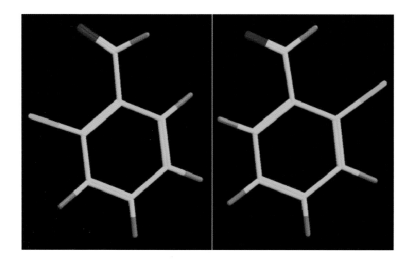

Plate 6   *Optimized cis and trans o-chlorobenzaldehyde resulting from optimization starting from different starting geometries (see Figure 9.3).*

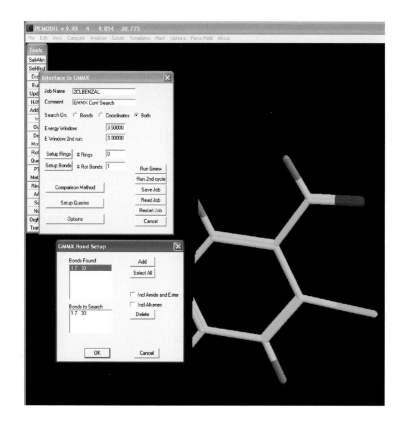

Plate 7   *Setting up rotatable bonds in GMMX (see Figure 9.4).*

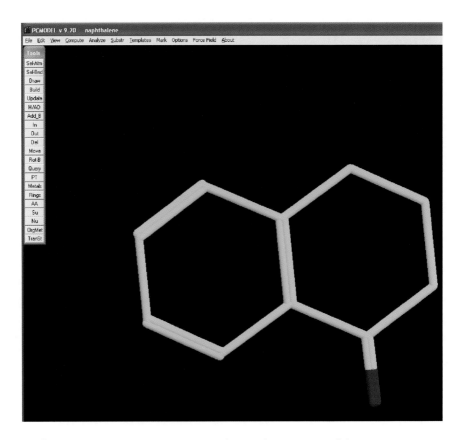

Plate 8 *Heteroatom representation of α-tetralone in PCModel (see Figure 9.6).*

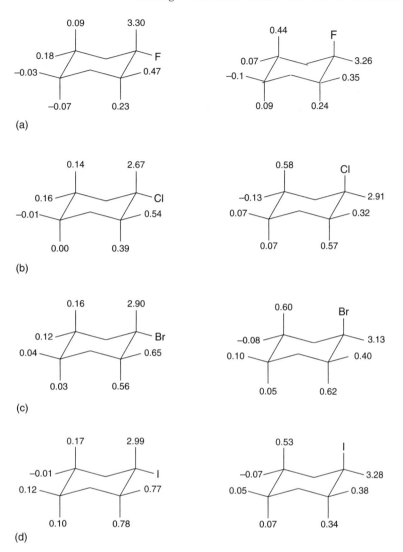

Figure 6.3 *Observed halo SCSs in (a) fluoro-,(b) chloro-, (c) bromo- and (d) iodo-cyclohexanes.*

bond was modelled by four magnetic dipoles parallel to but placed perpendicular to the C—X bond at a given distance (d_m) from X (model B). The result is a shielding cone as shown in Figure 6.4 (model B) characteristic of an anisotropy term. Thus the parameters to be fitted for each halogen are two shielding coefficients a_s (Equation (3.11)) and the push–pull coefficients in model A and one shielding plus push–pull coefficient and the anisotropy ($\Delta\chi$(C—X)) and distance d_m (Figure 6.4) in model B. Both models gave improved results compared to the original steric term.

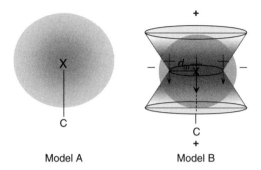

Model A   Model B

Figure 6.4 *Models of C—X shielding: model A, only steric term; model B, steric term plus an anisotropy term.*

6.3.2   Aromatic Halides

The chemical shift data for H-5 in the 4-halophenanthrenes provides a definitive test of the two models and these results are given in Table 6.8. It is clear that model B gives much better agreement with the observed data and it will be used in CHARGE henceforth.

Table 6.8 *Observed[45] versus calculated[43] SCSs of H-5 in 4-halophenanthrenes*

Halogen	Observed SCS	Calculated, model A	Calculated, model B
—Cl	0.98	0.65	0.77
—Br	1.21	0.59	1.14
—I	1.28	0.53	1.39

The meta' effect The failure of the classic steric, π-electron and electric field effects to explain the SCS on the *meta* protons in halobenzenes was noted by Abraham *et al.*[46] who suggested that this may be due to a through bond W effect, since the 3,5eq proton chemical shifts of the 1-eq-halocyclohexanes were also not calculated very well by CHARGE. However, the current model of CHARGE accurately predicts the 1-eq-halocyclohexanes chemical shifts. In the 1-halonaphthalenes (Table 6.10) the magnitude of the SCS on the *meta* protons is larger than in the corresponding halobenzenes and this suggests that this effect may be due to the aromaticity of the system.

In order to obtain more detailed data on this effect, a set of *ortho*-substituted bromo benzenes were selected and the bromo SCSs determined.[43] Bromobenzenes were used since the SCS of the halogen on the *meta* proton is quite large and also the geometry of the bromobenzenes is more reliable than that of the iodo-compounds when modelled. The geometries used in these calculations were minimized at the B3LYP/6-31G** level except the substituted bromobenzenes and the bromo- and iodo-alkanes which were minimized using the MMFF94 force field. The recommended[47] basis set (Lanl2DZ) for iodo benzenes produced bond lengths, which were in poor agreement with experimental and experimental geometries were used. Full details are given elsewhere.[41,43]

Table 6.9 *Observed*[48] *versus calculated ¹H chemical shifts for ortho-substituted bromobenzenes*

X	Observed/calculated	H3	H4	H5	H6
—H	Observed	7.24	7.30	7.24	7.50
	Calculated	7.24	7.30	7.24	7.49
—CH₃	Observed	7.18	7.18	7.01	7.50
	Calculated	7.16	7.23	7.05	7.44
—OH	Observed	7.02	7.22	6.80	7.46
	Calculated	6.91	7.19	6.82	7.37
—CN	Observed	7.65	7.45	7.45	7.65
	Calculated	7.72	7.50	7.46	7.72
—Cl	Observed	7.40	7.20	7.10	7.60
	Calculated	7.35	7.21	7.14	7.51
—Br	Observed	7.55	7.15	7.15	7.55
	Calculated	7.52	7.16	7.16	7.52
—I	Observed	7.85	6.95	7.15	7.60
	Calculated	7.75	7.00	7.17	7.49

The observed data (Table 6.9) shows clearly that the *meta* effect is present in all cases for H-5 but not for H-3. The effect practically disappears for H-3 if there is any atom other than hydrogen in the *ortho* position. Using additive tables[49] the SCS of the H-3 proton of o-methyl bromobenzene is −0.28 (obs. −0.16) whilst the H-5 proton is −0.30 (obs. −0.33). The large discrepancy at the H-3 proton disappears if the *meta* SCS of the halogen (−0.08) is removed. Similar observations that the additive SCS tables poorly represent the observed data for *ortho*-substituted benzenes have been made previously but no attempt has been made to resolve this problem.[50,51] This remarkable *meta* effect was reproduced in CHARGE by simply introducing a γ effect from C-1 to the *meta* proton provided that no alpha substituent was present. The three bond effect of C-1 in naphthalene to H-8 remains at zero. This was required since H-8 is affected by steric, anisotropic and electric field effects and thus this contribution cannot be isolated and determined. Support for this treatment comes from the ³J_CH coupling. A ¹³C experiment with gated ¹H decoupling allowed this coupling to be determined. In bromobenzene the ³J_CH coupling (C1—H3) was 10.6 Hz whereas in 9-bromoanthracene the ³J_CH (C9-H1) coupling was only 6.3 Hz. This large reduction is most likely due to the W orientation of C-1 and H-3 and suggests that the *meta* and peri interactions differ and should be handled separately. This model gives the calculated chemical shifts in Table 6.9 in excellent agreement with the observed shifts.

The extension to the halonaphthalenes can also be tested and Table 6.10 gives the observed vs. calculated shifts for the halonaphthalenes examined. This table also shows the results of GIAO calculations for these molecules. The charge calculations give in general very good agreement with an overall rms error of 0.06 ppm. The GIAO calculations are not as accurate. For the fluoro, chloro and bromo compounds the calculated shifts are all

Table 6.10   *Observed versus calculated ¹H chemical shifts in 1-halonaphthalenes*[43]

Halogen		H2	H3	H4	H5	H6	H7	H8
−F	CDCl₃	7.134	7.381	7.613	7.845	7.535	7.511	8.106
	CHARGE	7.078	7.482	7.495	7.812	7.523	7.451	7.995
	GIAO	7.305	7.687	7.756	7.990	7.787	7.800	8.444
−Cl	CDCl₃	7.545	7.343	7.725	7.820	7.505	7.567	8.257
	CHARGE	7.427	7.408	7.710	7.828	7.503	7.517	8.304
	GIAO	7.743	7.668	7.887	8.015	7.811	7.885	8.582
−Br	CDCl₃	7.772	7.234	7.797	7.831	7.519	7.586	8.229
	CHARGE	7.613	7.369	7.755	7.839	7.505	7.533	8.291
	GIAO	7.861	7.650	7.904	8.016	7.827	7.902	8.655
−I	CDCl₃	8.080	7.164	7.821	7.753	7.502	7.564	8.080
	CHARGE	7.871	7.225	7.787	7.834	7.496	7.536	8.167
	GIAO	7.312	6.980	7.091	7.205	7.133	7.222	7.621

ca. 0.2–0.3 ppm larger than observed, whilst for iodonaphthalene the calculated shifts are ca. 0.2–0.7 ppm less than observed.

These results together with those given earlier for substituted benzenes (Chapter 5, Section 5.1 and Appendix A) which use the present theory show that the CHARGE model now can be used with confidence to predict the ¹H chemical shifts of any halogenated benzenoid compound.

These investigations included the halocyclohexanes and gave the following values for the parameters. The shielding coefficients a_s and anisotropy distance d_m (Figure 6.4) for Cl, Br and I were 90.0, 95.0 and 120.0 Å[6] and 0.50, 1.00 and 1.20 Å, respectively. The values of $\Delta\chi$ (C−X) for Cl, Br and I were 12,17 and 22 ppm Å³ mol⁻¹ and the push–pull coefficients for methylene protons were 0.75 (Cl), 0.50 (Br) and 0.40(I), cf. 0.53 (Me) to give the observed shielding effects.

6.3.3   Alkyl Halides

The chemical shifts of the methine protons beta to the substituent in the bromo and iodo alkanes show an enhanced deshielding effect from that given by Equationss (3.2) and (3.3), in particular where the proton is beta to two carbon atoms. Thus the Br beta effect was increased by 19% for RCHXBr (X ≠ C) and 47% for RR′CHBr (R, R′ = alkyl) methine protons. Similarly for iodo-alkanes the beta effect on methine protons was increased by 47% for RCHXI and by 105% for RR′CHI protons. For the highly polarizable iodine atom, even the beta methylene protons were more deshielded than calculated in the presence of a beta carbon, and the iodine beta effect was increased by 28% to compensate.

The observed and calculated ¹H chemical shifts for the acyclic haloalkanes and SCSs for the cyclic compounds studied are given in Tables 6.11 to 6.18. In Table 6.11 the values for both the *trans* and *gauche* conformers of 1-halo- and 2-methyl-1-halo-propanes are given but in the cases of the 1,2-dihalo-, 1,1,2-trihalo- and 1,1,2,2-tetrahalo-ethanes the shifts for both conformers are the same, due to the non-orientational nature of the through bond H.C.C.X γ effect.

Table 6.11   *Observed versus calculated ^{1}H chemical shifts (δ) for acyclic haloalkanes*

Molecule	Group	X=Cl Obs.[a]	X=Cl Calc.	X=Br Obs.[a]	X=Br Calc.	X=I Obs.[a]	X=I Calc.
CH$_3$X	CH$_3$	3.05	3.12	2.68	2.79	2.16	2.21
CH$_2$X$_2$	CH$_2$	5.33	5.27	4.94	4.74	3.90	3.79
CHX$_3$	CH	7.27	7.28	6.82	6.34	4.91[c]	5.12
CH$_3$CH$_2$X	CH$_2$	3.57[b]	3.58	3.36	3.19	3.15	3.08
	CH$_3$	1.49	1.49	1.65	1.65	1.86	1.87
CH$_3$CHX$_2$	CH	5.87[c]	5.72	5.86[c]	5.68	5.24[d]	5.36
	CH$_3$	2.23	2.12	2.47	2.42	2.96	2.83
CH$_3$CX$_3$	CH$_3$	2.75	2.72				
CH$_2$XCH$_2$X	CH$_2$	3.69	3.85	3.63	3.64	3.64[e]	3.65
CH$_2$XCHX$_2$	CH	5.74	5.99	5.61[f]	5.85	—	—
	CH$_2$	3.97	4.18	4.04	4.04	—	—
CHX$_2$CHX$_2$	CH	5.94	5.99	6.03	6.01	—	—
CX$_3$CHX$_2$	CH	6.12[c]	6.12	—	—	—	—
n-PrX[i]	CH$_2$Cl	3.47[c]	3.64 (g)	3.36	3.25 (g)	3.16	3.14 (g)
			3.44 (t)		3.13 (t)		3.02 (t)
	CH$_2$	1.81	1.74 (g)	1.89	1.84 (g)	1.86	1.99 (g)
			1.74 (t)		1.84 (t)		1.99 (t)
	CH$_3$	1.06	1.00 (g)	1.04	0.94 (g)	1.04	0.96 (g)
			0.95 (t)		0.98 (t)		0.93 (t)
Me$_2$CHX	CH	4.13	4.31	4.20	4.46	4.24	4.55
	CH$_3$	1.54	1.56	1.71	1.69	1.88	1.91
CMe$_3$X	CH$_3$	1.58	1.54	1.76	1.71	1.95	1.93
Me$_2$.CHCH$_2$X[i]	CH$_2$X	—	—	3.31[c]	3.16 (g)	3.15[h]	3.05 (g)
					3.31 (t)		3.21 (g)
	CH	—	—	1.98	2.03 (g)	1.73	2.10 (g)
					2.03 (t)		2.10 (t)
	CH$_3$	—	—	1.03	0.99 (g)	1.01	0.97 (g)
					1.01 (t)		0.99 (t)
Me$_3$.C.CHX$_2$	CH	—	—	—	—	5.21[g]	5.10
	CH$_3$	—	—	—	—	1.19	0.99

[a] Data from Tiers[52], unless stated otherwise.
[b] Data from Altona *et al.*[53].
[c] Data from Gasteiger and Marsili[54].
[d] Pure liquid, data from Neumann and Rahm[55].
[e] Data from Pachler and Wessels[56].
[f] Data from Stepanyants *et al.*[57].
[g] Data from Puchert and Behnke[48].
[h] Data from Abraham *et al.*[1].
[i] (g) *gauche* conformer; (t) *trans* conformer.

The observed and calculated (CHARGE and GIAO) ^{1}H chemical shifts in the halocyclohexanes are shown in Table 6.12. The GIAO calculations use the same formulation as Table 6.10. The CHARGE calculations show general agreement with the observed shifts. The GIAO calculations also reproduce the observed trends but are in general less accurate than CHARGE. The calculated ^{1}H shifts in CHARGE of the 2,6 axial and equatorial hydrogens reproduce the observed values even though the calculated values are obtained from the non-orientational γ effect. The observed shifts for axial iodine are less than the calculated values (but much greater than the GIAO values) and also much less than the

Table 6.12   *Observed versus calculated ¹H chemical shifts (δ) for halocyclohexanes*

Halogen		1-ax	1-eq	2,6-ax	2,6-eq	3,5-ax	3,5-eq	4-ax	4-eq
Eq-Cl	Obs.	*3.88*		1.58	2.22	1.33	1.84	1.18	1.68
	CHARGE	3.96		1.61	2.07	1.24	1.75	1.27	1.70
	GIAO	4.14		1.80	2.23	1.58	1.91	1.48	1.72
Ax-Cl	Obs.		4.59	1.76	2.00	1.77	1.55	1.26	1.75
	CHARGE		4.38	1.62	2.07	1.63	1.63	1.18	1.73
	GIAO		4.66	1.83	2.11	2.23	1.57	1.50	1.87
Eq-Br	Obs.	4.09		1.75	2.33	1.35	1.8	1.22	1.72
	CHARGE	3.96		1.71	2.17	1.22	1.73	1.25	1.69
	GIAO	4.38		1.97	2.33	1.60	1.91	1.53	1.73
Ax-Br	Obs.		4.81	1.81	2.08	1.79	1.60	1.24	1.78
	CHARGE		4.37	1.73	2.17	1.73	1.64	1.17	1.72
	GIAO		4.98	1.93	2.17	2.42	1.58	1.50	1.88
Eq-I	Obs.	4.18		1.97	2.45	1.36	1.67	1.30	1.80
	CHARGE	4.14		1.71	2.33	1.18	1.67	1.21	1.64
	GIAO	3.27		1.33	1.12	0.76	0.80	0.79	0.60
Ax-I	Obs.		4.96	1.53	2.06	1.72	1.62	1.26	1.73
	CHARGE		4.56	1.87	2.33	1.63	1.60	1.14	1.68
	GIAO		3.87	1.14	1.06	1.46	0.65	0.67	0.94

bromo value, in contrast to the shifts for equatorial iodine. This may well be due to steric perturbation of the axial iodine atom (see later). However the steric calculation for the 3,5 axial hydrogens do replicate the observed trends.

For all the 3,5-axial hydrogens the rms error of observed and calculated shifts is ca. 0.10 ppm. The separation of the SCS into the constituent parts will be deferred until the next section.

Table 6.13   *Observed versus calculated ¹H SCSs (ppm) for 1,1-dichloro-[a] and trans-1,2-dibromo[b] cyclohexanes*

	1,1-Dichloro-			trans-1,2-Dibromo			
	Obs.	Calc.		Diequatorial		Diaxial	
Proton			Proton	Obs.	Calc.	Obs.	Calc.
1a (CH)	—	—	1,2ax (CH)	2.89	3.18	—	—
1e (CH	—	—	1,2eq (CH)	—	—	3.02	3.14
2,6a	0.93	0.81	3,6ax	0.74	0.57	1.26	1.07
2,6e	0.86	0.83	3,6eq	0.86	0.62	0.25	0.47
3,5a	0.56	0.53	4,5ax	0.22	0.10	0.63	0.59
3,5e	0.07	−0.02	4,5eq	0.13	0.16	−0.04	0.00
4a	0.05	0.07					
4e	0.07	0.10					

[a] Data from Abraham *et al.*[41].
[b] Data from Abraham *et al.*[42].

Table 6.14 *Observed[a] versus calculated[b] ¹H chemical shifts (δ) for 9-chloro-trans-decalin and SCSs (ppm) for 9-chloro, 9-bromo and 9-iodo trans-decalin*

	9-Chloro,		SCS					
	chemical shift		Chloro		Bromo		Iodo	
Proton	Obs.	Calc.	Obs.	Calc.	Obs.	Calc.	Obs.	Calc.
1,8a	1.46	1.43	0.53	0.42	0.54	0.40	0.46	0.51
1,8e	1.92	2.04	0.38	0.42	0.52	0.55	0.68	0.70
2,7a	1.87	1.85	0.62	0.68	0.70	0.69	0.82	0.72
2,7e	1.56	1.57	−0.11	−0.18	−0.08	−0.10	−0.01	−0.11
3,6a	1.27	1.18	0.02	0.01	0.05	−0.01	0.11	−0.02
3,6e	1.73	1.81	0.06	0.06	0.05	0.05	0.04	0.04
4,5a	1.46	1.73	0.53	0.71	0.48	0.75	0.44	0.77
4,5e	1.30	1.43	−0.08	−0.20	−0.23	−0.12	−0.25	−0.13
10 (CH)	1.26	1.07	0.38	0.20	−0.04	0.10	−1.20	0.14

[a] Data from Schrumpf *et al.*[40].
[b] Data from Abraham *et al.*[41].

Table 6.13 shows that the SCS from multiple halo substitution on either the same or vicinal carbon atoms is also well reproduced with only a few values differing by more than 0.15 ppm.

In 9-substituted-trans-decalins (Table 6.14) the calculated SCSs show slightly greater deviation from the observed, although problems with the geometry were found for 9-chloro-*trans*-decalin. The initial calculated shifts for the 1,8a (and 10) protons of 9-chloro-*trans*-decalin using RHF/6-31G* optimized geometry were inconsistent with the observed shifts and also those for the axial-chloro-cyclohexane analogue.[41] The same calculations were therefore performed using the RHF/6-31G* optimized unsubstituted *trans*-decalin geometry with a C—H bridge proton replaced by a C—Cl bond. This unstrained structure gave more realistic results for both $H_{1,8a}$ and H_{10} and the results of these calculations are given in Table 6.14. Since the unstrained structure gives closer agreement with the experimental shifts, which are also consistent with the cyclohexane value, it would appear this is the more valid geometry. The distortion of the *trans*-decalin ring caused by the 9-chloro substituent, presumably interacting with four parallel and close through space C—H bond protons, appears to be over-emphasized at the RHF/6-31G* level. The consequential reduction in the H..H distances to $H_{1,8a}$ and H_{10} lead to a calculated increase in the shielding steric interactions, highlighting the sensitivity of the r^{-6} term to the chosen geometry

Schneider[20] reported anomalous shielding effects in 9-bromo and 9-iodo-*trans*-decalins on H-10, and this is supported by the CHARGE calculations. In the case of iodine the shielding effect is even greater than the deshielding effect on H-1,8-ax, and this data point was thus excluded from the parameterization of the scheme. Clearly, there is some additional effect occurring through the C—C bridge than is calculated by the CHARGE non-orientational γ effect (see later).

The observed and calculated SCSs in halo-bornanes and norbornanes are given in Table 6.15. It should be noted that the SCS for a single substituent derived from multi-functional compounds, as in the case of the iodo-camphor, is dependent upon the non-interaction of the substituents. Since the C=O group is adjacent to the C—I bond in 3-substituted camphors some interaction is possible. The 2-iodo norbornane was not available for comparison, but the SCS effect of the β proton in the O=C—C̲H̲—Cl fragment in 2-endo-chlorobornane is +2.68 ppm, yet in 3-endo-camphor it is only +2.04 ppm. The observed iodo-camphor SCS in Table 6.15 may be a less definite representation of the iodobornane SCS effects.

The steric shielding on H-6n in endo-bromobornane (calc. 0.84 vs. obs. 0.91 ppm) and the 7s proton in exo-halonorbornanes reflect the observed effects. The smaller size of the push–pull coefficient for iodine than bromine can be seen experimentally in the SCS effects on the 7-anti protons. In 2-exo-iodobornane the H-7a SCS is +0.17 ppm, greater than for the bromo-norbornane (+0.11 ppm), yet the 7-syn protons are similarly shielded (+0.74 and +0.68 ppm) and the linear electric field effect is less (cf. CHARGE partial atomic charge, I = −0.089 electrons and Br = −0.125 electrons). The relative deshielding of H-7anti may be attributed to a smaller iodine push–pull shielding effect.

The halo γ SCSs in the bicycloheptanes in Table 6.15 appear anomalous around $\theta(\angle HCCX) \sim 120°$. The SCS of H_{3x} in 2-endo-chlorobornane is. +0.75 ppm (obs.) vs. +0.37 (calc.) ppm and in the 2-endo bromo +0.82 (obs.) vs. 0.48 (calc.) ppm. The 2-endo iodo has not been determined but in the 2-exo-iodo H_{3n} is +0.97 (obs.) vs. +0.67 (calc.) ppm, thus it is a general effect. The experimental halo SCSs for the 2-endo- and 2-exo- bicycloheptanes in Table 6.15 are all >0.75 ppm for $\theta \sim 120°$ and are nearly always larger than the SCSs for $\theta \sim 0°$ of 0.2–0.3 ppm (Cl), 0.6–0.9 ppm (Br) and 0.76 ppm (I). This effect is also observed[58] in acenaphthenes where the Cl SCSs = +0.46 ppm ($\theta \sim 120°$) and +0.16 ppm ($\theta \sim 0°$) and hexachlorobicyclo [2.2.1.]heptanes, SCS = +0.59 ppm ($\theta \sim 120°$) and +0.22 ppm ($\theta \sim 0°$). Clearly, the non-orientational halo γ effect in the CHARGE scheme would not be expected to reproduce these effects.

In the case of the haloandrostane data (Table 6.16) derived from 3-substituted-androstan-17-ones the interaction between the substituents would be expected to be minimal. The validity of this assumption can be verified by the good agreement between the SCS data derived from 3α-chloroandrostane and from 3α-chloro-androstan-17-one in Table 6.16.

One exception is H-1α and H-1β in 3α-chloro-androstane which were noted as overlapping signals with reported SCSs both as +0.23 ppm[1]. The chemical shifts of 1α and 1β thus obtained are (0.89 + 0.23) 1.02 and (1.66 + 0.23) 1.89 ppm, respectively, with a separation of 0.87 ppm between the protons. Clearly, the reported SCSs are incorrect. The analogous SCSs from 3α-chloro-androstan-17-one again reported[16] as overlapping signals give chemical shifts for 1α at (0.89 + 0.60) 1.49 and for 1β also at (1.67 − 0.18) 1.49 ppm, thus these that values are more consistent.

Apart from the above the reported chlorine SCSs from 3α-chloro-androstane and androstan-17-one are consistent. The halo SCSs on the C and D ring protons are ≤ ±0.05 ppm and have not been shown in table 6.16. For H-9 the alkane SCSs for 3α substitution of +0.06 are less than half that from the ketone (+0.15, +0.18), but is in agreement with the CHARGE calculations (+0.05 ppm). Considering the uncertainties in the

Table 6.15 *Observed versus calculated[a] SCSs (ppm) for 2-halo-bicyclo[2.2.1]heptanes*

Proton	X=Cl 2-endo-[c] Obs.	X=Cl 2-endo-[c] Calc.	X=Cl 2-exo-[d] Obs.	X=Cl 2-exo-[d] Calc.	X=Br 2-endo-[c] Obs.	X=Br 2-endo-[c] Calc.	X=Br 2-exo-[b] Obs.	X=Br 2-exo-[b] Calc.	X=I 2-exo-[d] Obs.	X=I 2-exo-[d] Calc.
1 (CH)			0.20	0.21			0.32	0.26	0.41	0.33
2n	—	—	2.71	2.40	—	—	2.83	2.84	2.82	2.91
2x	2.68	2.43	—	—	2.83	2.78	—	—	—	—
3n	0.22	0.46	0.75	0.44	0.87	0.54	0.87	0.52	0.97	0.67
3x	0.75	0.37	0.33	0.41	0.82	0.48	0.60	0.53	0.76	0.65
4 (CH)	0.08	0.10	0.12	0.11	0.06	0.08	0.12	0.03	0.06	0.08
5n	0.14	0.04	-0.09	0.09	0.14	0.02	-0.08	0.06	-0.06	0.05
5x	0.04	0.10	0.00	0.07	0.04	0.08	0.01	0.07	0.02	0.05
6n	0.84	0.92	-0.02	0.01	0.84	0.91	0.02	0.02	0.07	-0.01
6x	-0.15	-0.21	0.14	0.11	-0.07	-0.14	0.17	0.09	0.10	0.06
7a			0.06	-0.02			0.11	-0.04	0.17	-0.04
7s			0.59	0.48			0.68	0.54	0.74	0.48
8-Me	0.05	0.05			0.05	0.04				
9-Me	0.05	0.04			0.04	0.03				
10-Me	0.10	0.11			0.14	0.09				

[a] Calculated SCSs, cf. bornane or cf. norbornane (Chapter 4, Table 4.2); data from Abraham *et al.*[41].
[b] Data from Abraham and Fisher[22].
[c] Observed SCSs, cf. 3-endo- and 3-exo-chloro, -bromo and -iodo camphors; data from Kaiser *et al.*[23].
[d] Data from Abraham *et al.*[21].

experimental SCSs due to signal overlap at the 1,2,4 and 6 positions in the 3-haloandrostanes the calculated SCSs are encouraging. For H-5 where the experimental data are more reliable in the 3α-halo substituent the scheme predicts the SCSs extremely well.

The calculated γ effect on H-2,8,9 in the 1-halo- and H-1,3 in 2-halo-adamantanes (Table 6.17) is in complete agreement with the observed, and even the SCS on H-3 in the 2,4-dihalo-adamantanes (Table 6.18) with two γ effects appears additive and in agreement with the calculated values. Also the long-range effect in the 2-halo-adamantanes (Table 6.17) on the deshielded sterically perturbed H-4,9ax and correspondingly shielded geminal H-4,9-eq shows the general applicability of the steric plus push–pull term to these systems. In 2-bromoadamantane the calculated steric effect on H-4,9ax (calc. +0.64 ppm, obs. +0.59 ppm) and push–pull on H-4,9eq (calc. −0.12 ppm, obs. −0.13 ppm) demonstrate the applicability of these terms to the caged structures. Similar effects are noted (Table 6.18) in *trans*-2,4-dichloro-adamantane on H-10syn/anti, H-6syn/anti and H-9ax/eq, and in the *cis*-2,4-conformer for H-6syn/anti. The calculated electric field effects on H-8,10 are also in good agreement with the observed SCSs.

Table 6.16   *Observed*[18] *versus calculated*[c,42] *SCSs (ppm) for 3-haloandrostanes*

		3α-Chloro-		3β-Chloro-	
Proton	Obs.[b]	Obs.[a]	Calc.	Obs.[a]	Calc.
1α	0.23*	0.60*	0.49	0.13	0.07
1β	0.23*	−0.18*	−0.03	0.09	0.12
2α	0.36	0.42*	0.42	0.57*	0.43
2β	0.45	0.42*	0.43	0.31*	0.40
3α	2.81	—	—	2.63	2.51
3β	—	2.83	2.54	—	—
4α	0.37*	0.38	0.44	0.51	0.44
4β	0.51*	0.46	0.40	0.36	0.39
5 (CH)	0.64	0.67	0.61	0.10	0.03
9 (CH)	0.06	0.15	0.05	−0.03	0.00
19-Me	0.00	0.00	0.01	0.05	0.04

	X=Br				X=I			
	3α-		3β-		3α-		3β-	
Proton	Obs.[a]	Calc.	Obs.[a]	Calc.	Obs.[a]	Calc.	Obs.[a]	Calc.
1α	0.65*	0.55	0.16	0.05	0.64*	0.52	0.16	0.03
1β	−0.13*	−0.05	0.06	0.11	−0.14*	−0.05	−0.09	0.09
2α	0.49*	0.51	0.69*	0.52	0.47*	0.66	0.68*	0.66
2β	0.49*	0.55	0.51*	0.54	0.24*	0.70	0.79*	0.69
3α	—	—	2.80	2.95	—	—	2.93	3.03
3β	3.05	2.94	—	—	3.26	3.04	—	—
4α	0.43*	0.51	0.58*	0.51	0.48*	0.65	0.72*	0.65
4β	0.43*	0.52	0.58*	0.52	0.21*	0.67	0.72*	0.66
5 (CH)	0.72	0.69	0.12	0.04	0.68	0.67	0.11	0.02
9 (CH)	0.18	0.04	−0.02	0.02	0.18	0.02	−0.03	0.02
19-Me	0.01	0.01	0.07	0.03	0.02	0.01	0.07	0.02

* Unresolved.
[a] SCSs, cf. 3α- and 3β- halo-androstan-17-ones.
[b] SCS, cf. 3α- chloro-androstane vs. 5α-androstane.
[c] Calculated SCSs, cf. 3α- and 3β-haloandrostanes vs. 5α-androstane.

Table 6.17 *Observed versus calculated SCSs for chloro-, bromo- and iodoadamantanes*

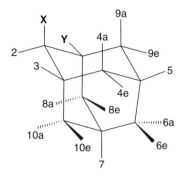

	1-Chloro- (X=H, Y=Cl)			2-Chloro- (X=Cl, Y=H)	
Proton	Obs.[a,41]	Calc.	Proton	Obs.[a,41]	Calc.
2,8,9	0.39	0.41	1,3 (CH)	0.20	0.21
3,5,7 (CH)	0.27	0.11	2 (CH)	2.65	2.24
4,6,10-ax	−0.08	0.07	4,9-ax	0.52	0.58
4,6,10-eq	−0.08	0.09	4,9-eq	−0.18	−0.09
			5 (CH)	−0.01	0.07
			6[B]	0.01	0.05
			7 (CH)	−0.01	0.08
			8,10-ax	0.05	0.01
			8,10-eq	0.20	0.10

	1-Bromo X=H, Y=Br		1-Iodo X=H, Y=I			2-Bromo X=Br, Y=H		2-Iodo X=I, Y=H	
Proton	Obs.[22]	Calc.[42]	Obs.[26]	Calc.	Proton	Obs.[22]	Calc.	Obs.[61]	Calc.
2,8,9	0.62	0.51	0.88	0.65	1,3 (CH)	0.28	0.26	0.29	0.33
3,5,7 (CH)	0.23	0.09	0.10	0.07	2 (CH)	2.92	2.67	3.25	2.73
4,6,10-ax	−0.02	0.06	0.11*	0.04	4,9-ax	0.59	0.64	0.63	0.62
4,6,10-eq	−0.02	0.06	0.11*	0.05	4,9-eq	−0.13	−0.12	−0.04	−0.10
					5 (CH)	0.01	0.06	0.03*	0.05
					6	−0.01	0.03	0.04	0.03
					7 (CH)	0.01	0.06	0.03*	0.04
					8,10-ax	0.22	0.19	0.20*	−0.02
					8,10-eq	0.11	0.07	0.20*	0.04

* Unresolved.
[a] SCSs, cf. adamantane (Chapter 4, Table 4.2).
[b] γ/δ and ε-ax/ε-eq unresolved.

6.3.4 Contributions to the ¹H SCSs in Halocyclohexanes

The generally good agreement between the observed and calculated halo SCSs is encouraging and demonstrates the general applicability of the scheme to the calculation of ¹H chemical shifts in halo-substituted alkanes. For the cyclic chloro-alkanes studied with 70 chemical shifts spanning a range of ca. 1.1 to 7.3δ the CHARGE scheme fits the

Table 6.18 *Observed*[60] *versus calculated*[42] *SCSs for dichloro-, dibromo- and diiodo-adamantanes*

	2(ax),4(eq)-Dihalo-(X=H; Y, Z=Cl)			2(eq),4(eq)-Dihalo-(X, Z=Cl, Y=H)	
Proton	Obs.	Calc.	Proton	Obs.	Calc.
1 (CH)	0.18	0.28	1,5 (CH)	0.16	0.28
2-eq (CH)	2.76	2.33	2,4-ax (CH)	2.51	2.22
3 (CH)	0.45	0.41	3 (CH)	0.39	0.41
4-ax (CH)	3.07	2.94	6,8-anti	−0.18	−0.05
5 (CH)	0.18	0.28	6,8-syn	0.51	0.63
6-anti	−0.17	−0.05	7 (CH)	−0.02	0.14
6-syn	0.52	0.63	9-ax	0.01	0.01
7 (CH)	—	0.14	9-eq	0.36	0.20
8-anti	0.11*	0.15	10	0.33	0.51
8-syn	0.11*	0.05			
9-ax	0.60	0.58			
9-eq	0.05	0.01			
10-anti	−0.17	−0.11			
10-syn	0.71	0.71			

	X=H, Y/Z=Br		X=H, Y/Z=I			X/Z=Br, =H		X/Z=I, Y=H	
Proton	Obs.	Calc.	Obs.	Calc.	Proton	Obs.	Calc.	Obs.	Calc.
1 (CH)	0.28	0.32	0.23	0.37	1,5 (CH)	0.28	0.32	0.31	0.37
2-eq (CH)	2.98	2.74	3.21	2.76	2,4-ax (CH)	2.75	2.67	3.14	2.67
3 (CH)	0.54	0.51	0.43	0.65	3 (CH)	0.51	0.51	0.49	0.65
4-ax (CH)	3.40	3.41	3.73	3.48	6,8-anti	−0.12	−0.08	−0.02	−0.08
5 (CH)	0.28	0.32	0.27	0.38	6,8-syn	0.60	0.69	0.65	0.66
6-anti	−0.13	−0.08	−0.07	−0.08	7 (CH)	−0.02	0.12	−0.02	0.09
6-syn	0.49	0.69	0.61	0.66	9-ax	0.16	0.03	0.40*	−0.04
7 (CH)	—	0.12	—	0.09	9-eq	0.45	0.14	0.40*	0.08
8-anti	0.18*	0.10	0.20*	0.06	10	0.50	0.57	0.61	0.55
8-syn	0.18*	0.05	0.20*	0.01					
9-ax	0.71	0.63	0.76	0.59					
9-eq	0.12	−0.02	0.10	−0.04					
10-anti	−0.02	−0.11	0.15	−0.14					
10-syn	0.69	0.75	0.80	0.70					

* Unresolved.

experimental data to an rms of 0.15 ppm, which is not much more than the combined errors in the observed data and in the calculations. The corresponding values for the bromo compounds are 118 shifts and an rms of 0.10 ppm and for the iodo compounds 96 shifts and an rms of 0.14 ppm.

On the basis of this good agreement it is possible to examine the relative size of the various contributions to the ¹H chemical shifts of the halocyclohexanes (Figure 6.5). These

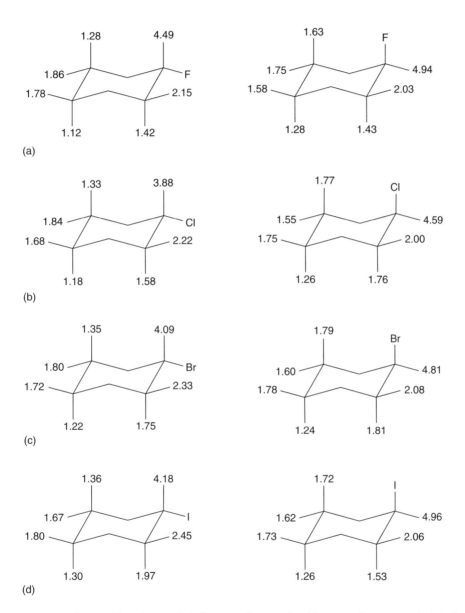

Figure 6.5 *Observed ¹H chemical shifts in (a) fluoro-, (b) chloro-, (c) bromo- and (d) iodo-cyclohexanes.*

should be considered with respect to the contributions for cyclohexane itself (Chapter 4, Figure 4.8).

Charge contributions. The through bond contributions to the SCS in the halocyclohexanes are only for the β (CHX) and vicinal (HCCX) protons. For the β (CHX) protons, i.e. 1ax and 1eq, the chemical shifts are not in the order of the electronegativity of the halogen atom but for both the axial and equatorial substituents the order is F > I > Br > Cl (Figure 6.5). This was noted earlier for the average chemical shifts of the β methine protons in halocyclohexanes at room temperature and attributed to an increase in the contribution of possible resonance forms.[35] In the CHARGE scheme this order is obtained by two opposing through bond effects. The β charge term is given by a general electronegativity term (Chapter 3, Equation (3.3)) with an explicit correction for the heavy halogen atoms in the order I > Br > Cl. In contrast the deshielding effect of the halogen at the γ protons (H-2ax and H-2eq) increases in the order F < Cl < Br < I for the equatorial conformer, but in the axial form H-2eq is essentially constant at 2.04δ and H-2ax is anomalous in axial iodo-cyclohexane (Table 6.12). There is also no obvious orientational dependence of the γ SCS supporting the CHARGE treatment. In CHARGE the γ effect is proportional to the polarizability of the substituent and this simple relationship gives generally good agreement with the observed SCS. For β and γ protons this is the only factor in the halogen SCSs. There are small indirect ring deformation contributions amounting to ca. −0.03 to 0.02 ppm. For the more distant protons there is no charge effect of the substituent in CHARGE and the SCSs on these protons are thus due to the other contributions.

Steric contributions. The steric effect (Chapter 3, Equation (3.11)) does not operate on the β and γ protons, thus the major steric effect of the halo-substituent is on H-3,5ax in axial halo-cyclohexanes and this is shown in Figure 6.6. For all the other axial protons the steric term stems largely from the cyclohexane H..H shielding at ca. −0.2 ppm (Chapter 4, Figure 4.8). For the equatorial protons the steric contribution is much less. Steric deshielding

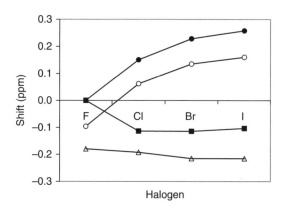

Figure 6.6 *The steric and push–pull contributions to the chemical shifts of halocyclohexanes. Steric contributions: (●) δ X..H steric [3ax]; (o) δ total steric [3ax]; (■) δ X..H push–pull [3eq]; (Δ)δ total steric [4ax] (reproduced from Abraham et al.[42] with permission of the Royal Society of Chemistry).*

of H-3ax and push–pull shielding on H-3eq results in δeq < δax for H-3ax and H-3eq in axial chloro-, bromo- and iodo-cyclohexanes in contrast to the usual situation.

The X..H steric contribution in the 1,3 syn di-axial orientation increases with halogen size (Figure 6.6) from 0.00 ppm for fluorine to +0.26 ppm for iodine, but the total steric term on H-3ax also includes one H..H shielding effect as one of the 1,3-syn diaxial H..H interactions in cyclohexane is replaced by an X..H deshielding term. Thus the overall effect for fluorine is shielding, but deshielding for the other halogens. H-3eq experiences a 'push–pull' effect which is a function of the deshielding effect on H-3ax. Consequently, for fluorine this is zero, but for Cl, Br and I the proportionality constants vary to give an approximately constant contribution at ca. −0.1 ppm.

Electric field and CX anisotropy contributions. A detailed account of the contributions to the SCSs of H-3ax and H-3eq in the halocyclohexanes is given in Table 6.19. In Equation (3.12) the component of the electric field along the bond (E_z) is dependent upon the relative orientation of the C—X and H—C bonds, as well as the halogen atom charge and X.H distance. The charge on the halogen decreases in the order F > Cl > Br > I, and this should be the determining effect for the electric field contribution for any particular proton. In contrast the C—X anisotropy as parameterized is zero for F and increases in the order Cl < B < I. However the angular dependence of this term ($3\cos^2\varphi - 1$), Equation (3.14) does have important implications. For example the anisotropy term for H-3ax in the axial conformer actually changes sign in going from Cl, Br to I (Table 6.19) due to the change in the angle φ as the C—X bond length increases. In both axial and equatorial conformers the SCSs are multifunctional with both the anisotropy and electric field terms present in varying amounts (cf. H-3ax and H-3eq, Table 6.19). The electric field is more often larger but this is not always the case. For the axial conformer the calculated values for

Table 6.19 *Contributions to the SCSs of the 3-ax and 3-eq protons in halocyclohexanes*

Conformer	Axial				Equatorial			
	F	Cl	Br	I	F	Cl	Br	I
3-ax H								
C—X electric	0.237	0.205	0.178	0.136	0.112	0.097	0.087	0.065
X..H steric	0.000	0.132	0.228	0.258	0.00	0.011	0.009	0.010
C—X anisotropic	0.000	0.089	0.074	−0.026	0.00	−0.074	−0.089	−0.101
Δ(C—H electric)	0.043	0.043	0.045	0.045	0.01	0.090	0.100	0.008
Δ(H..H steric)	0.092	0.099	0.111	0.107	0.01	0.020	0.020	0.010
Total calculated	0.37	0.58	0.65	0.53	0.13	0.14	0.13	−0.07
Observed	0.44	0.58	0.60	0.53	0.09	0.14	0.16	0.17
3-eq H								
C—X electric	0.054	0.067	0.062	0.053	0.139	0.120	0.108	0.083
X..H push–pull	0.000	−0.113	−0.114	−0.103	0.000*	0.006*	0.006*	0.006*
C—X anisotropic	0.000	−0.008	−0.022	−0.045	0.000	−0.041	−0.055	−0.066
Δ(C—H electric)	0.025	0.090	0.100	0.008	0.006	0.025	0.020	0.025
Total calculated	0.08	−0.07	−0.07	−0.09	0.15	0.11	0.08	0.05
Observed	0.06	−0.13	−0.08	−0.07	0.18	0.16	0.12	−0.01

* Steric shift.

H-3ax F < Cl < Br > I and H-3eq (F > Cl ≈ Br ≈ I) reproduce the observed trends very well. In the equatorial conformer the observed and the calculated SCSs on H-3eq are in the expected order F > Cl > Br > I (Table 6.19) though the observed decrease is rather larger than that calculated. The calculated SCSs for H-3ax (F ≈ Cl ≈ Br > I) is in reasonable agreement with the observed trends (F < Cl < Br < I) except for iodine which is reversed from that calculated. It was noted previously that iodine is anomalous in this orientation and it is probable that other long-range mechanisms may influence the SCSs. The calculated SCSs for H-4 (Table 6.12) are in general agreement with the observed data except for fluorine though here the SCSs are small and subject to larger errors due to the complexity of the spectra in this region.

Because the SCSs are multifunctional it is only possible to examine the calculated contributions and Figure 6.7 shows the electric field contributions to H-3 and H-4. The closest 'long-range' proton is H-3ax and a steady decrease in the electric field contribution

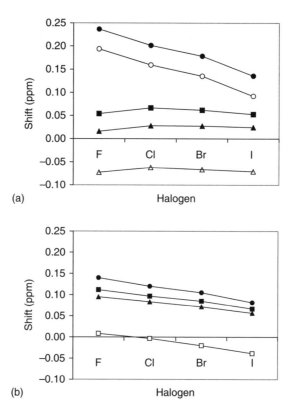

Figure 6.7   *The linear electric field contributions to the chemical shift of (a) axial and (b) equatorial halocyclohexanes. (a) Electric field contributions: (•) δ C—X electric [3ax], (o) δ total electric [3ax], (■) δ C—X electric [3eq], (▲) δ C—X electric [4ax] and (Δ) δ total electric [4ax]. (b) Electric field contributions: (•) δ C—X electric [3eq], (■) δ C—X electric [3ax], (▲) δ C—X electric [4eq] and (□) δ total electric [3ax] (reproduced from Abraham et al.[42] with permission of the Royal Society of Chemistry).*

with decreasing charge is observed from 0.24 ppm for fluorine ($q = -0.213$ electrons) to 0.14 ppm for iodine ($q = -0.098$ electrons) (Figure 6.7(a)). For H-3ax and H-4ax the SCSs also includes one C—H electric field contribution from the missing proton. This contribution is shielding at ca. -0.04 ppm per 1,3-syn C—H bond. For the equatorial protons the cut-off eliminates all C—H linear electric field contributions. For the more distant protons the C—X electric field contribution is much smaller. The contribution to H-3eq is virtually identical to H-4eq and about twice that of H-4ax. The increase in the C—X bond length down the group and variations in ring deformation due to the 1,3-syn diaxial halogen interactions, particularly for Cl vs. F, compensate for the charge decrease. Hence, the contribution for H-3eq is ca. 0.05 to 0.06 ppm, and for H-4ax ca. 0.02 to 0.03 ppm for all halogen substituents.

It can be seen that the removal of the proton by the axial substituent contributes a significant proportion of the SCS (ca. 0.12 ppm). The electric field term is larger than the steric term for fluorine and chlorine, but for bromine and iodine this is reversed.

For H-3eq in Table 6.19 the C—X linear electric field is of the opposite sign to the observed SCSs for all but fluorine. However, the larger shielding push–pull contribution for the heavier halogen atoms gives the overall observed shielding.

The chloro-, bromo- and iodo-steric terms are compared in Figure 6.8. As expected the steric contribution has I > Br > Cl for any given distance. Some illustrative points are marked on each curve representing increasing distances in the order: H-5-endo in 3-endo-halo-bornanes, H-3axial in axial halocyclohexanes and H-7syn in exo-halo-norbornanes. The calculated steric effect on H-3ax is similar to that on H-7syn, and this matches with the observed SCSs. The chloro SCSs for H-3ax is +0.54 ppm and for H-7syn +0.59 ppm.

In contrast H-5endo is much closer to the halogen substituent with a larger calculated steric contribution, and the observed SCS is greater at +0.84 ppm. The distances for similar systems such as halo-*trans*-decalins and androstanes lie within the range shown above.

For the syn 1,3-diaxial protons the major contribution to the halo (X = Cl, Br and I) SCSs is thus the steric contribution, although the C—X electric field contribution is not negligible. From Table 6.19 it can amount to over half the value of the steric term, and the contribution from the removal of the proton needs also to be accounted for. The effect on close through space protons in Figure 6.8 (<2.6 Å) is calculated to increase rapidly, although this remains to be substantiated by experimental data.

The importance of the linear electric field is thus evident for the other 'long-range' protons. This agrees with the analyses of Schneider on *trans*-decalins and steroids.[16,20] Whether the electric field term is subject to some shielding effect due to obstructing bonds would require a more extensive analysis. H-4 values in fluoro-cyclohexane (Table 6.12) are in poor agreement with the calculated SCSs and neglecting the electric field term would improve the fit. There may be solvent effects in this case as reaction field effects would be greatest for the substituent with the largest halogen charge but in general solvent effects are expected to be minimal as all samples were run in low concentration in non-polar solvents (50:50 CDCl₃:CF.Cl₃).

6.3.5 Steric Coefficients for Halogens

Bothner-By and Naar-Colin[35] noted unusual SCSs on β protons in isopropyl and cyclohexyl halides (RR′CHX) which were inconsistent with the halogen electronegativity alone. They

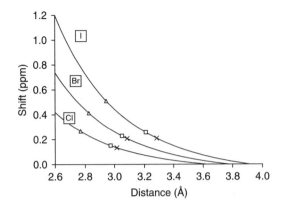

Figure 6.8   *The steric contributions to the chemical shifts as a function of the X..H distance for chlorine (bottom line), bromine (middle line) and iodine (top line). Protons marked as follows: (Δ) H_{5-endo} in 3-endo-halo-bornanes, (□) $H_{3-axial}$ in axial-halo-cyclohexanes and (×) H_{7-syn} in exo-halo-norbornanes (reproduced from Abraham et al.[42] with permission of the Royal Society of Chemistry).*

attributed this in part to the increase of the contribution of possible resonance forms. Further, the importance of these forms would increase 'with increasing atomic number of the halogen and with increasing branching on the α-carbon'. This effect is reproduced in this study, and accounted for by an enhancement of the bromine and iodine β effects on methine protons, and for iodine on methylene protons. Such effects are greater for iodine than for bromine, and the effect of two beta carbons (RR'CHX) greater than for one (RCHXY).

This argument may also be applied to the γ protons, although only the unusual H-10 SCSs in 9-bromo- and 9-iodo-*trans*-decalins show obvious shielding. Here, the α- and β-carbons across the bridge are surrounded by the maximum number of carbons, i.e. X—C(R)(R')—C(R'')(R''')—H increasing steric compression and possibly these contributions.

The latest values of the Cl, Br and I steric shielding coefficients (a_s) obtained[43] are 90.0, 95.0 and 120.0 ppm $Å^{-6}$ which are much less than obtained previously without the C—X anisotropy contribution (200.0 ppm $Å^{-6}$)[1] These values allow the origins of the so called 'steric term' to be examined, as this has been considered either as a van der Waals' term or as a quadratic electric field effect. The latter as given by Buckingham[5] on the δ scale is the BE^2 term in Equation (6.1) which is also a function of the electric field squared (r^{-6}). The observed SCS of the chloro substituent in axial-chlorocyclohexane on H-3ax (0.58 ppm) is made up of on the CHARGE scheme 0.21 ppm due to the electric field contribution, 0.13 ppm steric term plus other small contributions (total 0.58 ppm, Table 6.19). Using the B value calculated by Grayson and Raynes[12,13] the quadratic electric field effect using a chlorine charge of −0.155 electrons (cf. CHARGE) and H..Cl distance of 2.774 Å (cf. HF/6-31G* geometry) is only 0.002 ppm. For protons it would appear that the quadratic electric field effect even from polar groups is negligible and that the steric term is entirely due to van der Waals' interactions.

This conclusion is supported by previous work. Comparison of the Cl, Br and I steric coefficients with the hydrogen and carbon values (Chapter 4, Section 4.2) allows comparison with previously derived van der Waals' terms. Abraham and Holker[61] calculated the intramolecular van der Waals' effect from a methyl group in 2-bromo-3-oxo-steroids using the intermolecular van der Waals' equation of Raynes *et al.*[59] (Equation (6.2)):

$$\sigma_{VDW} = -3B\alpha I/r^6 \tag{6.2}$$

where α and I are the polarizability and ionization potential of the substituent. They derived a value of 163 ppm $\mathring{A}^{-6}$ for the methyl group using a B value of 1.0×10^{-18} esu. Later determinations of the dispersion constant (B) by Tribble *et al.*[62] empirically, and the shielding hyperpolarizability by Grayson and Raynes[13] using finite-field SCS calculations, suggested values of 0.27 and 0.23×10^{-18} esu. These results suggest a more realistic value of $3B\alpha I$ of ca. 42 ppm $\mathring{A}^{-6}$.

For the polar substituents (Cl, Br and I) the a_s coefficients are 90.0, 95.0 and 120.0 ppm $\mathring{A}^{-6}$ compared to ca. 100 ppm $\mathring{A}^{-6}$ calculated by Equation (6.2). These are very similar, in contrast to the quadratic electric field term above which produced results two orders of magnitude too small to account for the CHARGE steric term.

Further the ratios of the Cl, Br and I a_s coefficients of 1:1.06:1.33 is in the same sense as the ratios of the polarizabilities[62] (1:1.54:2.55), but not the first ionization energies[63] (1:0.91:0.82) again supporting the origin of this term as due to van der Waals' interactions and not to the quadratic electric field.

In contrast to the results of Equation (6.2) the shielding coefficient for carbon has an average value of 206.0 ppm $\mathring{A}^{-6}$. One problem with this comparison is the interdependence of the carbon and hydrogen coefficients, since both terms are invariably involved in methyl group effects. In the CHARGE scheme the H..H steric interactions are *shielding* in contrast to a *deshielding* effect predicted from Equation (6.2). Thus the carbon a_s value would be expected to be more deshielding than the sum of the coefficients, $3B\alpha I$ from Equation (6.2).

The decrease in the size of the push–pull coefficient for the halogen atoms (Cl = 75%, Br = 50% and I = 40%) may be a consequence of the increased polarizability of the atom. Alternatively, the angle between the halogen atom and the affected H—C—H bond may be important. For instance, in 2-exo-halo-norbornanes the X—H$_{7s}$—C$_7$ angle increases from 97.6° for the chloro substituent to 100.0° for bromine and 101.6° for iodine, partly as a consequence of the increased C—X (X=Cl, Br and I) bond length. However, more data would be needed to substantiate this hypothesis.

6.4 Alcohols and Phenols

6.4.1 Introduction

The effect of the electronegative oxygen atom on ^{1}H chemical shifts has been known for ca. five decades[64] and a number of attempts to analyse OH SCSs have been made.[8,16,19,65] Zürcher[8] was limited to observing only the methyl groups in steroids in his pioneering studies but concluded that the C—O bond anisotropy was not important for the OH group SCSs. Schneider *et al.*[16] regarded the electric field term as the dominant term but Hall[65] suggested that the chemical shift difference between the anomeric protons of the C$_2$—O axial and C_2—O equatorial sugars could be accounted for by C—O anisotropy alone. Yang *et al.*[19] concluded that electric field, anisotropy and a constant term were necessary to reflect the observed ether SCSs in oxasteroids but did not consider any steric contributions. Abraham *et al.*[66] in a study of acyclic and cyclic ethers (Chapter 7, Section 7.5) predicted the ¹H chemical shifts in these systems using the CHARGE model including both oxygen steric and electrostatic terms but no C—O bond anisotropy.

The biologically important polyhydroxy compounds (sugars, etc.) are insoluble in CDCl₃, thus for a general prediction of the ¹H spectra of polyhydroxy compounds it is necessary to extend the analysis to include other solvents. Abraham *et al.*[67] reported the ¹H chemical shifts of alcohols, diols and inositols using CDCl₃, D₂O and DMSO. We will consider the problem of solvation further in Chapter 8, thus will confine this section to CDCl₃ solvent (and some D₂O measurements) and hence to alcohols, diols and phenols soluble in this solvent. A major problem in such analyses is the conformational isomerism about the oxygen atom. For example the SCS of the hydroxyl group may well be dependent on the position of the OH proton. The value of the CH.OH coupling in alcohols in solution shows that the OH proton is usually not in a single orientation but may have a preferred conformation.[68] However it is difficult to estimate accurately the populations of the different OH conformers, especially in D₂O solution.

Inter- and intramolecular hydrogen bonding in alcohols has been studied in the past through NMR and IR spectroscopy. It is well known that both the OH proton chemical shift and the OH stretch absorption frequency are affected by hydrogen bonding. Farrar *et al.*[69] studied the OH chemical shift of ethanol at a range of concentrations in six solvents, including CCl₄, CDCl₃ and DMSO-d_6. The chemical shift of pure ethanol is 5.4 ppm whereas the infinite dilution shifts ($\delta_{OH,\infty}$) in CCl₄, CDCl₃ and DMSO are 0.7, 1.1 and 4.45 ppm, respectively. The large deshielding of the OH protons in ethanol and DMSO compared to CCl₄ and CDCl₃ solvent is due to intermolecular hydrogen bonding.

The ¹H spectra of the simple alcohols have been well documented but the conformations of the 1,2-diols and derivatives in solution has been the subject of some controversy. The presence or absence of a hydrogen bond in the *gauche* conformer of ethane-1,2-diol (**13**), Figure 6.9 has been questioned. Recent NMR investigations[70,71] found that the *gauche* form predominates with ca. 10–20% of the *trans* form in solution and other investigations on related diols have obtained similar results.[72,73] Sanders *et al.*[74] from the observed OH chemical shifts concluded that (**13**) and propane-1,3-diol (**16**) are intramolecularly H-bonded in chloroform solution.

The conformation of *cis*-cyclohexane-1,3-diol (**23**) was determined in a variety of solvents from the ³J$_{HH}$ couplings.[75] In polar solvents the diequatorial form predominates but in CDCl₃ the dieq and diax conformers are of ca. equal energy due to an intra-molecular hydrogen bond between the 1,3-diaxial hydroxy groups. Intramolecular hydrogen bonds have also been observed in both *cis* and *trans* cyclohexane-1,2-diol[76] (**20**) and cyclo-pentanol (**12**) and *cis* and *trans* cyclopentane-1,2-diols are interconverting between several conformations in CDCl₃ solution.[77]

6.4.2   Alcohols and Diols

In order to apply the theory of Chapter 3 to alcohols both the short-range electronic effects of the OH group and the long-range contributions need to be obtained. The short-range effects include the γ (H.C.C.OH) parameters and the 1,2-dioxy compounds need to be treated separately in this respect so that there is another set of parameters for the H.C(OR).C(OR) fragment.[67]

The long-range contributions include in principle steric, anisotropic and electric field effects. The steric effect of the oxygen atom was shown to vary considerably in ethers,[66] phenols[78] and esters,[79] thus it is necessary to obtain the value of a_s (Chapter 3, Equation (3.11)) for the OH function. Again it was found that the 1,2-diols required a separate, smaller value of a_s than the simple alcohols (see later). Also it was assumed that the C—O and O—H bonds of the alcohols are not magnetically anisotropic.

The electric field term requires some consideration. The O—H, O—C and C—H bond dipole moments are ca. 1.53, 0.86 and 0.3D.[67] These bond dipoles are opposed in the HC.OH system to give a resultant dipole moment along the C.O bond of ca. 0.3D. The major component of the C.OH dipole is perpendicular to the C—O bond. However the OH proton is rapidly rotating between three different possible conformations. Thus the time average electric field produced along a given C—H bond by the rapidly rotating C.OH group will be very small. It was therefore decided to ignore any electric field effects due to the HC.OH group in the CHARGE chemical shift calculations.

Observed versus calculated shifts. The compounds investigated are shown in Figure 6.9. The ¹H chemical shifts of methanol, ethanol, n-propanol, i-propanol, t-butanol, ethylene glycol and 2-methoxyethanol are from Gottlieb *et al.*,[80] the data for *endo* and *exo* norborneol from Abraham *et al.*,[21] that for *cis* and *trans* 4-t-butyl cyclohexanols and cis 1,3-cyclohexanediol from Abraham *et al.*[75] and that for cyclopentanol and *cis* and *trans* 1,2-cyclopentanediol from Abraham *et al.*[77] The molecular geometries were obtained using the MMFF94 force field.

The observed and calculated ¹H chemical shifts are given in Tables 6.20 and 6.21. In order to avoid H-bonding effects on the OH chemical shift due to inter-molecular hydrogen bonding, the OH chemical shift was extrapolated to ∞ dilution for compounds **1-6, 13, 16, 17, 18, 20** and **23**.[81] The data in the tables can now be evaluated. Inspection of the ¹H chemical shifts of the compounds in CDCl₃ and D₂O shows a remarkable consistency. There are 21 compounds which are soluble in both solvents and in which the conformational profile is unchanged in the two solvents, giving 79 distinct chemical shifts (excluding the OH). The rms difference $\Delta\delta$ (D₂O—CDCl₃) of these 79 chemical shifts is 0.034 ppm, a value not much greater than the experimental errors. The largest difference, not surprisingly is for the methyl group in methanol (0.12 ppm). Thus for these compounds *the ¹H chemical shifts in D₂O are essentially identical to those in CDCl₃.*

Inspection of other tabulated shifts[80] shows that this identity applies to other common functional groups, e.g. acetone 2.17 vs. 2.22; acetonitrile 2.10 vs. 2.06 and nitromethane 4.33 vs. 4.40.

This identity only applies to compounds with the same conformational profile in the two solvents and this is illustrated by the results in the tables. For those compounds in which intra-molecular hydrogen bonding is possible the chemical shifts in D₂O and CDCl₃ differ considerably. The clearest example is *cis*-1,3-cyclohexanediol (**23**) which has been shown to exist solely in the di-equatorial conformation in D₂O but as a 1:1 mixture of the di-equatorial and di-axial conformers in CDCl₃.[75] As expected the chemical shifts differ considerably in the two solvents. Another possible example is 1,3-propanediol (**16**) in which

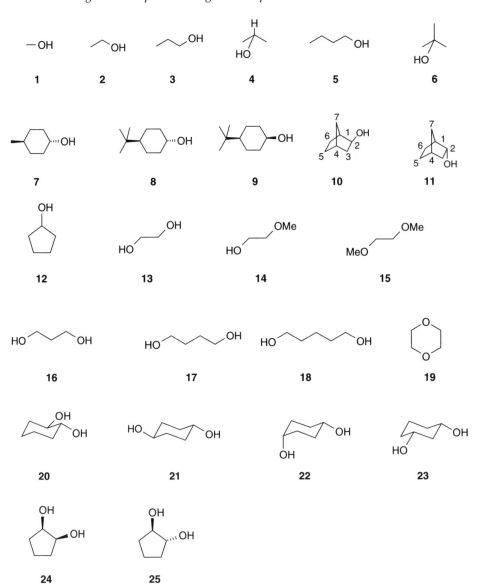

Figure 6.9 *Molecules investigated with numbering.*

an intra-molecular hydrogen bond is likely in the gauche conformer. 1,4-Butanediol and 1,5-pentanediol do not show any evidence of different conformational profiles in the two solvents and this is consistent with the higher energy of the seven- and eight-membered rings needed to form an intra-molecular hydrogen bond in these compounds.

This identity of the ^{1}H shifts in the two solvents for these compounds may be rationalized by noting that both these solvents are not magnetically anisotropic and they have similar dipole moments. Neither of these factors is true for DMSO and as a result the ^{1}H shifts in

Table 6.20   *Observed versus calculated ¹H chemical shifts (δ) of alcohols[a]*

Molecule		Observed		Calculated
		CDCl$_3$	D$_2$O	
Methanol[b] (**1**)	Me	3.49	3.34	3.37
	OH	0.85	—	0.89
Ethanol[b] (**2**)				
	CH$_2$	3.71	3.65	3.74
	Me	1.24	1.17	1.18
	OH	1.10	—	1.20
n Propanol[b] (**3**)				
	H-1	3.59	3.61	3.57
	H-2	1.59	1.57	1.48
	Me	0.94	0.89	0.85
	OH	1.22	—	1.21
i Propanol[b] (**4**)				
	CH	4.04	4.02	3.98
	Me	1.22	1.17	1.25
	OH	1.23	—	1.21
n Butanol (**5**)				
	H-1	3.64	3.61	3.59[f], 3.62[g]
	H-2	1.56	1.51	1.30, 1.47
	H-3	1.39	1.35	1.16, 1.34
	Me	0.94	0.91	0.84, 0.82
	OH	1.17	—	1.21
t Butanol[b] (**6**)				
	t-Bu	1.28	1.24	1.30
	OH	1.22	—	1.22
trans-4-Methylcyclohexanol (**7**)				
	1ax	3.52	3.58	3.59
	2,6eq	1.94	1.90	1.84
	2,6ax	1.25	1.25	1.38
	3,5eq	1.70	1.70	1.63
	3,5ax	0.96	0.97	0.86
	4ax	1.33	1.33	1.36
	Me	0.88	0.86	0.87
	OH	2.30	—	1.26
trans-4-t-Butylcyclohexanol[c] (**8**)				
	1ax	3.52	—	3.58
	2,6eq	2.01	—	1.84
	2,6ax	1.22	—	1.38
	3,5eq	1.78	—	1.64
	3,5ax	1.05	—	0.86
	4ax	0.97	—	1.01
	t-Bu	0.85	—	0.88
	OH	1.26	—	1.31
cis-4-t-Butylcyclohexanol[c] (**9**)				
	1eq	4.03		4.02
	2,6eq	1.83	—	1.83
	2,6ax	1.49	—	1.53
	3,5eq	1.54	—	1.60
	3,5ax	1.35	—	1.34
	4ax	0.99	—	0.98
	t-Bu	0.86	—	0.89
	OH	1.25	—	1.29

Table 6.20   (*Continued*)

Molecule		Observed		Calculated
		CDCl$_3$	D$_2$O	
2-exo-Norborneol^d (**10**)				
	H-1	2.14	2.05	2.07
	2en	3.77	3.70	3.61
	3ex	1.29	1.38	1.25
	3en	1.67	1.67	1.66
	H-4	2.26	2.26	2.21
	5ex	1.43	1.41	1.55
	5en	1.02	1.00	1.18
	6ex	1.46	1.46	1.55
	6en	1.02	1.03	1.25
	7s	1.57	1.50	1.60
	7a	1.12	1.14	1.07
	OH	—	—	1.32
2-endo-Norborneol^d (**11**)				
	H-1	2.25	2.24	2.36
	2ex	4.23	4.12	4.01
	3ex	1.96	1.92	1.99
	3en	0.84	0.83	0.90
	H-4	2.17	2.15	2.19
	5ex	1.57	1.56	1.53
	5en	1.34	1.38	1.27
	6ex	1.36	1.40	1.39
	6en	1.88	1.71	1.77
	7s	1.34	1.35	1.17
	7a	1.29	1.30	1.17
	OH	1.44	—	1.24
Cyclopentanol^e (**12**)				
	H-1	4.32	4.30	3.94
	2,5cis	1.56	1.58	1.41
	2,5tr	1.76	1.79	1.79
	3,4cis	1.76	1.79	1.68
	3,4tr	1.56	1.58	1.50
	OH	1.28	—	1.25

^{a} Data from Abraham *et al.*[67], unless stated otherwise.
^{b} Data from Gottlieb *et al.*[80].
^{c} Data from Abraham *et al.*[75].
^{d} Data from Abraham *et al.*[21].
^{e} Data from Abraham *et al.*[77].
^{f} All-*trans* conformer.
^{g} C.C.C.O gauche.

this solvent differ considerably from those in CDCl$_3$ (cf. Chapter 8). There is one caveat to the above conclusion. Charged molecules would not be expected to give the same ^{1}H shifts in CDCl$_3$ and D$_2$O. Both the species present (ions, ion-pairs, etc.) and the effect of the integral charge would be expected to differ considerably in the two solvents.

The OH SCSs. The conformational dependence of the OH SCSs and comparison with the corresponding fluoro and iodo SCSs was given by Schneider *et al.*[16] in their analysis of

Table 6.21   *Observed versus calculated* ¹ *H chemical shifts (δ) of diols and related ethers[a]*

Molecule		Observed		Calculated
		CDCl₃	D₂O	
Ethylene glycol[b] (**13**)	CH₂	3.73	3.65	3.72
	OH	1.78	—	1.49
2-Methoxyethanol[b] (**14**)				
	α-CH₂	3.70	3.71	3.85
	β-CH₂	3.51	3.56	3.48
	Me	3.40	3.38	3.35
	OH	3.10	—	1.35
1,2-Dimethoxyethane[b] (**15**)				
	Me	3.40	3.37	3.34
	CH₂	3.55	3.60	3.62
1,3-Propanediol (**16**)				
	α-CH₂	3.85	3.69	3.60[t], 3.77[g]
	β-CH₂	1.82	1.80	1.65, 1.75
	OH	1.89	—	1.40, 3.25
1,4-Butanediol (**17**)				
	α-CH₂	3.68	3.63	3.60[t], 3.77[g]
	β-CH₂	1.68	1.60	1.34, 1.53
	OH	1.83	—	1.40, 2.60
1,5-Pentanediol (**18**)				
	α-CH₂	3.67	3.62	3.60[t], 3.62[g]
	β-CH₂	1.61	1.58	1.33, 1.44
	γ-CH₂	1.47	1.41	1.01, 1.19
	OH	1.24	—	1.40, 1.40
Dioxan[b] (**19**)				
	CH₂	3.71	3.75	3.74
trans-1,2-Cyclohexanediol (**20**)				
	1,2ax	3.33	3.37	3.49
	3,6eq	1.95	1.92	1.86
	3,6ax	1.24	1.25	1.38
	4,5eq	1.69	1.67	1.67
	4,5ax	1.24	1.25	1.26
	OH	2.07	—	1.40
trans-1,4-Cyclohexanediol (**21**)				
	1,4ax	3.68	3.66	3.61
	2,3eq	1.97	1.93	1.86
	2,3ax	1.36	1.34	1.39
	OH	—	—	1.26
cis-1,4-Cyclohexanediol (**22**)				
	1,4	3.83	3.81	3.80
	2,3 *cis*	1.75	1.66	1.74
	2,3 *trans*	1.66	1.66	1.67
	OH	—	—	1.41
cis-1,3-Cyclohexanediol[c] (**23**)				
	1,3	3.82	3.65	3.61
	2eq	1.97	2.22	2.03
	2ax	1.36	1.21	1.49
	4,6eq	2.01	1.89	1.84
	4,6ax	1.22	1.13	1.40
	5eq	1.78	1.78	1.70
	5ax	1.05	1.25	1.28
	OH	1.90	—	1.27

Table 6.21   (*Continued*)

Molecule		Observed		Calculated
		CDCl$_3$	D$_2$O	
cis-1,2-Cyclopentanediol[d] (**24**)				
	1,2	4.00	4.00	3.65
	3,5 *cis*	1.66	1.64	1.52
	3,5 *trans*	1.86	1.84	1.90
	4 *cis*	1.80	1.77	1.63
	4 *trans*	1.52	1.53	1.54
	OH	—	—	1.38
trans-1,2-Cyclopentanediol[d] (**25**)				
	1,2	4.00	4.00	3.50
	3,5 *cis*	1.53	1.55	1.34
	3,5 *trans*	2.01	2.00	2.07
	4	1.71	1.72	1.57
	OH	—	—	1.35

[a] Data from Abraham *et al.*[67], unless stated otherwise.
[b] Data from Gottlieb *et al.*[80].
[c] Data from Abraham *et al.*[75].
[d] Data from Abraham *et al.*[77].
[f] All-*trans* conformer.
[g] C.C.C.O gauche.

steroid spectra and shown in Figure 6.10. It can be seen that there are general similarities with the halogen SCSs, despite the very different dipole moments of the substituents. The data in Tables 6.20 and 6.21 allow a more detailed investigation of the SCS of the OH group on ^1H chemical shifts and also provide a critical test of the application of the CHARGE model to these compounds. Both short-range and long-range effects need to be considered.

The short-range effects include the γ (H.C.C.OH) effect and the dependence of the H.C.C.OH SCS on the H.C.C.O dihedral angle is shown in Figure 6.11 for the five compounds in Table 6.20 containing the C.CH$_2$.CH(OH).C fragment in a fixed conformation. The curve in Figure 6.11 was simulated by a cos $n\phi$ ($n = 1, 2, 3$) series. The best fit was given by Equation (6.3) with only one-fold and two-fold potentials.

$$\gamma(SCS) = 0.228 - 0.290 \cos \phi - 0.212 \cos 2\phi \qquad (6.3)$$

The coefficient of the three-fold term was zero. It is of interest that although the minimum SCS occurs at 0° dihedral angle the largest SCS occurs for a dihedral angle of ca. 120°. This angle dependence had not been recognized previously and it would be of some interest to determine the theoretical basis for this result. It was also found that the gem diols needed to be treated separately so the H.C(OH).C(OH) fragment was separately parameterized using a similar equation. Following the treatment given earlier the only long-range effect is the oxygen steric effect which was parameterized together with the γ effects. The oxygen steric coefficient (Chapter 3, Equation (3.11)) was found to differ in the alcohols, ethers and 1,2-diols with values of a_s of 85, 136 and 26 ppm Å^6 res.

The result of this parameterization is shown in the calculated shifts in the tables. For the alcohols and diols the 150 chemical shifts are reproduced with an rms error of 0.076 ppm. This excludes the OH chemical shifts which have larger errors due to possible inter- and

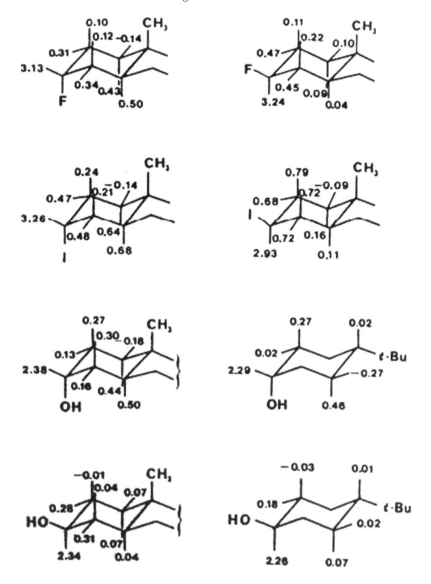

Figure 6.10 OH, F and I SCSs in the A ring of androstane and t-Bu-cyclohexane (reproduced from Schneider et al.[16] with permission of the American Chemical Society).

intra-molecular hydrogen bonding (see later). Note that in D_2O the OH chemical shift for any molecule is set for convenience in CHARGE equal to the water value of 4.70 ppm. The good agreement of the observed vs. calculated shifts is strong support for the model used in the calculations and for the assumptions made of no magnetic anisotropy and no electric field effects for the OH group. Previous theoretical and experimental investigations support these assumptions. Pople[82] concluded there was no theoretical evidence for C—C bond anisotropy and Williamson et al.[83] have successfully analysed ¹H chemical shifts in

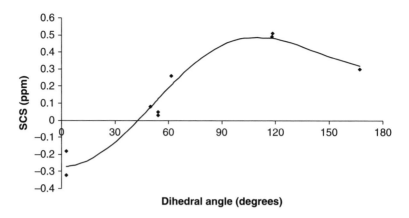

Figure 6.11 *SCSs of the OH group in the C.CH$_2$.CH(OH).C fragment versus the H.C.C.O dihedral angle.*[67]

proteins without including any magnetic anisotropy for the single bonds, though there was significant anisotropy from the C=O double bond and also from the N—CO partial double bond. Indirect support for the absence of any electric field contribution from the OH groups comes from the identical chemical shifts of the compounds investigated in CDCl$_3$ and D$_2$O. A significant electric field effect from the OH group would have been expected to produce different shifts in the two solvents, as any electric field in D$_2$O solution would be much reduced from the corresponding field in CDCl$_3$ due to the much greater solvent permitivity in D$_2$O (80 vs. 4.6).

DMSO solvent effects. ^{1}H shifts in DMSO differ considerably from those in the other solvents. This is true not only for the OH proton, which is found at a different region in the spectrum in CDCl$_3$ (ca. 1.6 ppm) and DMSO (ca. 4.4 ppm) but also for other protons. Thus the effect of DMSO on the ^{1}H chemical shifts needs to be considered explicitly and this is detailed in Chapter 8, Section 8.6.

The chemical shift range of the OH protons studied here in CDCl$_3$ solution is ca. 1–4 ppm, due to both inter- and intra-molecular O—H..O hydrogen bonding. The inter-molecular H-bonding can be removed by measurements at sufficiently low concentration. Sanders *et al.* noted that 10^{-2} molar was sufficient to remove H$_2$O/OH exchange and give essentially ∞ dilution OH shifts.[74] This means that the OH group *in dilute solution* can be treated in CHARGE in the same manner as any other proton and this has been performed with the OH data in the tables for those compounds examined at ∞ dilution. It was found that the short-range electronic effects (e.g. HOC.C) were not needed for these compounds, probably due to the rotational averaging around the C—O bond. For the simple alcohols, the OH chemical shift is almost constant at 1.2δ. In the diols, the major factor influencing the OH shift is steric deshielding due to the effect of a near non-bonded oxygen (O—H. . .O). For example, the OH chemical shifts in ethylene glycol (**13**), 1,3-propanediol (**16**) and *trans* 1,2-cyclohexanediol (**20**) are all ca. 2.0δ due to intra-molecular hydrogen bonding. To reproduce these shifts it is necessary to determine the proportions of the free and H-bonded conformers in solution and this has been attempted.[81] For stronger hydrogen bonds with

H-bond distances (O..H < 2.0 Å) Equation (3.16) is used (see next section), but none of the aliphatic alcohols studied has such strong hydrogen bonds.

6.4.3 Phenols

Both inter- and intramolecular H-bonding in phenols has been investigated for many years with IR and NMR being the major tools for investigating these H-bonds in solution.[84–87] The OH chemical shift of phenol has been measured in several solvents. It varies from 4.6δ in dilute CDCl₃ to 9.2δ in DMSO-d_6[88] due to intermolecular H-bonding in DMSO. In CCl₄ the OH shift increases linearly with concentration from 5.36δ at infinite dilution to 6.98δ for the saturated solution[89] (ca. 120 mg ml⁻¹) at which point the chemical shift is constant and independent of concentration. This concentration dependence was attributed to a monomer–trimer equilibrium.[89,90] In contrast the chemical shift of the OH proton of 2-hydroxy-benzophenone is insensitive to changes in both temperature and concentration in CCl₄ due to a strong intramolecular H-bond impeding the formation of intermolecular H-bonds.[91] The OH bond weakens and lengthens.[92–96] on hydrogen bonding and this produces a decrease in the OH stretching frequency and a large deshielding of the OH chemical shift.[97,98] Indeed these effects were found to be correlated for intramolecularly hydrogen bonded molecules.[89,93,99] The majority of the early studies were aimed at determining the hydrogen bond energy for the weakly H-bonded systems such as *ortho* halophenols (i.e. excluding stronger interactions such as in 2-hydroxyacetophenone). Abraham and Mobli have reported an NMR, IR and theoretical investigation on the ¹H chemical shifts and hydrogen bonding in phenols.[78] Their *ab initio* calculations on the changes in the OH chemical shifts with H-bonding are given in Chapter 3, Section 3.5.4. Here we discuss their NMR data and the application of these calculations to the determination of the ¹H shifts of phenols in CDCl₃ solvent.

Concentration effects. The ¹H spectrum of phenol, 2- and 4-cyanophenol and 4-nitrophenol was measured from 1 mg ml⁻¹ to 100 mg ml⁻¹ (≈ 0.01–1M) in CDCl₃. The concentration dependence of the OH shifts in these compounds is accurately linear with concentration (correlation coefficient, $R^2 = 0.99$) and given by the equations below, where c is the concentration in M/1. For phenol, the OH chemical shift only changes by 0.1 ppm from a concentration of 10 mg ml⁻¹ to ∞ dilution. As 0.1 ppm is the tolerated error in chemical shift calculations this concentration may be safely used in such predictions for such weakly intermolecular H-bonded systems. The concentration limit is much greater for compounds with intramolecular H-bonds as they inhibit any intermolecular H-bonding. However the gradients for the substituted phenols are all much larger than phenol illustrating the much stronger intermolecular H-bonding in these compounds.

Phenol $\delta_{(OH)} = 4.60 + 0.940c$
4-Nitro $\delta_{(OH)} = 5.356 + 3.96c$
2-Cyano $\delta_{(OH)} = 5.659 + 5.71c$
4-Cyano $\delta_{(OH)} = 5.266 + 7.85c$

The intermolecular hydrogen bonding of the OH proton is determined by the acidity of the OH proton, the H-bonding acceptor strength of the substituent, steric hindrance of

Table 6.22 *Ortho and para substituted phenols used and their nomenclature*

No.	−X	−Y
[1]	H	H
[2]	F	H
[3]	Cl	H
[4]	Br	H
[5]	I	H
[6]	Me	H
[7]	OMe	H
[8]	CN	H
[9]	CF$_3$	H
[10]	NO$_2$	H
[11]	COOCH$_3$	H
[12]	H	CN
[13]	H	F
[14]	H	t-Bu
[15]	H	NO$_2$
[16]	H	CF$_3$
[17]	H	OMe
[18]	Cl	Cl
[19]	CHO	H
[20]	COCH$_3$	H

the OH proton by the substituent (for *ortho* substituted phenols) and the strength of any intramolecular hydrogen bond. For example, in 2-nitrophenol the intramolecular H-bond dominates thus there is little concentration dependence of the OH shift. In contrast in 4-cyanophenol there is no intramolecular H-bond and this OH shows the strongest concentration dependence. Abraham and Mobli[78] measured the ¹H shifts of a number of *ortho* and *para* substituted phenols (Table 6.22) in CDCl$_3$ and DMSO solvents. The major conformational problem in these compounds is the possible existence of two conformers in CDCl$_3$ solution for the *ortho* compounds (Figure 6.12).

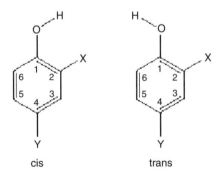

Figure 6.12 *Cis and trans conformers of 2,4-substituted phenols.*

Ab initio calculations and IR spectra were used to determine the proportions of the *cis* and *trans* conformers. Their results agreed with earlier investigations in that the *cis* conformer predominates for all substituents capable of forming a hydrogen bond to the OH, that is, all *ortho* compounds in Table 6.22 except the *ortho* methyl (**6**). In this compound the *trans* conformer is favoured ($\Delta E(cis–trans) = -0.80 \, \text{kcal mol}^{-1}$).[78]

Their data for CDCl₃ solvent are given in Tables 6.23 and 6.24 together with the calculated shifts from CHARGE and GIAO calculations. In the GIAO calculations CH₄ is taken as the reference using B3LYP-6311++g(d,p), for both optimization and GIAO calculation.

The shifts due to DMSO solvent are considered in Section 8.6. In Tables 6.23 and 6.24 the infinite dilution value of the OH shift is used to compare with the calculated values.

Chemical shift calculations. In order to calculate the chemical shifts of the phenols by CHARGE, as well as the effects considered in the last section, for the phenols there are effects due to the π-electrons and also the H-bonding shifts of the OH in the intramolecular H-bonded phenols investigated.

π -Shift contribution through oxygen. In the CHARGE program the π shift contribution to the ¹H chemical shift of an aromatic proton (i.e. H.C_{AR}) is given by Equation (3.18) as $10 \, \text{ppm el}^{-1}$.[100] The effect of the π density on the oxygen atom on the OH chemical shifts can be obtained from the data in Table 6.24 for the *para* substituted phenols. *Para* substituted phenols are ideal as *meta* substituents have little π effect on the OH group and the *ortho* substituents have more contributions than merely π effects. Table 6.25 gives the OH chemical shift vs. the excess π-electron density on the oxygen atom as calculated by CHARGE. There is a linear correspondence with a coefficient of 42.3 ppm per π electron. This when inserted into the CHARGE program gives the calculated shifts in Table 6.25. The coefficient for oxygen is ca. four times that for carbon. Interestingly this value is identical to that for the nitrogen atom in anilines (42.7 ppm/el) from a similar correlation for *para* substituted anilines (Section 6.5).[101] Thus one 2p-electron pair on the X atom in a X—H bond is sufficient to produce this effect.

H-bonding contributions to the OH chemical shift. In previous investigations on ortho substituted phenols with strong intramolecular H-bonds (e.g. 2-hydroxybenzaldehyde) with OH· · ·O=C distances < 2.0 Å, the large deshielding effect of the ortho substituent on the OH proton was modelled in CHARGE by the function $((2/r)^{12} - 1)/2$ which when added to the other contributions gave good agreement with the observed shifts.[102] This function only applies for r <2.0 Å, since the CHARGE calculations beyond this point are represented well with the current model. The ab initio calculations (Chapter 3, Section 3.5.4) show clearly that the OH shielding is a linear function of the H· · ·O distance for r < 2.1 Å and Equation (3.16) is the correct form for these calculations. The slope of the plot (B) is given by the ab initio calculations as $-7.6 \, \text{ppm Å}^{-1}$. It is of interest to note that if the two observed OH shifts for 2-hydroxybenzaldehyde **19** and 2-hydroxyacetophenone **20** (Table 6.23) are used with the ab initio H· · ·O distances the value for the slope B is $-7.8 \, \text{ppm Å}^{-1}$, in complete agreement with the theoretical value. Thus this value was adopted for all functional groups considered. In all the compounds considered the H· · ·O=C—R entity is planar, thus φ=0 or 180° and the $\cos^2 \varphi$ dependence cannot be experimentally verified.

The values of *A* and r_0 were found to be zero and 2.05 Å for the strong intramolecular hydrogen bonds studied, that is with *ortho* nitro, aldehyde, ketone and ester groups. In

Table 6.23 ^{1}H chemical shifts (δ) of ortho substituted phenols in CDCl$_3$ with CHARGE and GAIO calculated chemical shifts

Phenol(s)	Solvent	H2	H3	H4	H5	H6	OH	Substituent
[1] Phenol	CDCl$_3$	6.826	7.240	6.928	7.240	6.826	4.692	
	CHARGE	6.843	7.212	6.917	7.212	6.843	4.704	
	GIAO[a]	6.897	7.388	7.014	7.388	6.897	3.781	
[2] 2-Fluorophenol	CDCl$_3$		7.060	6.845	7.010	7.010	5.077	
	CHARGE		6.959	6.931	6.980	6.858	4.957	
	GIAO[a]		7.231	6.882	7.180	7.254	4.715	
[3] 2-Chlorophenol	CDCl$_3$		7.310	6.864	7.175	7.014	5.508	
	CHARGE		7.207	6.877	7.178	6.913	5.616	
	GIAO[a]		7.393	6.924	7.307	7.191	5.018	
[4] 2-Bromophenol	CDCl$_3$		7.455	6.804	7.217	7.019	5.470	
	CHARGE		7.372	6.832	7.193	6.918	5.459	
	GIAO[a]		7.504	6.926	7.344	7.249	5.156	
[5] 2-Iodophenol	CDCl$_3$		7.658	6.680	7.247	6.999	5.281	
	CHARGE		7.551	6.624	7.130	6.843	5.005	
	GIAO[a]		6.942	6.467	6.807	6.904	4.407	
[6] 2-Methylphenol	CDCl$_3$		7.114	6.842	7.076	6.761	4.604	2.250
	CHARGE		7.019	6.860	7.023	6.743	4.525	2.294
	GIAO[a]		7.306	6.958	7.161	6.594	3.827	2.325
[7] 2-Methoxyphenol	CDCl$_3$		6.850	6.850	6.850	6.911	5.590	3.886
	CHARGE		6.686	6.811	6.802	6.709	5.570	3.782
	GIAO[a]		6.670	6.900	7.051	7.115	5.310	3.908
[8] 2-Cyanophenol	CDCl$_3$		7.511	7.000	7.478	6.998	5.659	
	CHARGE		7.539	7.175	7.496	7.119	5.567	
	GIAO[a]		7.601	7.033	7.595	7.233	5.441	
[9] 2CF$_3$-phenol	CDCl$_3$		7.510	7.008	7.422	6.955	5.440	
	CHARGE		7.746	7.113	7.432	7.089	5.295	
	GIAO[a]		7.705	7.259	7.652	7.087	5.649	
[10] 2-Nitrophenol	CDCl$_3$		8.094	6.973	7.565	7.144	10.555	
	CHARGE		8.090	7.233	7.577	7.176	10.663	
	GIAO[a]		8.423	6.925	7.634	7.322	11.498	
[11] Methyl salicylate	CDCl$_3$		7.833	6.873	7.449	6.978	10.727	3.949
	CHARGE		7.898	7.135	7.388	7.047	10.997	3.894
	GIAO[a]		8.085	6.877	7.547	7.150	11.078	3.914
[19] 2-Hydroxy-benzaldehyde	CDCl$_3$		7.567	7.027	7.535	6.997	11.024	9.903
	CHARGE		7.678	7.161	7.520	7.066	11.056	9.893
[20] 2-Hydroxy-acetophenone	CDCl$_3$		7.730	6.896	7.466	6.972	12.242	2.627
	CHARGE		7.617	7.165	7.511	7.083	12.292	2.665

[a] CH$_4$ as reference using B3LYP-6311++g(d,p), for both optimization and NMR calculations.

methyl salicylate the large steric coefficient of the CO oxygen was reduced for these short distances, but no other change was needed in the calculation. Thus this equation may be applied to any functionality capable of forming such hydrogen bonds.

Table 6.24   *¹H chemical shifts (δ) of para substituted phenols in CDCl₃ with CHARGE and GAIO calculated chemical shifts*

Phenol(s)	Solvent	H2,6	H3,5	OH	Substituent
[12] 4-Cyanophenol	CDCl₃	6.917	7.556	5.266	
	CHARGE	7.116	7.592	5.095	
	GIAO[a]	6.867	7.663	4.203	
	DMSO	6.903	7.630	10.583	
[13] 4-Fluorophenol	CDCl₃	6.763	6.921	4.604	
	CHARGE	6.878	7.013	4.450	
	GIAO[a]	6.783	7.056	3.688	
	DMSO	6.736	6.969	9.313	
[14] 4-t-Bu-phenol	CDCl₃	6.749	7.252	4.557	1.290
	CHARGE	6.817	7.082	4.540	1.380
	GIAO[a]	6.764	7.337	3.652	1.291
	DMSO	6.668	7.159	9.064	1.224
[15] 4-Nitrophenol	CDCl₃	6.898	8.157	5.356	
	CHARGE	7.183	8.083	5.334	
	GIAO[a]	6.829	8.476	4.428	
	DMSO	6.948	8.117	10.997	
[16] 4-CF₃-phenol	CDCl₃	6.896	7.509	5.050	
	CHARGE	7.024	7.683	4.842	
	GIAO[a]	6.910	7.713	4.185	
	DMSO	6.921	7.519	10.243	
[17] 4-Methoxyphenol	CDCl₃	6.758	6.791	4.521	3.759
	CHARGE	6.748	6.757	4.364	3.766
	GIAO[a]	6.587	6.986	3.397	3.741
	DMSO	6.670	6.738	8.848	3.652
	Solvent	**H3**	**H5**	**H6**	**OH**
[18] 2,4-Chlorophenol	CDCl₃	7.324	7.150	6.950	5.466
	CHARGE	7.227	7.195	6.890	5.562
	GIAO[a]	7.458	7.209	7.101	5.065
	DMSO	7.432	7.199	6.967	10.442

[a] CH₄ as reference using B3LYP-6311++g(d,p), for both optimization and NMR calculations.

Table 6.25   *The OH chemical shift of para substituted phenols versus the OH π-electron density (me)*

Para substituent	Experimental	Electron density	Calculated
NO₂	5.356	120.4	5.334
CN	5.266	115.0	5.095
CF.₃	5.050	109.5	4.842
H	4.692	107.2	4.704
F	4.604	100.1	4.561
t-Bu	4.557	103.1	4.540
OMe	4.521	98.5	4.364

The chemical shifts for all phenols in Tables 6.23 and 6.24 were calculated using the CHARGE program modified as described and with *ab initio* GIAO calculations. These were plotted against the observed chemical shifts and the slope, intercept and the correlation coefficients of these plots are given below:

$$\delta_{observed} = 0.993\delta_{calculated} + 0.059, \quad R^2 = 0.994, \quad \text{CHARGE}$$

$$\delta_{observed} = 0.869\delta_{calculated} + 1.043, \quad R^2 = 0.965, \quad \text{GIAO}$$

The CHARGE calculations reproduce the experimental data well. The *ortho* substituted compounds not involved in strong intramolecular hydrogen bonds (OH···O distance > 2 Å) have not been used in the parameterization of the CHARGE program and serve as an objective basis for comparison with other chemical shift calculation methods and to determine the general accuracy of the program.

There are a number of discrepancies around the OH region (4—6 ppm) from the GIAO calculations and this may be due to the exclusion of solvent effects in these calculations, and an improved correlation is found when the OH protons are removed ($R^2 = 0.982$). The CHARGE data correlate well in this region and when the OH data are removed the correlation coefficient is hardly affected ($R^2 = 0.996$). The GIAO data contain some outliers, but follows a trend which is slightly tilted compared to the diagonal. This tilted trend had been observed before.[43,103] and may be due to inaccuracies in the DFT calculations or the exclusion of solvent effects.

The results show that atoms in the vicinity of hydroxyl protons have a large effect on their chemical shifts. By elucidating the OH oxygen π shift the CHARGE program can accurately model the non-bonded effects for atoms well removed from the OH proton. The classical models of anisotropy, steric, and electric field incorporated in the program fail to describe the chemical shifts of OH protons closer than 2 Å to H-bonding oxygen atom. This interaction has been characterized and reproduced by the introduction of an additional term obtained from *ab initio* DFT calculations.

The above analysis was performed on phenols but since the effect of conjugation is treated and parameterized separately it should not affect calculations of non-conjugated systems. However as detailed elsewhere[104] the effects of an oxygen atom which is not sp² hybridized (such as the oxygen atom in the DMSO molecule) on OH-bonded chemical shifts differs from the calculations in Chapter 3, Section 3.5.4 thus these shift calculations are only strictly applicable to cases where the acceptor atom is a sp² hybridized oxygen atom. The H-bonding Equation (3.16) would need to be separately parameterized for systems not containing the C=O group. Fortunately in organic chemistry and particularly in biologically relevant systems the carbonyl oxygen is by far the most common hydrogen bond acceptor in strong hydrogen bonds.

The authors noted that combined molecular mechanics/ CHARGE calculations to predict ¹H chemical shifts should be used with caution in cases where strong hydrogen bonding occurs, since the force field would have to predict H-bond distances to an accuracy of fractions of an angstrom, to obtain accurate chemical shifts. Unless the force fields used can reproduce such high accuracy the resulting chemical shift calculations should be used as estimates (indeed for binding studies such estimates may still be of great interest).

6.5   Amines

6.5.1   Introduction

Amines are one of the most important classes of organic compounds, being constituents of alkaloids, amino acids, proteins, etc. and many other natural products. Also many drugs are amines (e.g. amphetamines, morphine, nicotine, etc.). Despite this common occurrence there are few studies on the ^{1}H NMR chemical shifts of the amino group. Zurcher's[8] pioneering investigation did not consider amines, probably due to lack of good data. Pretsch[105] reported literature data for some amines, mainly of attached methyl groups. Alkorta and Elguero[106] calculated (GIAO) ^{1}H shieldings in amines and gave literature data for the amines they investigated (methyl and ethylamine and some heterocyclic amines). They calculated the influence of the nitrogen lone pair on the ^{1}H chemical shifts and noted that it was a function of ring size and the nitrogen substitution.

There are possible reasons for the lack of investigations into amino chemical shifts. One is the strong basicity of alkyl amines. Abraham *et al.*[107] found that the ^{1}H shifts of the NH protons of aliphatic amines showed little consistency in either CDCl$_3$ or DMSO solvent and often the signals were too broad to observe. They note that the NH$_2^+$ protons of dimethylamine hydrochloride occur at 9.2 ppm in CDCl$_3$ compared to the free base at ca. 1 ppm[48] hence traces of acid will produce large shifts for these protons and also broaden the signal due to exchange. They suggested that this was the most probable reason for the lack of consistency of these shifts in either solvent and concluded that these protons are not suitable for diagnostic purposes. Other near protons may be expected to experience similar but smaller effects due to protonation. Aryl amines are much less basic and the chemical shifts of the NH protons in these compounds can be satisfactorily predicted once all the factors affecting their chemical shifts, including π contributions, are evaluated.

The other factor is the trivalent nature of the nitrogen atom. This gives additional degrees of freedom when compared to e.g. halo substituents. For example, the SCS of the amino group is dependent on the position of the lone pair.[106] In solution the NH$_2$ group is usually not in a single orientation but the precise populations of the different conformers are often unknown. For example, ethylamine has two distinct conformers with the lone pair anti or gauche to the C—C bond but the conformer energies in solution have not been measured accurately. Gas phase *ab initio* calculations by Alkorta and Elguero[106] gave ethylamine as a 50:50 mixture but solvent effects were not considered. In more complex acyclic amines the number of conformers is prohibitive for any quantitative calculation.

For these reason Basso *et al.*[108] investigated a variety of cyclic amines in which the carbon framework is in a rigid conformation (Figure 6.13) and determined the orientation of the amino group by *ab initio* calculations.

The ^{1}H NMR spectra of these molecules were complex but completely assigned to give a database of ^{1}H chemical shifts and this database was used to parameterize the CHARGE model for the amino group.

Alkorta and Elguero[106] calculated amine shieldings at the GIAO/B3LYP/6-311++G** level but noted that their ^{1}H calculations did not correlate as well with the observed shifts as the analogous ^{13}C calculations. Basso *et al.*[108] used a similar approach to Alkorta and Elguero in that the same density function theory and basis set was used for the geometry optimization

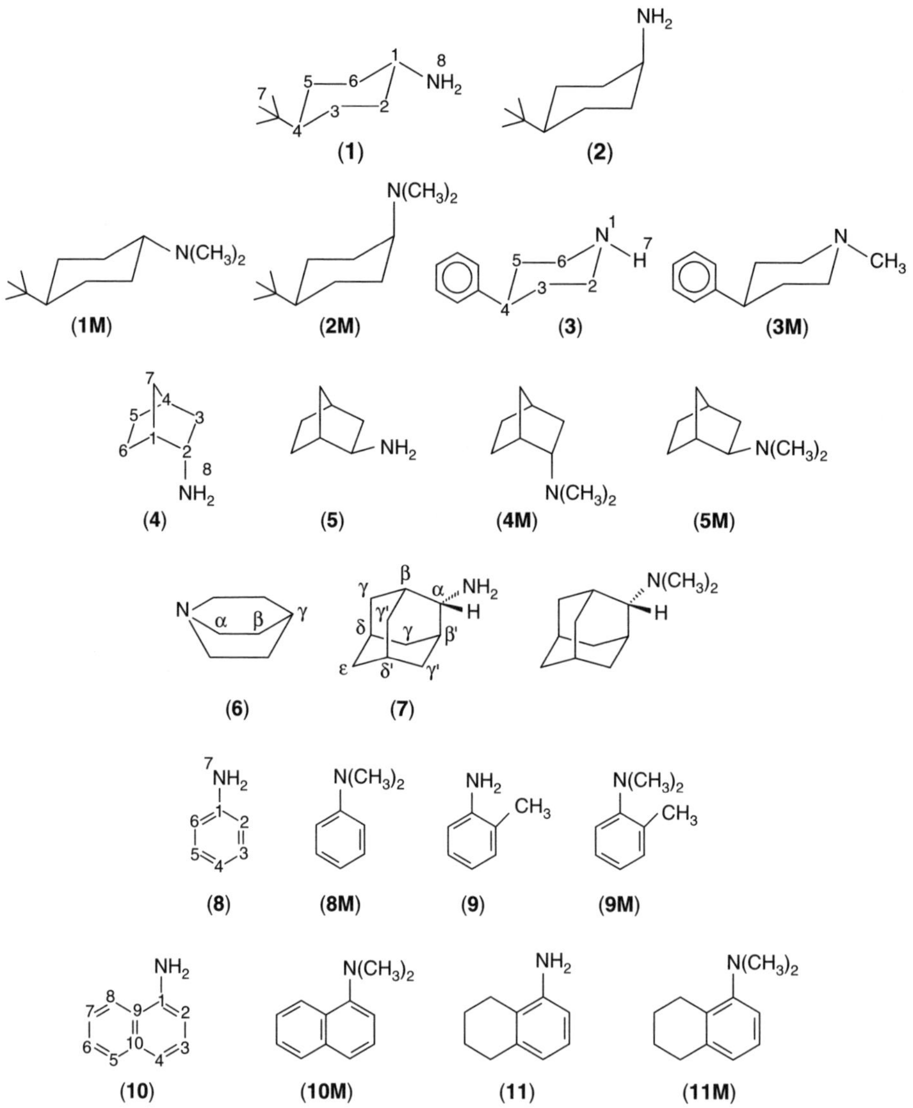

Figure 6.13   *Amino compounds investigated.*

and GIAO calculations. The GIAO derived chemical shifts were then compared with the shifts from the CHARGE parameterization.

6.5.2   Theory and Application to Amines

In the CHARGE routine, single bonds are considered to be isotropic and this assumption was made for the C—N and N—H bonds, thus there is no anisotropic term for amines. The other long-range effects in the CHARGE formulation (electric field, steric effects, π effects and ring currents) need to be considered. The electric field shift is given directly from the calculated partial atomic charges and these can be checked by comparison of the observed and calculated dipole moments. For methylamine, dimethylamine and trimethylamine the observed[109] vs. calculated dipole moments are 1.33, 1.01 and 0.63D vs. 1.32, 1.10 and 0.86D. The agreement is such that the electric field term was used unchanged in the calculations. The nitrogen steric effect needs to be determined. The aromatic ring current in substituted benzenes is the same as in benzene (Chapter 5, Section 5.1) and this was assumed for the amino benzenes. The effect of the π-electron density on the ring protons is given by Equation (3.18) but the effect of the π density on the nitrogen atom on the chemical shifts of the NH proton requires a different coefficient. The influence of the nitrogen lone-pair on the shielding of neighbouring protons has to be considered. This was not included explicitly for the case of the oxygen atom in alcohols and ethers.[110] (Section 6.3 and Chapter 7, Section 7.4) as the steric effect of the oxygen atom plus the α, β and γ effects of the oxygen atom reproduced the observed shifts. The same procedure was used for the nitrogen atom of the amines considered.

Thus it was necessary to evaluate the steric coefficient a_s (Equation (3.11)), π-electron coefficient (Equation (3.18)) and the α, β and γ effects of the nitrogen atom on the chemical shifts of the near protons. These parameters were obtained from an iterative calculation on the observed shifts.

6.5.3   Observed versus Calculated Shifts

The primary amines used were obtained commercially, as well as the dimethylamino compounds **8M**, **9M** and **10M** (Figure 6.13). The remaining dimethylamino derivatives were synthesized according to literature procedures. ¹H spectra were obtained at 400.13 MHz. Compounds **1M**, **2M**, **3M**, **4M** and **10** gave complex overlapping ¹H spectra at 400 MHz and the ¹H spectra were obtained at 700.13 MHz. Full assignments are given in Tables 6.27 and 6.28.

The geometries of all the compounds were minimized using *ab initio* calculations. For all compounds the potential energy surfaces were constructed at the HF/3-21G level in order to determined the preferred amino group orientation. The stable conformers were then minimized at the B3LYP/6-311++G(d,p) level and the GIAO calculations performed using the B3LYP/6-31G(d,p) level, referenced to methane and converted to TMS using $\delta = 0.23$ ppm for methane.[111]

6.5.4   Conformational Analysis

The hydrocarbon fragments in the molecules studied are rigid structures, thus only the rotational isomerism about the C—N bond needed to be determined. In order to determine the most stable rotamer of the amino group, potential energy surface (PES) calculations were performed, varying the C2—C1—N—H(C) dihedral angle with increments of 10°. The nitrogen lone-pair is not defined in the compound specification, but it is convenient to describe the rotational isomerism about the H—C—N—R (R=H, Me) bond in terms of the H—C—N lone-pair dihedral angle (θ). In *trans* 4-t-butylcyclohexylamine (**1**) the symmetrical conformer ($\theta = 180°$) was calculated to be slightly more stable (0.3 kcal mol^{-1}) than the gauche ($\theta = 60°$) conformer. The statistical weight of two for the gauche conformer results in almost equal populations of the two forms (anti:gauche, 45:55). In the dimethyl derivative (**1M**) the symmetric conformer is much more stable than the gauche by 1.7 kcal mol^{-1}. In *cis* 4-t-butylcyclohexylamine (**2**) the symmetric form ($\theta = 180°$) is more stable than the gauche conformer with one hydrogen atom pointing into the ring ($\Delta E = 1.0$ kcal mol^{-1}) and is again the more populated form (77%). In the dimethyl derivative (**2M**) the methyl groups are too bulky to point into the ring and only the symmetric form is present. The conformations of piperidine derivatives analogous to (**3**) and (**3M**) have been determined previously.[112] The N—H equatorial conformer in piperidine is favoured over the axial NH conformer by 0.4 kcal mol^{-1} and in *N*-methyl piperidine the equatorial *N*-methyl is favoured over the axial conformer by 3.0 kcal mol^{-1}. The remaining compounds were found to have one major conformer. In both endo aminonorbornane (**4**) and the dimethyl derivative (**4M**) the stable conformer has the lone-pair anti to the CH proton but in the corresponding exo compounds (**5**) and (**5M**) the lone-pair is anti to the C$_2$—C$_3$ bond. Both 2-adamantanamine (**7**) and its dimethyl derivative (**7M**) exist in a symmetric conformation with the lone-pair anti to the CH bond ($\theta = 180°$).

The *ab initio* calculations iterate to a planar nitrogen atom for aniline (**8**), *N,N*-dimethylaniline (**8M**) and *ortho*-toluidine (**9**) but in (**9M**) the nitrogen atom is pyramidal with one methyl group at ca. 90° to the phenyl ring. Interestingly the *ab initio* calculations give a non-planar nitrogen atom for both the naphthalene derivatives (**10**) and (**10M**) and the tetrahydronapthalenes (**11**) and (**11M**). Where the contribution of each rotamer was significant as in compounds **1, 1M, 2** and **3** the chemical shifts were calculated for each conformer and the final chemical shifts obtained from the weighted average.

Effect of the π density on the nitrogen atom on NH chemical shifts. The effect of the excess π- electron density at an aromatic carbon atom on the shifts of the attached proton is given by Equation (3.18) in which the coefficient was determined as 10 ppm per π electron.[100] Basso *et al.*[108] determined the analogous value for an NH proton by comparing the NH shifts of *para* substituted anilines with the calculated π densities at the nitrogen atom (Table 6.26).

Table 6.26 NH$_2$ *chemical shifts versus nitrogen π-electron excess for para-substituted anilines*

Substituent	π-Electron excess (me)[a]	NH$_2$ Chemical shift (ppm)	
		Obs.[b]	Calc.[c]
NMe$_2$	135	3.28	3.15
OEt	140	3.40	3.37
F	141	3.51	3.48
Me	145	3.48	3.59
Cl	149	3.63	3.74
Br	148	3.64	3.73
I	149	3.57	3.72
H	149	3.56	3.77
CF$_3$	152	3.90	3.92
CO$_2$Me	153	4.19	3.99
CN	158	4.30	4.20

[a] Millielectrons.
[b] Data from Puchert and Behnke[48].
[c] CHARGE program.

The *para* substituent is sufficiently distant from the nitrogen atom that the only factor affecting the NH shift is the π electron density. There is a linear correlation between the shifts and the π densities with a coefficient of 42.8 ppm per π electron and rms error of 0.142. The shifts calculated by CHARGE with the above coefficient inserted are also shown in the table. The agreement clearly demonstrates the π dependence of the shifts, but it is important to note that the coefficient depends on the π calculation. The modified Huckel program detailed previously[110] was used. This coefficient was included in the calculations of the NH chemical shifts of the aromatic amines in Table 6.28. It has no effect on the NH shifts of the aliphatic amines as none of them are conjugated. The remaining parameters necessary for the calculation of the ¹H chemical shifts were obtained from an iterative calculation. The coefficients A and B of the γ effects (H—C—C—N and H—C—N—C) were determined as well as the nitrogen steric coefficient (a_S, Equation (3.11)) of 66.6 ppm Å^{-6} .

CHARGE model versus GIAO calculations. The observed and calculated ¹H chemical shifts of the aliphatic amines are shown in Table 6.27. The observed chemical shifts range from ca. 0.60 to 3.30δ and were predicted with the CHARGE model with an rms error of 0.108 ppm. The majority of the shifts show deviations of less than 0.1 ppm, showing the agreement between observed and calculated chemical shifts.

One of largest deviations of the calculated shifts was associated with the gamma substituent effect (N—C—C—H). This has a large dihedral angle dependence. The norbornane derivatives and the 4-t-butylcyclohexane derivatives have very different dihedral angles (N—C—C—H), see Figure 6.14, and this may be the reason for such deviations. The γ effect has been parameterized using a simple cos ϕ term and this may not be adequate to fully account for this dihedral angle dependence.

The observed versus calculated ¹H chemical shifts for the aromatic amines are given in Table 6.28. The majority of the shifts were predicted to within 0.05 ppm, showing even better agreement than the aliphatic amines.

The CHARGE and GIAO calculated shifts were compared from the data in Tables 6.27 and 6.28. The GIAO calculations were, in general, less accurate than CHARGE calculations.

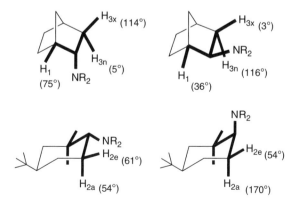

Figure 6.14 *Structure and dihedral angles (N—C—C—H) of derivatives (R=H, CH₃) of 2-amino-norbornane and 4-t-butylcyclohexylamine.*

Table 6.27 Experimental and calculated ^1H chemical shifts of aliphatic amines

Compound	Method	H1a	H1e	H2a/6a	H2e/6e	H3a/5a	H3e/5e	H4a	NR$_2$	t-Bu	rms
trans-4-t-Butylcyclohexylamine (**1**)	CDCl$_3$	2.551	—	1.033	1.885	1.033	1.755	0.962	1.100	0.842	—
	CHARGE	2.750	—	1.153	1.772	0.908	1.653	1.065	1.081	0.904	0.124
	GIAO	2.688	—	1.058	1.729	1.206	1.742	1.087	0.338	0.869	0.113
cis-4-t-Butylcyclohexylamine (**2**)	CDCl$_3$	—	3.148	1.535	1.654	1.274	1.525	0.962	1.100	0.859	—
	CHARGE	—	3.211	1.517	1.611	1.288	1.514	1.019	1.183	0.907	0.041
	GIAO	—	3.151	1.423	1.546	1.851	1.376	1.104	0.269	0.889	0.239
trans-4-t-Butyl- (**1M**) N,N-dimethylcyclohexylamine	CDCl$_3$	2.068	—	1.161	1.931	1.006	1.820	0.938	2.267	0.843	—
	CHARGE	2.109	—	1.063	1.637	0.893	1.650	1.056	2.335	0.905	0.142
	GIAO	2.074	—	1.227	1.822	1.169	1.801	1.069	2.115	0.872	0.103
cis-4-t-Butyl- (**2M**) N,N-dimethylcyclohexylamine	CDCl$_3$	—	1.945	1.296	2.005	1.324	1.431	1.032	2.205	0.847	—
	CHARGE	—	2.030	1.139	1.854	1.329	1.411	1.008	2.279	0.904	0.090
	GIAO	—	1.930	1.207	2.006	1.828	1.358	1.210	2.126	0.895	0.196
		H2a/6a	H2e/6e	H3a/5a	H3e/5e	H4a	NR	*ortho*	*meta*	*para*	
trans-4-Phenylpiperidine (**3**)	CDCl$_3$	2.737	3.180	1.636	1.827	2.609	1.100	7.222	7.290	7.199	—
	CHARGE	2.720	3.155	1.680	1.921	2.589	1.217	7.212	7.313	7.186	0.040
	GIAO	2.407	2.582	1.435	1.192	1.845	0.385	6.971	6.994	6.913	0.465
trans-N-Methyl- 4-phenylpiperidine (**3M**)	CDCl$_3$	2.040	2.969	1.820	1.820	2.465	2.317	7.209	7.271	7.166	—
	CHARGE	2.098	2.952	1.864	1.798	2.599	2.277	7.212	7.312	7.187	0.056
	GIAO	2.028	2.628	1.831	1.524	2.113	2.092	7.234	7.270	7.194	0.205

Compound		H1	H2	H3x	H3n	H4	H5x	H5n	H6x	H6n	H7a	H7s	NR$_2$	rms
endo-2-Aminonorbornane (**4**)	CDCl$_3$	2.071	3.256	1.956	0.588	2.142	1.547	1.221	1.366	1.728	1.296	1.388	1.100	—
	CHARGE	2.287	3.192	1.983	0.564	2.186	1.541	1.269	1.349	1.780	1.187	1.149	1.088	0.109
	GIAO	1.920	3.401	1.667	0.700	2.055	1.482	1.289	1.129	2.494	1.313	1.344	0.266	0.270
exo-2-Aminonorbornane (**5**)	CDCl$_3$	1.927	2.806	1.026	1.635	2.208	1.419	1.046	1.464	1.079	1.097	1.469	1.100	—
	CHARGE	1.975	2.769	0.985	1.640	2.205	1.56	1.234	1.509	1.243	1.090	1.475	1.138	0.090
	GIAO	1.785	3.001	1.073	1.369	2.146	1.447	1.016	1.442	1.017	1.047	2.039	0.153	0.206
endo-2-*N,N*-Dimethylamino norbornane (**4M**)	CDCl$_3$	2.228	2.087	1.709	0.898	2.165	1.497	1.270	1.280	1.748	1.288	1.353	2.165	—
	CHARGE	2.484	1.977	1.569	0.709	2.199	1.530	1.278	1.118	2.065	1.177	1.158	2.267	0.164
	GIAO	1.863	2.019	1.363	0.649	1.795	1.222	1.007	0.976	2.012	1.026	1.090	1.768	0.320
exo-2-*N,N*-dimethylamino norbornane (**5M**)	CDCl$_3$	2.346	1.811	1.378	1.354	2.242	1.445	1.077	1.495	1.077	1.053	1.470	2.171	—
	CHARGE	2.343	1.955	1.510	1.176	2.199	1.554	1.062	1.485	1.231	1.062	1.496	2.215	0.106
	GIAO	2.325	1.758	1.465	1.319	2.181	1.436	1.097	1.482	1.045	1.045	1.717	1.976	0.081

Quinulidine (**6**)		Hα	Hβ	Hγ										rms
	CDCl$_3$	2.854	1.531	1.735										—
	CHARGE	2.836	1.663	1.802										0.086
	GIAO	2.615	1.281	1.580										0.229

Compound		Hα	Hβ/β'	Hγ'a	Hγ'e	Hγa	Hγe	H8'	H8	Hε	NR$_2$			rms
2-Adamantanamine (**7**)	CDCl$_3$	2.982	1.720	1.731	1.847	1.979	1.527	1.786	1.822	1.717	1.419			—
	CHARGE	3.082	1.881	1.592	1.631	2.123	1.459	1.923	1.929	1.664	1.294			0.191
	GIAO	3.186	1.453	1.617	1.775	2.591	1.279	1.643	1.645	1.645	0.238			0.236
2-*N,N*-Dimethyl adamantanamine (**7M**)	CDCl$_3$	1.870	2.051	1.667	1.848	2.043	1.435	1.809	1.790	1.709	2.216			—
	CHARGE	2.263	2.029	1.601	1.615	2.226	1.336	1.920	1.927	1.664	2.288			0.172
	GIAO	1.939	1.965	1.627	1.796	2.320	1.324	1.676	1.655	1.663	2.085			0.127

Table 6.28 Experimental and calculated ${}^1H$ chemical shifts of aromatic amines

Compound	Method	H2/6	H3/5	H4	NR$_2$			rms
Aniline (**8**)	CDCl$_3$	6.635	7.127	6.733	3.500			—
	CHARGE	6.661	7.134	6.678	3.538			0.035
	GIAO	6.159	7.028	6.474	2.837			0.318
N,N-Dimethylaniline (**8M**)	CDCl$_3$	6.730	7.231	6.713	2.926			—
	CHARGE	6.651	7.124	6.669	2.726			0.122
	GIAO	6.402	7.172	6.662	2.923			0.169
		H3	H4	H5	H6	CH$_3$	NR$_2$	
o-Toluidine (**9**)	CDCl$_3$	7.021	6.683	7.010	6.625	2.128	3.488	—
	CHARGE	6.929	6.612	6.946	6.585	2.376	3.933	0.060
	GIAO	6.918	6.471	6.960	6.168	2.924	1.860	0.260
o – N,N-dimethyltoluidine (**9M**)	CDCl$_3$	7.147	6.934	7.142	7.023	2.322	2.689	
	CHARGE	7.090	6.990	7.076	6.980	2.325	2.715	0.047
	GIAO	7.189	6.959	7.121	6.927	2.192	2.507	0.102

		H2	H3	H4	H5	H6	H7	H8	NR$_2$	
1-Aminonaphthalene (**10**)	CDCl$_3$	6.676	7.272	7.308	7.800	7.441	7.433	7.788	3.966	—
	CHARGE	6.530	7.212	6.937	7.718	7.465	7.333	8.107	3.973	0.201
	GIAO	6.503	7.257	7.125	7.647	7.425	7.400	7.732	3.236	0.114
1-*N,N*-dimethylaminonaphthalene (**10M**)	CDCl$_3$	7.023	7.344	7.474	7.775	7.413	7.443	8.223	2.853	—
	CHARGE	7.047	7.384	7.433	7.807	7.493	7.431	8.422	2.848	0.080
	GIAO	7.024	7.365	7.398	7.702	7.466	7.509	8.369	2.729	0.083
5-Amino-1,2,3,4-tetrahydronaphthalene (**11**)	CDCl$_3$	6.449	6.901	6.512	2.704	1.723	1.826	2.398	3.475	—
	CHARGE	6.463	6.931	6.467	2.679	1.816	1.808	2.550	3.995	0.072
	GIAO	6.217	6.913	6.451	2.626	1.618	1.713	2.229	2.540	0.129
5-*N,N*–Dimethylamino-1,2,3,4-tetrahydronaphthalene (**11M**)	CDCl$_3$	6.889	7.074	6.798	2.794	1.760	1.780	2.735	2.669	—
	CHARGE	6.898	7.080	6.89	2.694	1.808	1.800	2.828	2.679	0.061
	GIAO	6.836	7.083	6.801	2.737	1.661	1.646	2.706	2.464	0.098

In particular the NH chemical shifts are calculated much more shielded than the observed value. This is very likely due to the absence of solvation effects in the GIAO calculations. The differences between theory and experiment for these protons were so large that they were not included in the rms error calculations. The theoretical predictions for protons attached to the carbon chains do not shown a regular behaviour. Some of the predicted shifts were shielded and others were deshielded. Nevertheless the rms error values are only 0.096 for CHARGE and 0.187 for GIAO.

In conclusion the GIAO calculations yield reasonable results and may be improved by using alternative basis sets or solvation theory. Overall, the semiempirical calculations produce more reliable results and provide a rapid and useful tool for routine use in chemical shifts prediction.

6.6   Cyanides

6.6.1   Introduction

Nitriles are of considerable importance in all branches of chemistry. They are both versatile synthetic intermediates and important compounds[113] and in consequence the ¹H spectra of nitriles has been studied since the beginning of NMR spectroscopy. Despite this, there is still some controversy and uncertainty over the causes of the SCSs of the cyano group[114]. The cyano group is both strongly polar and also anisotropic and both of these factors were proposed to account for cyano SCSs. Early workers suggested that the CN magnetic anisotropy should be similar to that of the analogous C≡C bond and Reddy and Goldstein[115] used a correlation between C^{13}—H couplings and the ¹H chemical shift to estimate $\Delta\chi$ as $-16.5 \times 10^{-6}\,cm^3\,mol^{-1}$ for both the CN and the C≡C bond. Cross and Harrison[116] used the value of the CN anisotropy obtained by Reddy and Goldstein to calculate the shifts of the C-19 methyl groups in some 5α and 5β-cyano steroids. They found that the shifts were opposite to those predicted from the anisotropy and suggested that the CN electric field could be responsible.

Subsequently Zurcher[8] and ApSimon et al.[117] conducted more detailed analyses of the CN SCSs. They used the McConnell equation[118] to calculate the magnetic anisotropy of the cyano group and the CN dipole to calculate the electric field but did not consider any steric effects of the CN group. They also assumed that the CN anisotropy could be calculated from the centre of the triple bond, although the π-electron system may be more or less displaced towards the more electronegative atom. Both studies concluded that the electric field effect was predominant. However both these studies used mainly the methyl groups of steroids to determine the SCSs. When they extended their calculations to include nearer protons large differences between the observed and calculated shifts were found.

Abraham and Reid[119] then presented an extensive data set of CN SCSs using conformationally rigid molecules with fully assigned ¹H spectra. The aliphatic nitriles analysed were *trans-* and cis-4-t-butylcyclohexanecarbonitrile (**1a**, **1b**), axial and equatorial cyclohexanecarbonitrile (**2a**, **2b**) and ax–ax and eq–eq *trans*-1,4-dicyano cyclohexane (**3a**, **3b**). Included also in the analysis were the ¹H spectra of 2-exo and 2-endo norbornanecarbonitrile (**4a**, **4b**) and 1-adamantane carbonitrile (**5**) recorded previously[120] and the ¹H shifts of acetonitrile (**6**), propionitrile (**7**), isobutyrocarbonitrile (**8**) and trimethylacetonitrile (**9**) from the

Aldrich catalogue.[48] The aromatic nitriles recorded were benzonitrile (**10**), *o*-, *m*- and *p*-di-cyanobenzene (**11**, **12**, **13**), 1- and 2-cyanonaphthalene (**14**,**15**) and 9-cyanoanthracene (**16**). The ^{1}H shifts of acrylonitrile (**17**) were obtained from the Aldrich catalogue.[48].

These data provided a sufficient data set for an analysis of cyano SCSs using the CHARGE model.[46,110]

6.6.2   Theory and Application to the Cyano Group

The cyano group has in principle steric, electric field and anisotropic effects on protons more than three bonds away plus for aromatics a large effect on the π-electron densities. All these were incorporated into the model. The electric field of the cyano group was calculated as being due to the charge on the nitrogen atom of the CN and an equal and opposite charge on the carbon atom of the CN bond. The charge on the nitrogen atom is already calculated in CHARGE and the coefficient in Equation (3.12) is known so the electric field is given directly.

This of course assumes that the charges used in Equation (3.12) provide a reasonable measure of the electric field of the cyano group. The partial atomic charges obtained in the CHARGE program were derived from the observed molecular dipole moments and the extent of the agreement provides a check on the electric field calculation. The calculated versus observed (in parenthesis) dipole moments[109] (in debye) of acetonitrile, propionitrile, t-butylcarbonitrile, (**1a**), (**1b**), acrylonitrile and benzonitrile were 3.81 (3.97), 3.77 (4.02), 3.82 (3.95), 3.87 (3.82), 3.65 (3.76) , 4.11 (3.89) and 4.15 (4.14) and the good agreement provides strong support for the electric field calculation.

The CN group has cylindrical symmetry and Equation (3.14) may be used to calculate the contribution of the anisotropy to the ^{1}H chemical shifts. The steric effects of the CN group are calculated by use of Equation (3.11). Thus the unknowns to be determined are $\Delta\chi$, the molar anisotropy of the CN bond and the steric coefficient a_s.

For protons three bonds or less from the CN group it is necessary to determine the γ effect (GSEF) of the cyano carbon and nitrogen on the neighbouring protons, following Equation (3.5), in which the coefficients A and B may differ for the CN group in aromatic versus saturated compounds. The γ effect from the nitrogen atom has no orientation dependence and was considered as dependent only on the polarizability of the nitrogen atom.

For the aromatic cyanides it was necessary to obtain the appropriate values of the factors h_r and k_{rs} for the Huckel integrals for the CN group (Equation (3.7)) and these were obtained by simulating π-electron densities obtained from *ab initio* calculations. The π-electron densities and dipole moments from these *ab initio* calculations were very dependent on the basis set used. The 3-21G basis set gave the best agreement with the observed dipole moment and the π densities from this basis set were used to parameterize the Huckel calculations. Values of h_r of -0.12 and 0.19 for C(sp) and N(sp) and of k_{rs} of 1.05 for (Csp^2–Csp) and 1.20 for (Csp–Nsp) gave π-electron densities for the aromatic nitriles in reasonable agreement with those from the *ab initio* calculations. The electron densities (total and π) and dipole moments for benzonitrile using CHARGE and *ab initio* calculations are given in Table 6.29.

Table 6.29 *Total and π (in parenthesis) charges (me) and dipole moments for benzonitrile*

Atom	Method				
	STO-3G	3-21G	6-31G	CHARGE	Observed
N(sp)	−200 (−49)	−504 (−87)	−273 (−63)	−484 (−137)	
C(sp)	73 (26)	338 (31)	21 (52)	390 (109)	
C_1	2 (−56)	−58 (−77)	10 (−76)	13 (−9)	
C_o	−42 (24)	−194 (37)	−148 (37)	−47 (14)	
C_m	−58 (2)	−232 (0)	−212 (1)	−72 (−1)	
C_p	−49 (28)	−227 (36)	−180 (34)	−66 (11)	
μ (D)	3.65	4.55	4.82	4.15	4.14

6.6.3   Observed versus Calculated Shifts

^{1}H NMR were obtained at 400.13 MHz and the complex spectra for **1a** and **1b** at 750 MHz. The spectral assignments of the compounds examined are given in Tables 6.31–6.35 along with the calculated values from the CHARGE model. Full details of the spectral assignmnts are given elsewhere.[119,121]

The geometry of the compounds investigated were optimized using the molecular mechanics force fields and also the *ab initio* calculations at the RHF/6-31G* and MP2/6-31G* levels.

The spectra of the separate conformers (**2a, 2b**) and (**3a, 3b**) were obtained from the ^{1}H spectrum at −60 °C. For (**2**) the equatorial conformer was more favoured (ΔE (ax–eq) = 0.27 kcal mol^{-1}) in agreement with literature values (0.2 kcal mol^{-1}).[112] The commercial sample of 1,4-dicyano cyclohexane was identified as the *trans* isomer from the m.pt. of 140–141 °C. (cf. literature 139–140 °C[122]) and confirmed by the ^{1}H spectrum The separate conformers (**3a, 3b**) were assigned from the −60 °C spectra. The di-equatorial conformer was the more stable (1.5/1.0 ratio) with ΔE (ax–eq) = 0.17 kcal mol^{-1}.

As the ^{1}H chemical shifts of the individual conformers of compounds (**2**) and (**3**) were measured at low temperatures (−60 °C), it was of interest to determine whether the ^{1}H chemical shifts had an intrinsic temperature dependence. This was achieved by measuring the spectra of **1a** and **1b** at various temperatures and the results are shown in Table 6.30. The only protons experiencing a significant (> 0.05 ppm) change in the chemical shift on

Table 6.30 ^{1}H *chemical shifts (δ) of trans- and cis-4-t-butyl-cyclohexanecarbonitrile (**1a, 1b**) as a function of temperature*

^{1}H number	Trans (**1a**)			Cis (**1b**)		
	RT	−20 °C	−60 °C	RT	−20 °C	−60 °C
1e	—	—	—	2.921	2.973	3.019
1a	2.314	2.347	2.388	—	—	—
2e	2.161	2.179	2.192	2.037	2.059	2.077
2a	1.529	1.535	1.550	1.516	1.520	1.528
3e	1.855	1.856	1.862	1.771	1.782	1.794
3a	0.981	0.985	0.990	1.367	1.341	1.324
4a	1.023	1.025	1.030	0.986	0.986	0.987

going from RT to $-60\,^\circ$C are the H-1 protons in both **1a** and **1b**. δ (H1eq) changes by 0.098 ppm and δ (H1ax) changes by 0.072 ppm. The corresponding protons in compounds **2** and **3** were corrected by these amounts subsequently.

Aromatic Nitriles. The assignment of the aromatic nitriles is given in Tables 6.34 and 6.35. This agrees with previous data for (**10**), (**11**) and (**12**)[123–126] and (**16**).[127] The ^{1}H chemical shifts for propionitrile (**7**), iso-butyronitrile (**8**), trimethylacetonitrile (**9**) and acrylonitrile (**17**) were measured directly from the Aldrich ^{1}H NMR catalogue.[48]

The data for the aromatic nitriles obtained in dilute CDCl$_3$ solution are in excellent agreement with earlier data obtained in CCl$_4$ solution.[124–126] For example, the *ortho*, *meta* and *para* ^{1}H shifts in benzonitrile in CDCl$_3$ and in CCl$_4$ solution (in parentheses) are 7.660 (7.631), 7.482 (7.452) and 7.559 (7.552). As found previously for the aromatic hydrocarbons[46] there is a small almost constant shift to higher δ values in CDCl$_3$ compared to CCl$_4$.

The data obtained for the cyano compounds were combined with the ^{1}H chemical shifts of the parent compounds[46,66] to give the cyano SCSs in these compounds. These are shown in Figure 6.15 for (**1a**, **1b**) and (**14**, **15**), together with the SCSs obtained previously for (**4a**, **4b**).[120] The SCSs are invariably deshielding. The SCS on the β protons (H.C.CN) is almost constant at 1.24 ($\pm$0.04) ppm. The γ effect of the CN group (i.e. H.C.C.CN) is also

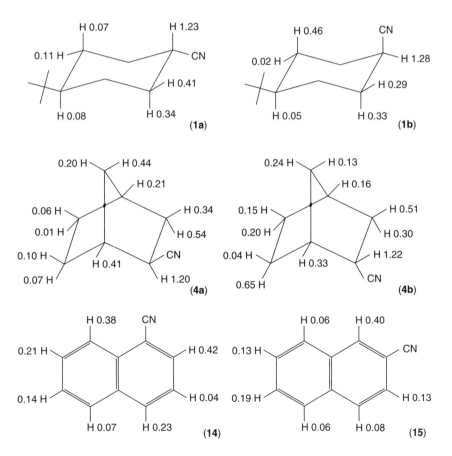

Figure 6.15 *Cyano SCSs in aliphatic and aromatic molecules.*

deshielding with for the saturated nitriles little orientational dependance. For example, the 2ax and 2eq protons in **1a** and **1b** and the 3-exo and 3-endo protons in **4a** and **4b** all give almost identical SCSs of 0.41 (±0.02) ppm.

The long-range (> 3 bonds) effects of the cyano group are also large and extend over both the cyclohexane and bicycloheptene system. For **1a** the CN SCS decreases with increasing distance of the proton from the CN, with the equatorial protons generally displaying a greater CN SCS than the axial protons. However for **1b** the SCS of H-3a is very large. Similar large effects are observed for the 7syn protons in **4a** and the 6-endo protons in **4b**. All these protons are in a similar environment to the cyano group, i.e. essentially orthogonal to the CN bond. Although these SCSs can be due to either the CN anisotropy or electric field, the CN SCSs at protons situated along the CN bond (e.g. the 3ax and 3eq protons in **1a**, the 7syn protons in **4b**, etc.) is also deshielding which would not be the case if the SCSs were primarily due to the CN anisotropy. This suggestion was confirmed by the detailed analysis in terms of the CHARGE model (see later). Similar CN SCSs are observed for the aromatic nitriles **14** and **15** though in these compounds π-electron effects will be present. Again all the SCSs are deshielding and they are considerable even for the protons in the non-substituted aromatic ring.

The data collected in Tables 6.31–6.35 provide a rigorous test of the application of both the CHARGE model and also of theories of cyano SCSs. All the molecules considered are of fixed conformation and the geometries calculated by *ab initio* calculations, thus the only empirical parameters to be determined are those required for the model. These are the anisotropy and steric coefficient of the cyano group and the factors involved in the γ effect (Equation (3.5)). The anisotropy of the CN bond $\Delta\chi^{CN}$ was taken from the centre of the CN bond and the steric effect of the sp carbon atom from the atom considered. The nitrogen atom was considered to be of a sufficient distance from the protons of the molecules investigated to have no noticeable steric interaction with them. There is however a possible γ effect from the nitrogen of the CN group (i.e. H.C.CN) which was considered as a polarizability effect (see theory).

Thus the data set of Tables 6.31–6.35 was calculated with a total of ten possible parameters which were the anisotropy of the CN bond, the carbon steric effect, the γ effect of the sp

Table 6.31 *Observed versus calculated ^{1}H chemical shifts (δ) for trans and cis 4-t-butyl-cyclohexanecarbonitrile (**1a**, **1b**), axial and equatorial cyclohexanecarbonitrile (**2a**, **2b**) and ax–ax and eq–eq trans-1,4-dicyanocyclohexane (**3a**, **3b**)*

H no.	(**1a**)		(**1b**)		(**2a**)[a]		(**2b**)[a]		(**3a**)[a]		(**3b**)[a]	
	Obs.	Calc.	Obs.	Calc.	Obs.	Calc.	Obs.	Calc.	Obs.	Calc.	Obs.	Calc.
1e	—	—	2.921	2.886	2.960	2.859	—	—	3.040	2.999	—	—
1a	2.314	2.416	—	—	—	—	2.386	2.342	—	—	2.445	2.440
2e	2.161	2.067	2.037	2.076	2.000	2.035	2.076	2.034	2.009	2.196	2.208	2.184
2a	1.529	1.646	1.516	1.641	1.538	1.587	1.521	1.591	1.918	1.990	1.582	1.695
3e	1.855	1.807	1.771	1.824	1.700	1.788	1.760	1.776	—	—	—	—
3a	0.981	0.985	1.367	1.290	1.500	1.575	1.220	1.284	—	—	—	—
4e	—	—	—	—	1.700	1.763	1.700	1.730	—	—	—	—
4a	1.023	1.095	0.986	1.078	1.200	1.254	1.220	1.277	—	—	—	—

[a] Measured at −60 °C; H-1e and H-1a have been corrected by 0.098 and 0.072 ppm, respectively.

Table 6.32   *Observed versus calculated ¹H chemical shifts (δ)for 2-exo (**4a**) and 2-endo (**4b**) norbornanecarbonitriles*

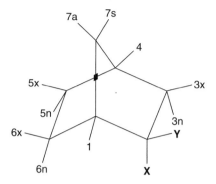

(**4a**) X=H, Y=CN (exo)
(**4b**) X=CN, Y=H (endo)

¹H number	(4a)		(4b)	
	Observed[120]	Calculated	Observed[120]	Calculated
1	2.599	2.402	2.520	2.373
2x	—	—	2.694	2.873
2n	2.360	2.539	—	—
3x	1.810	1.947	1.982	1.928
3n	1.697	1.664	1.458	1.631
4	2.397	2.204	2.348	2.182
5x	1.528	1.643	1.619	1.641
5n	1.171	1.328	1.356	1.400
6x	1.570	1.620	1.505	1.639
6n	1.225	1.402	1.814	1.835
7s	1.621	1.533	1.308	1.290
7a	1.381	1.356	1.417	1.335

Table 6.33   *Observed versus calculated ¹H chemical shifts (δ) for 1-adamantanecarbonitrile (**5**) and the acyclic nitriles*

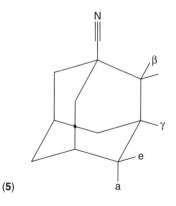

(**5**)

Table 6.33 *(Continued)*

Compound		Obs.[120]	Calc.	Compound		Obs.[48]	Calc.
(5)	β	2.04	1.97	CH_3CN		2.03	2.07
	γ	2.04	2.12	CH_3CH_2CN	Me	1.30	1.22
	e	1.74	1.76		CH_2	2.47	2.44
	a	1.74	1.77	Me_2CHCN	Me	1.35	1.28
Acrylonitrile	gem	5.66	5.86		CH	2.78	2.80
	cis	6.24	6.09	t-BuCN	Me	1.40	1.33
	trans	6.10	5.94				

Table 6.34 *Observed[119] versus calculated ¹H chemical shifts (δ) of benzonitrile (10), o-, m- and p- dicyanobenzenes (11, 12, 13)*

Proton	(10)		(11)		(12)		(13)	
	Obs.	Calc.	Obs.	Calc.	Obs.	Calc.	Obs.	Calc.
H2	7.660	7.684	—	—	7.971	8.042	7.806	7.876
H3	7.482	7.550	7.850	7.888	—	—	7.806	7.876
H4	7.615	7.576	7.782	7.775	7.916	7.916	—	—
H5	7.482	7.550	7.782	7.775	7.671	7.760	7.806	7.876
H6	7.660	7.684	7.850	7.888	7.916	7.916	7.806	7.876

Table 6.35 *Observed[119] versus calculated ¹H chemical shifts (δ) for 1- and 2-naphthalenecarbonitriles (14, 15) and 9-anthracenecarbonitrile (16)*

Proton	(14)		(15)		(16)	
	Observed	Calculated	Observed	Calculated	Observed	Calculated
H1	—	—	8.245	8.245	8.431	8.362
H2	7.900	7.897	—	—	7.728	7.779
H3	7.512	7.721	7.611	7.779	7.596	7.699
H4	8.069	8.112	7.925	8.012	8.089	8.181
H5	7.916	7.928	7.908	7.895	—	—
H6	7.612	7.564	7.663	7.566	—	—
H7	7.685	7.624	7.610	7.548	—	—
H8	8.226	8.133	7.907	7.935	—	—
H10	—	—	—	—	8.691	8.956

carbon atom (coefficients *A* and *B*, Equation (3.5)) which may differ for aliphatic and aromatic nitriles and the nitrogen polarizability.

Iterative calculations were used to determine the best fit values of all these parameters using all the above data, a total of 93 shifts. Iterations were carried out including both the steric and anisotropy terms, the anisotropy alone and the steric term alone. All the iterations performed yielded little or no improvement of the calculated chemical shifts than calculations with no steric or anisotropic terms present. It was therefore concluded that the steric and anisotropic terms of the cyano group were negligible and the major factor influencing the long-range proton chemical shifts was the electric field effect. The final parameterization of the cyano group therefore included electronic effects for protons two or three bonds removed from the carbon atom and the electric field effect for protons more than three bonds away. It was found that Equation (3.5) could be further simplified with $B_1 = B_2$. Thus the entire data set was reproduced with only five parameters. The values of the coefficients A and B in Equation (3.5) were $+0.110$ and -0.047 for the saturated nitriles and -0.185 and $+0.030$ for the unsaturated nitriles. The orientation dependance of the γ CN effect (H.C.C.CN) was very small in both the saturated and unsaturated compounds. The nitrogen polarizability was obtained as 0.19 somewhat less than the value used previously (0.44).

6.6.4   Cyano SCSs

Aliphatic nitriles. The 62 ¹H chemical shifts of the saturated nitriles in Tables 6.31–6.33 range from ca. 0.70 to 3.50δ and were predicted with an rms error of 0.087 ppm and generally good agreement between the observed and calculated shifts. The agreement for the norbornanes (Table 6.32) was not as good as the cyclohexanes (Table 6.31) and this was suggested to be due to the larger errors in the observed versus calculated shifts in norbornane than for cyclohexane. Some support for this was the good agreement between the observed and calculated SCSs for the norbornanes (Table 6.36).

In particular the calculated SCSs for H-1, H-2 and H-4 were in good agreement with the observed SCSs confirming that the calculated CN SCSs were accurate even for these systems. The large deshielding of the H-6endo in **4b** is particularly well reproduced showing that the simple electric field model gives excellent agreement with the observed SCSs.

The calculated chemical shifts for the acyclic molecules (Table 6.33) were also in good agreement with the observed shifts. Zurcher[8] could not predict the α proton chemical shifts in these compounds (H.C.CN) or the chemical shifts of protons three bonds from the CN group (H.C.C.CN) in 2-endo/exo-norbornenecarbonitriles from his linear electric field model. ApSimon *et al.*[117] also failed to calculate CN SCSs in their investigations on the

Table 6.36   *Observed versus calculated SCSs for 2-exo (**4a**) and 2-endo (**4b**) norbornanecarbonitriles*

¹H number	(4a)		(4b)	
	Observed	Calculated	Observed	Calculated
1	0.41	0.43	0.33	0.40
2x	—	—	1.22	1.34
2n	1.20	1.30	—	—
3x	0.40	0.41	0.51	0.39
3n	0.54	0.42	0.30	0.45
4	0.21	0.23	0.16	0.21
5x	0.06	0.11	0.15	0.11
5n	0.01	0.09	0.19	0.16
6x	0.10	0.09	0.09	0.10
6n	0.06	0.16	0.65	0.60
7s	0.44	0.30	0.13	0.06
7a	0.20	0.12	0.24	0.10

Table 6.37 Observed versus calculated CN SCSs with the C—CN/C—H electric field and H-steric contributions for trans (**1a**) and cis (**1b**) 4-t-butylcyclohexanecarbonitriles

Compound	Proton no.	Obs. SCS	Calc. SCS	C—CN electric field	C—H electric field	H-steric
(**1a**)	2e	0.411	0.413	0.332	−0.001	0.000
	2a	0.339	0.413	0.336	−0.001	0.000
	3e	0.105	0.153	0.120	0.027	0.006
	3a	0.071	0.108	0.079	0.017	0.012
	4a	0.083	0.090	0.061	0.022	0.007
(**1b**)	2e	0.287	0.408	0.344	−0.001	0.000
	2a	0.326	0.422	0.262	−0.001	0.000
	3e	0.021	0.170	0.153	0.005	0.012
	3a	0.457	0.413	0.270	0.040	0.103
	4a	0.046	0.073	0.070	−0.005	0.009

long-range shielding effects of the CN group on methyl protons in several cyano-steroids and also on the ring protons in 2-endo/exo-norbornenecarbonitriles. They obtained a poor correlation between the observed and calculated shifts and concluded that a modification of the solvent–solute interaction may be responsible for the poor correlation of some protons.

In the CHARGE program these protons are short-range and therefore the SCSs of the CN group is given by Equation (3.5), not by the electric field effect. The actual magnitudes of the contributions to the cyano SCSs are given in Table 6.37 with the observed versus calculated CN SCSs for **1a** and **1b**. The contributions to the CN SCSs include effects due to the removal of the hydrogen in forming the CN derivative. These are the C—H electric field and the steric effect of the hydrogen. However the dominant effect for all long- range protons can be seen to be the CN electric field effect.

For protons that are > three bonds away from the cyano group the sum of the components gives the total calculated SCS. For the H-2e/H-2a protons the components do not add up to give the calculated value of the CN SCS as these protons experience γ electronic effects (Equation (3.5)). Even in these cases the electric field effect is the major effect.

Aromatic Nitriles. There are other mechanisms which affect the ^{1}H shifts in aromatic nitriles, in particular the ring current and π-electron effects. The ring currents in the aromatic hydrocarbons are calculated in CHARGE in Chapter 5, Section 5.1. In this treatment the ring current intensities of the naphthalene, anthracene and benzene rings all differ. The further assumption was made in this work that the introduction of the cyano group has no effect on the parent hydrocarbon ring current. Thus there are no ring current effects on the CN SCSs. In contrast the CN group does affect the π-electron densities and this has a significant effect on the CN SCSs.

The observed versus calculated ^{1}H chemical shifts for the aromatic nitriles are given in Tables 6.34 and 6.35 and the observed versus calculated SCSs for benzonitrile (**10**), 1- and 2-naphthalene carbonitrile (**14**, **15**) and 9-cyanoanthracene (**16**) in Table 6.38 together with the calculated contributions to the CN SCSs.

There is again generally good agreement between the observed and calculated shifts with the majority of shifts predicted to 0.1 ppm and the majority of SCSs to < 0.05 ppm. The large

Table 6.38 *Observed versus calculated CN SCSs with the electric field and π-electron contributions for benzonitrile (10) and 1- and 2-naphthalenecarbonitriles (14, 15) and 9-cyanoanthracene (16)*

Compound	Proton no.	Obs.	Calc.	CN electric field	C—H electric field	π shift
(10)	2,6	0.319	0.347	0.370	0.000	0.116
	3,5	0.141	0.213	0.127	0.046	0.044
	4	0.274	0.239	0.096	0.036	0.107
(14)	2	0.423	0.404	0.375	0.000	0.169
	3	0.035	0.228	0.126	0.046	0.059
	4	0.225	0.283	0.096	0.035	0.154
	5	0.072	0.099	0.058	0.014	0.028
	6	0.135	0.071	0.054	0.010	0.008
	7	0.208	0.131	0.089	0.012	0.032
	8	0.382	0.304	0.333	0.074	0.001
(15)	1	0.401	0.416	0.376	0.000	0.180
	3	0.134	0.286	0.367	0.000	0.059
	4	0.081	0.183	0.127	0.046	0.014
	5	0.064	0.066	0.040	0.013	0.012
	6	0.186	0.073	0.035	0.000	0.037
	7	0.133	0.055	0.039	0.000	0.015
	8	0.063	0.106	0.050	0.021	0.037
(16)	1	0.422	0.416	0.336	0.070	0.000
	2	0.261	0.226	0.092	0.010	0.055
	3	0.129	0.146	0.055	0.008	0.011
	4	0.080	0.235	0.059	0.010	0.048
	10	0.260	0.549	0.099	0.027	0.252

deshielding of the peri protons H-8 in **14** and H-1 in **16** is well predicted, again demonstrating the accuracy of the electric field calculation even at these short interatomic distances. There are also some discrepancies. The difference between the observed and calculated shift for H-3 in **14** is 0.21 ppm whereas the corresponding meta proton in benzonitrile is predicted quite well (7.48 vs. 7.55).

Table 6.38 shows that the observed SCS for H-3 in benzonitrile is 0.14 whereas the observed SCS for H-3 in **14** is 0.04. The calculated SCSs for these protons are very similar as would be expected. It would appear that the CN SCSs differ significantly in the naphthalene and benzene rings (cf. Section 6.3.2). The calculated shift of the H-10 proton in **16** is also too large by 0.27 ppm. and Table 6.38 shows that this error is due to the calculated SCSs for this proton. This was probably due to the approximations in the Huckel treatment which tends to overestimate the π-electron changes in substituted condensed aromatics such as anthracene.

6.7   Nitro Compounds

6.7.1   Introduction

The substantial and easily observed effect of a nitro substituent on the ¹H NMR spectrum of a molecule has been investigated ever since the discovery of this spectroscopic technique itself. The ¹H chemical shifts in nitrobenzene are in the order *ortho* > *para* > *meta* and this was initially explained as due to the electric field[128] or π-electron density.[129] Neither explanation alone gave a satisfactory account of the order. The magnetic anisotropy of the nitro group was invoked to explain the ¹H chemical shifts in *cis*- and trans-4-tert-butylnitrocyclohexane,[130] with a shielding cone orthogonal to the plane of the trigonal nitrogen atom.

Subsequent investigations on nitro-durenes and mesitylenes did not support these theories. The torsional twist of the nitro group from the ring plane had very little effect on the *meta* protons and the *para* methyl chemical shifts.[131] In contrast a similar ¹³C study found that inhibition of conjugation of the nitro group with the benzene ring due to *ortho*-methyl groups had a significant effect on these shifts[132] and good agreement with calculated π-electron densities was found. In *trans*-1-nitropropene H-1 is more shielded than the more distant H-2 and this cannot be explained solely by the inductive and anisotropic effects of the nitro group.[133] In a ¹³C study of nitro cyclohexanes a good correlation between the α carbon SCSs and Pauling electronegativity parameters was found.[133] The SCSs on the β and γ carbons did not correlate as well.

¹⁷O and ¹⁴N chemical shifts have been used to correlate π-electron densities to the out-of-plane torsion of the nitro group in aromatic compounds.[134-136] However a study of ¹H chemical shifts in such compounds revealed a significant anisotropy effect.[137] The significance of these findings was stressed as any additive scheme based on 2D (i.e. planar) structures cannot be applied to sterically crowded nitro aromatics.[138,139]

More recently the shielding effect of the nitro group was investigated by *ab initio* shielding calculations[140] using a methane molecule as a probe to monitor the change in nuclear shielding around the nitro group. The equation derived was used as a correction term in a phenanthrene compound where good agreement was found.

Although various contributions and models have been described above to explain the SCSs of the nitro group, none of them has given a quantitative explanation of the ¹H shifts of a variety of nitro compounds. Mobli *et al.*[141] obtained the ¹H chemical shifts of a number of aliphatic and aromatic nitro compounds and analysed the observed shifts using both the CHARGE model and GIAO calculations. The parameters in the CHARGE model were determined using aliphatic and aromatic nitro compounds of known fixed geometry (**1–7** in Figure 6.16). The parameters derived from these compounds were then used in the conformational analysis of some aromatic nitro compounds (**8–11** in Figure 6.16).

6.7.2   Theory and Application to Nitro Compounds

To apply the CHARGE theory to the nitro SCSs, both the short-range electronic effects and the long-range contributions of the nitro group need to be obtained. The nitrogen atom of the nitro group is specially identified in the CHARGE program but will be described simply as N here.

The 3 bond γ-substituent effects which need to be obtained are $N-C_{ar}-C_{ar}-H$, $N-C_{al}-C_{al}-H$ and $O-N-C_{al}-H$. The long-range terms include the nitrogen and oxygen steric (a_s) terms[142] (Equation (3.11)), the electric field and anisotropy as well as ring current and π contributions for the aromatic nitro compounds. The electric field of the nitro group was calculated as being due to the charge on each oxygen atom of the NO_2 group and an equal and opposite charge on the nitrogen atom. The charge on the oxygen atom is already

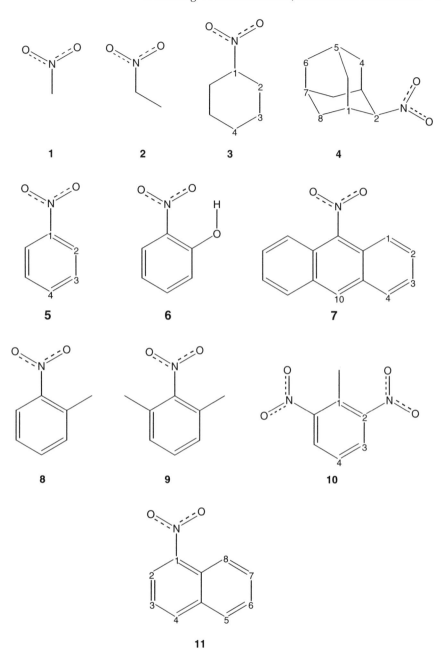

Figure 6.16   *Nitro compounds investigated.*

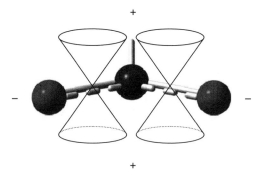

Figure 6.17   *The nitro anisotropy as modelled using classical shielding cones.*

calculated in CHARGE and the coefficient in Equation (3.12) is known so the electric field is given directly. This of course assumes that the charges used in Equation (3.12) provide a reasonable measure of the electric field of the nitro group. The partial atomic charges obtained in the CHARGE program were derived from the observed molecular dipole moments and the extent of the agreement provides a check on the electric field calculation. The calculated versus observed (in parenthesis) dipole moments[109] (in debye) of nitromethane, nitroethane, nitroethylene and nitrobenzene were 3.46 (3.35), 3.23 (3.13), 3.70 (3.95) and 3.99 (4.24) and the agreement provides strong support for the electric field calculation.

The nitro anisotropy was considered to be due to the circulation of electrons along the N=O bonds and therefore the origin of the anisotropy was set to the middle of the N=O bond, analogous to the carbonyl anisotropy[102] as shown in Figure 6.17. Note that this representation is in contrast to the classical models where the shielding cone is placed at the nitrogen atom.[130,137] Thus two anisotropy terms (parallel $\Delta\chi_1$ and perpendicular $\Delta\chi_2$) were required for the parameterization. As previously the ring currents of the nitro aromatics were assumed to be the same as the parent hydrocarbons (Chapter 5, Section 5.1.1) and the Hückel calculations were parameterized to reproduce *ab initio* derived π densities.

6.7.3   Observed versus Calculated Shifts

2-Nitroadamantane was prepared by a literature method,[143] the other compounds were commercial samples. ¹H spectra were run at 400 MHz and the complex ¹H spectra of **3–11** were also obtained at 700 MHz. The geometries of the compounds were obtained using ab initio calculations (B3LYP) with the 6-31G(d,p) basis set and the GIAO calculations used the B3LYP/6-311++G(2d,p) method. Theory and experiment[135] agreed with a non-planar geometry for 9-nitroanthracene (**7**). The geometries of compounds (**8**)–(**11**) are considered later. The nuclear shieldings were converted to chemical shifts using the chemical shift (0.23 ppm) and nuclear shielding (31.763 ppm) of methane using the same basis set. The π-electron densities of nitrobenzene were calculated using the B3LYP/3-21G method as this had been proved successful in previous investigations.[144]

The spectral assignments are given in Tables 6.39 and 6.41. Compound (**3**) was recorded at −85 °C to observe the separate conformers. The population of the axial conformer was very low as reported previously[133] (5% as measured by H-1) and the full assignment of the spectrum of this conformer could not be performed. The effect of solvent (50:50 mixture of

CDCl₃ and CFCl₃) and temperature on the 1eq-nitrocyclohexane spectrum was considered to be larger than the effect of 5% axial nitrocyclohexane at room temperature. Therefore the chemical shifts at room temperature were assigned to 1eq-nitrocyclohexane. Full details of the assignments are given elsewhere.[104,141]

6.7.4   SCSs of the Nitro Group

The determination of the unknown parameters was obtained by iterative methods from the shifts in Table 6.39. The three short-range γ substituent effects were directly obtained. The nitrogen and oxygen steric (aₛ) terms together with the anisotropy terms (parallel

Table 6.39   *Observed versus calculated chemical shifts for compounds (1)–(7)*

Compound	Method							
		CH₃						
(1) Nitromethane	CDCl₃	4.321						
	CHARGE	4.294						
	GIAO	4.319						
		CH₂	CH₃					
(2) Nitroethane	CDCl₃	4.418	1.581					
	CHARGE	4.407	1.505					
	GIAO	4.446	1.506					
		H1ax	H2eq	H2ax	H3eq	H3ax	H4eq	H4ax
(3) 1eq-Nitrocyclohexane	CDCl₃	4.372	2.231	1.851	1.866	1.367	1.667	1.280
	CHARGE	4.348	2.245	1.886	1.772	1.307	1.728	1.322
	GIAO	4.635	2.205	2.075	1.988	1.430	1.811	1.455
		H1	H2	H4eq	H4'ax	H5	H6	
(4) 2-Nitroadamantane	CDCl₃	2.755	4.284	1.682	1.879	1.871	1.772	
	CHARGE	2.844	4.292	1.755	1.889	1.988	1.720	
	GIAO	3.124	4.277	1.718	1.924	1.888	1.910	
		H7	H8ax	H8'eq				
	CDCl₃	1.927	1.786	2.000				
	CHARGE	2.004	1.685	1.761				
	GIAO	1.914	1.870	2.125				
		H2	H3	H4				
(5) Nitrobenzene	CDCl₃	8.236	7.549	7.700				
	CHARGE	8.147	7.611	7.681				
	GIAO	8.930	7.873	8.042				
			H3	H4	H5	H6	H7	
(6) 2-Hydroxynitrobenzene	CDCl₃		7.144	7.565	6.973	8.094	10.555	
	CHARGE		7.176	7.594	7.240	8.104	10.560	
	GIAO		7.322	7.634	6.925	8.423	11.498	
		H1	H2	H3	H4	H10		
(7) 9-Nitroanthracene	CDCl₃	7.944	7.654	7.563	8.073	8.619		
	CHARGE	7.960	7.638	7.630	8.090	8.654		
	GIAO	8.341	8.099	8.047	8.602	9.175		

$\Delta\chi_1$ and perpendicular $\Delta\chi_2$) were input for the parameterization and no significant nitrogen steric term could be established. The other parameters were re-optimized. The oxygen steric coefficient was $24\,\text{Å}^6$ and with the magnetic dipole placed in the middle of the N=O bond the parallel and perpendicular magnetic anisotropies were 2.5 and $-3.4 \times 10^{-30}\,\text{cm}^3$ molecule^{-1}. These values are much smaller than the comparable anisotropies for the carbonyl group (ca. 17 and $3.2\times10^{-30}\,\text{cm}^3$ molecule^{-1}, see (Section 7.2)). This is consistent with the O=N$^+$—O$^-$ structure for the nitro group. However as there are two NO bonds the overall anisotropy effect of the nitro group is comparable with that of the carbonyl group (cf. H-1 in 9-nitroanthracene, 7.94 ppm and in 9-acetylanthracene, 7.85 ppm).

This gave an overall rms error of 0.09 ppm for all protons included (note the OH proton of (**6**) was not included, see Section 6.4.3). The corresponding value for the GIAO calculations was 0.28 ppm. The correlation coefficients for both models were good (GIAO 0.997, CHARGE 0.999). In the GIAO calculations the calculated shifts are greater than observed for most aromatic protons but this does not apply to the aliphatic protons. An overestimate of proton deshielding in the vicinity of electronegative atoms has been observed previously in GIAO calculations using HF[111] and B3LYP theory.[43,102].

The contributions to the nitro SCSs for selected compounds are given in Table 6.40. The authors noted that as the parallel anisotropy is along both N=O bonds the anisotropies ($\Delta\chi_1$) of the two N=O bonds partially cancel when acting on a proton in the plane of the nitro group such as the *ortho* protons of nitrobenzene. This is due to the N=O bonds being almost orthogonal to each other and thus the parallel anisotropies partially cancel in the NO$_2$ plane. The large electric field of the polar N=O bond and π effects account for the majority of the large SCSs on the *ortho* protons of nitrobenzene. This large term can obscure the determination of the parallel component of the anisotropy and the steric contribution of the oxygens. However the iterative procedure used with the appropriate model compounds (**1–7**) gave well-determined values for these effects. It was also noted that since the dihedral angle of the nitro group in respect to the aromatic rings in 9-nitroanthracene is 90° the π orbitals are no longer overlapping leading to no conjugation and therefore no π shift on H-10.

Table 6.40   *Contributions of the nitro SCSs*

Compound/proton	$\delta_{(anis)}(\|\|)$	$\delta_{(anis)}(\perp)$	$\delta_{Osteric}$	δ_{el}	δ_π
9-Nitroanthracene					
H1/8	−0.206	−0.066	0.082	0.187	0
H10	−0.001	0.012	0.001	0.086	0
Nitrobenzene					
H2/6	−0.010	0.108	0.132	0.621	0.189
H3/5	0.000	0.019	0.003	0.141	0.064
H4	0.000	0.012	0.001	0.104	0.174
1-Nitrocyclohexane					
H2eq	−0.022	−0.052	0.038	0.314	0
1-Nitroadamantane					
H1	−0.001	0.099	0.11	0.448	0

6.7.5   Conformational Analysis

Initially compounds (**8**)–(**11**) were chosen for the parameterization of the nitro group. However the ab initio minimized geometry of 1-nitronaphthalene was calculated to be planar using the default ab initio method (B3LYP/6-31G(d,p)) which did not agree with dipole moment studies suggesting a non-planar conformation.[109,145] These molecules were then minimized using (a) a larger basis set 6-311++G(2d,p) with the B3LYP theory and (b) MP2 theory with the 6-31+G(d,p) basis set to observe the change in torsion angle of the nitro group. These geometries were then used for GIAO and CHARGE chemical shift calculations, the latter with the parameters derived for the nitro group. The GIAO calculations were performed with the B3LYP/6-311++G(2d,p) method.

Table 6.41 gives a comparison of the observed and calculated shifts and the calculated out-of-plane torsional angle of the nitro group with the rms error for each compound. In

Table 6.41 *Observed chemical shifts of (8)–(11) versus calculated chemical shifts (GIAO and CHARGE) based on different basis sets and geometries*

8	H3	H4	H5	H6	CH$_3$		rms error	Dihedral angle
Exp.	**7.342**	**7.498**	**7.340**	**7.965**	**2.605**			
GIAO^a	7.691	7.857	7.710	9.010	2.997		0.572	12.9
GIAO^b	7.595	7.757	7.605	8.611	2.787		0.361	24.9
GIAO^c	**7.610**	**7.800**	**7.617**	**8.126**	**2.517**		**0.234**	**44.5**
CHARGE^a	7.444	7.622	7.429	8.201	2.642		0.135	12.9
CHARGE^b	**7.446**	**7.619**	**7.431**	**8.077**	**2.615**		**0.096**	**24.9**
CHARGE^c	7.399	7.555	7.394	7.754	2.565		0.105	44.5
9	H3	H4	CH$_3$					
Exp.	**7.111**	**7.250**	**2.306**					
GIAO^a	7.505	7.594	2.603				0.347	44.6
GIAO^b	**7.377**	**7.467**	**2.420**				**0.209**	**56.6**
GIAO^c	7.413	7.574	2.332				0.256	66.4
CHARGE^a	7.209	7.496	2.581				0.220	44.6
CHARGE^b	7.197	7.471	2.552				0.197	56.6
CHARGE^c	**7.167**	**7.427**	**2.544**				**0.174**	**66.4**
10	H3	H4	CH$_3$					
Exp.	**8.003**	**7.529**	**2.588**					
GIAO^a	8.712	7.686	2.844				0.445	31.1
GIAO^b	8.327	7.579	2.593				0.190	38.5
GIAO^c	**8.075**	**7.622**	**2.353**				**0.152**	**50.5**
CHARGE^a	8.297	7.659	2.902				0.260	31.1
CHARGE^b	8.163	7.651	2.831				0.182	38.5
CHARGE^c	**7.916**	**7.606**	**2.741**				**0.111**	**50.5**

11	H2	H3	H4	H5	H6	H7	H8	rms error	Dihedral angle
Exp.	**8.242**	**7.555**	**8.131**	**7.967**	**7.633**	**7.731**	**8.575**		
GIAO^a	9.635	7.904	8.550	8.320	8.071	8.259	10.520	0.972	0
GIAO^b	8.795	7.805	8.403	8.234	7.986	8.100	9.389	0.453	30.4
GIAO^c	**8.196**	**7.787**	**8.235**	**8.133**	**7.939**	**8.028**	**8.722**	**0.206**	**46.9**
CHARGE^a	8.677	7.813	8.326	8.032	7.575	7.702	8.956	0.253	0
CHARGE^b	**8.345**	**7.789**	**8.276**	**8.020**	**7.588**	**7.682**	**8.560**	**0.116**	**30.4**
CHARGE^c	8.035	7.747	8.183	7.972	7.573	7.629	8.239	0.173	46.9

^{a} Calculated chemical shifts based on optimized geometry using B3LYP/6-31G(d,p).
^{b} Calculated chemical shifts based on optimized geometry using B3LYP/6-311++G(2d,p).
^{c} Calculated chemical shifts based on optimized geometry using MP2/6-31+G(d,p).

Table 6.42 *Methods used for the calculation of the chemical shifts of compounds (8)–(11)*

Method	rms error	Correlation coefficient
GIAO[a]	0.72	0.9823
GIAO[b]	0.36	0.9972
GIAO[c]	**0.22**	**0.9983**
CHARGE[a]	0.22	0.9977
CHARGE[b]	**0.14**	**0.9990**
CHARGE[c]	0.15	0.9977

[a] Chemical shifts calculated based on geometries optimized using B3LYP/6-31G(d,p).
[b] Chemical shifts calculated based on geometries optimized using B3LYP/6-311++G(2d,p).
[c] Chemical shifts calculated based on geometries optimized using MP2/6-31+G(d,p).

Table 6.42 the results are summarized and compared. The results are conclusive in that they all produce more accurate results when the dihedral angle is increased from that which is found using B3LYP/6-31G(d,p). However the correct answer is probably between the dihedral angles produced by the B3LYP/6-311++G(2d,p) and MP2/6-31+G(d,p) method.

The chemical shifts derived from the GIAO calculations using the B3LYP based geometries are without exception larger than the experimental value. The values produced using the MP2 based geometries are also mostly larger then the experimental values.

The geometries of (**8**)–(**11**) which gave the best agreement with experimental data (see Table 6.42) areMP2/6-31+G(d,p) for GIAO and B3LYP/6-311++G(2d,p) for CHARGE. There is excellent correlation for both methods and again there is a tendency for the values from GIAO calculations to be higher than the experimental data.

The sensitivity of the nitro group to steric influence from neighbouring atoms was evident in compounds (**8**)–(**11**) and the out of plane torsion of the nitro group in these compounds was monitored by the ¹H chemical shifts. The reduced conjugation leads to shielding of the *para* protons which is a simple method to estimate the extent of the out of plane bending of the nitro group.

The SCSs of the *para* protons in compounds **5**, **7** and **11** decreases from 0.36 for the planar nitrobenzene (**5**) to 0.29 for 1-nitronaphthalene (**11**) (ca. 45°) to 0.19 ppm for the orthogonal 9-nitroanthracene (**7**). The non-bonded effect on near protons is also very sensitive to the torsional angle, e.g. the SCSs on H-8 in 1-nitronaphthalene, 0.7 ppm compared to H-1 in 9-nitroanthracene −0.1 ppm.

Table 6.43 *Dihedral angles of the nitro group relative to the plane of the aromatic ring of compounds (8)–(11) using QM and MM methods*

Method	8	9	10	11
B3LYP/6-31G(d,p)	12.9	44.6	31.1	0.0
B3LYP/6-311++G(2d,p)	24.9	56.6	38.5	30.4
MP2/6-31+G(d,p)	44.5	66.4	50.5	46.9
MMX	7.8	53.7	41.3	34.8
MMFF94	48.8	78.6	54.0	58.9

Molecular mechanics calculations can provide realistic geometries[43] and Table 6.43 compares the geometries obtained using molecular mechanics force fields with the *ab initio* calculated geometries. The result of the MM calculations is comparable with the *ab initio* methods and the MMFF94 force field produced geometries often very similar to the MP2 calculations. These results are encouraging in the use of MM derived geometries for chemical shift calculations

6.8   Summary

In this chapter the observed ¹H SCSs for the monovalent substituents F, Cl, Br, I, OH, NH$_2$, CN and NO$_2$ have been given and analysed using the CHARGE model. In this model the SCSs over three bonds or less are determined by semi-empirical modelling of the electronic effects. The long-range (>3 bonds) SCSs of the fluoro substituent are determined solely by the electric field effect. For chloro-, bromo- and iodo-alkanes the largest contribution to the halogen SCSs is the steric term when the proton is in a syn-1,3-diaxial arrangement to the halogen. For other distant protons the electric field term is the major term, but a complete analysis requires a combination of the steric, anisotropic and electric field terms.

The ¹H chemical shifts of the alcohols and ethers investigated are *identical* for CDCl$_3$ and D$_2$O solutions, *provided* the conformational profile of the solute is the same in both solvents. The chemical shifts in both solvents were analysed using the CHARGE model. An accurate prediction of the observed shifts was obtained utilizing only the electronic (≤ 3 bonds) and steric effects. No anisotropy (C—O, O—H) or OH electric field effects were included.

There is a dramatic deshielding of the OH proton when an electronegative atom is within 2 Å of the proton and classical models fail to describe this effect. This effect was reproduced by *ab initio* calculations and is probably due to orbital overlap in the O—H..O=C system.

A correction was made for the effect of the π-electron excess on the oxygen atom on the chemical shift of the OH proton. The coefficient was much larger than that for a C—H atom. The chemical shifts calculated by the CHARGE model were compared to those from GIAO calculations. The results showed consistent chemical shift calculations using CHARGE whether the OH protons were included or not. The GIAO calculations are significantly improved when the OH protons are removed.

The amino SCSs followed the OH SCSs in that no CN anisotropy contribution was required, the SCSs were interpreted on the basis of steric and electric field contributions, with ring current and π-electron effects for the aromatic amines. A correction was made for the effect of the π-electron excess on the nitrogen atom on the chemical shift of the NH proton in *para*-substituted anilines. This correction was similar to that for the OH proton. Complicating factors for the aliphatic amine SCSs were the conformational profile of the amines and the possible presence of traces of acid which did not allow the ¹H shifts of the amino protons to be predicted accurately.

The CN SCSs over more than three bonds are determined by electric field effects, without the need to include any steric or anisotropic effects. The short-range CN SCSs (≤ 3 bonds) requires the inclusion of a γ effect from both the carbon and nitrogen of the CN substituent and these contributions plus the electric field effect for the γ protons (H.C.C.CN) were used to calculate the chemical shifts of the α and β protons respectively. The γ effect of the cyano carbon atom has a very small orientational dependence. The γ effect of the nitrogen (H.C.CN) cannot have an orientation effect and was modelled by adjusting the nitrogen polarizability.

The long-range (>3 bonds) nitro group SCSs were mainly due to the electric field with significant contributions from the NO anisotropy and steric effects. With the magnetic dipole placed in the middle of the N=O bond the parallel and perpendicular magnetic anisotropies were 2.5 and $-3.4 \times 10^{-30}\,\text{cm}^3$ molecule^{-1}. These values are much smaller than the comparable anisotropies for the carbonyl group and this is consistent with the $O{=}N^+{-}O^-$ structure for the nitro group. However as there are two NO bonds the overall anisotropic effect of the nitro group is comparable with that of the carbonyl group.

The electrical field term is the dominating effect on the SCSs of protons in the plane of the nitro group. This large term can obscure the determination of the parallel component of the anisotropy and the steric contribution of the oxygens.

The sensitivity of the nitro group to steric influence from neighbouring atoms was noted and the out of plane torsion of the nitro group was monitored by the ^{1}H chemical shifts. The reduced conjugation leads to shielding of the *para* protons which is a simple method to estimate the extent of the out of plane bending of the nitro group.

References

1. Abraham, R. J.;. Edgar, M.; Glover, R. P.; Warne, M. A.; Griffiths, L., *J. Chem. Soc. Perkin Trans. 2* 1996, 333.
2. Abraham, R. J.; Edgar, M.; Griffiths, L.; Powell, R. L., *J. Chem. Soc. Perkin Trans. 2* 1995, 561.
3. Abraham, R. J.; Griffiths, L.; Warne, M. A., *J. Chem. Soc. Perkin Trans. 2* 1997, 203.
4. Marshall, T. W.; Pople, J. A., *Mol. Phys.* 1958, **1**, 199.
5. Buckingham, A. D., *Can. J. Chem.* 1960, **38**, 300.
6. Petrakis, L.; Bernstein, H. J., *J. Chem. Phys.* 1962, **37**, 2731.
7. Petrakis, L.; Bernstein, H. J., *J. Chem. Phys.* 1963, **38**, 1562.
8. Zürcher, R. F., *Prog. Nucl. Magn. Reson. Spectrosc.* 1967, **2**, 205.
9. Aminova, R. and Gubaidullina, R. Z., *Zh. Strukt. Khim.* 1969, **10**, 253.
10. Fukui, H.; Kitamura Y. and Miura, K., *Mol. Phys.* 1977, **34**, 593.
11. Mukhomorov, V. K., *Zh. Strukt. Khim.* 1971, **12**, 326.
12. Grayson, M. and Raynes, W. T.,*Chem. Phys. Lett.* 1994, **34**, 270.
13. Grayson, M. and Raynes, W. T., *Magn. Reson. Chem.* 1995, **33**, 138.
14. Balalsubrahmanyam, S. N.; Narasimha Barathi S. and Usha, G., *Org. Magn. Reson.* 1983, **21**, 474.
15. Fukazawa, Y.; H. T., Yang Y. and Usui, S., *Tetrahedron Lett.* 1995, **36**, 3349.
16. Schneider, H. J.; Buchheit, U.; Becker, N.; Schmidt G. and Siehl, U., *J. Am. Chem. Soc.* 1985, **107**, 7027.
17. Davies, A. K.; Mathieson, D. W.; Nicklin, P. D.; Bell J. R. and Toyne, K. J., *Tetrahedron Lett.* 1973, **6**, 413.
18. Gschwendtner W. G. and Schneider, H. J., *J. Org. Chem.* 1980, **45**, 3507.
19. Yang, Y.; Haino, T.; Usui S. and Fukazawa, Y., *Tetrahedron* 1996, **52**, 2325.
20. Schneider H. J. and Jung, M., *Magn. Reson. Chem.* 1988, **26**, 679.
21. Abraham, R. J.; Barlow A. P. and Rowan, A. E., *Magn. Reson. Chem.* 1989, **27**, 1024.
22. Abraham, R. J. and Fisher, J., *Magn. Reson. Chem.* 1985, **23**, 856.

23. Kaiser, C. R.; Rittner R. and Basso, E. A., *Magn. Reson. Chem.* 1994, **3**, 503.
24. Frisch, M. J.; Trucks, G. W.; Schlegel, H. B.; Scuseria, G. E.; Robb, M. A.; Cheeseman, J. R.; Montgomery, J. A.; Jr., T. V.; Kudin, K. N.; Burant, J. C.; Millam, J. M.; Iyengar, S. S.; Tomasi, J.; Barone, V.; Mennucci, B.; Cossi, M.; Scalmani, G.; Rega, N.; Petersson, G. A.; Nakatsuji, H.; Hada, M.; Ehara, M.; Toyota, K.; Fukuda, R.; Hasegawa, J.; Ishida, M.; Nakajima, T.; Honda, Y.; Kitao,O.; Nakai, M. Klene, X. Li, J. E. Knox, H. P. Hratchian, J. B. Cross, V. Bakken, H.; Adamo, C.; Jaramillo, J.; Gomperts, R.; Stratmann, R. E.; Yazyev, O.; Austin, A. J.; Cammi, R.; Pomelli, C.; Ochtersk, J. W.; Ayala, P. Y.; Morokuma, K.; Voth, G. A.; Salvador, P.; Dannenberg, J. J.; Zakrzewski, V. G.; Dapprich, S.; Daniels,A. D.; Strain, M. C.; Farkas, O.; Malick, D. K.; Rabuck, A. D.; Raghavachari, K.; Foresman, J. B.; Ortiz, J. V.; Cui, Q.; Baboul, A. G.; Clifford, S.; Cioslowski, J.; Stefanov, B. B.; Liu, G.; Liashenko, T.;A.; Piskorz, P.; Komaromi, I.; Martin, R. L.; Fox, D. J.; Keith, M.; Al-Laham, A.; Peng, C. Y.; Nanayakkara, A.; Challacombe, M.; Gill, P. M. W.; Johnson, B.,Chen, W.; Wong, M. W.; Gonzalez C.; and Pople, J. A., *Gaussian* 03, Gaussian, D. Inc.: Wallingford, CT, 2006.
25. Landholt-Bornstein, *Structure Data of Free Polyatomic Molecules.* Springer-Verlag: New York, NY, 1976.
26. Fort, R. C. and Schleyer, P. R., *J. Org. Chem.* 1965, **30**, 789.
27. Bhandari, K. S. and Pincock, R. E., *Synth. Commun.* 1974, 655.
28. Abraham, R. J. and Grant, G. H., *J. Comput. Aid. Mol. Design* 1992, **6**, 273.
29. Bothner-By, A. A.; Castellano S.; and Gunther, H., *J. Am. Chem. Soc.* 1965, **87**, 2439.
30. Smith, S. L. and Ihrig, A. M., *J. Chem. Phys.* 1967, **46**, 1181.
31. Ihrig, A. M. and Smith, S. L., *J. Am. Chem. Soc.* 1972, **94**, 34.
32. DeWolf, M. Y. and Baldeschwieler, J. D., *J. Mol. Spectrosc.* 1964, **13**, 344.
33. Wheaton, G. A. and Burton, D. J., *J. Org. Chem.* 1983, **48**, 917.
34. Bothner-By A. A. and Naar-Colin, C., *Ann. NY Acad. Sci.* 1958, **70**, 833.
35. Bothner-By, A. A. and Naar-Colin, C., *J. Am. Chem. Soc.* 1958, **80**, 1728.
36. Spiesecke, H. and Schneider, W. G., *J. Chem. Phys.* 1961, **35**, 722.
37. Spiesecke, H. and Schneider,W. G., *J. Chem. Phys.* 1961, **35**, 731.
38. Hruska, F.; Hutton H. M. and Schaefer, T., *Can. J. Chem.* 1965, **43**, 2392.
39. Schaefer, F.; Hruska and Hutton, H. M., *Can. J. Chem.* 1965, **45**, 3143.
40. Schrumpf, G.; Sanweld W. and Machinek, R.; *Magn. Reson. Chem.* 1987, **25**, 11.
41. Abraham, R. J.; Griffiths L. and Warne, M. A.; *J. Chem. Soc. Perkin Trans. 2* 1997, 881.
42. Abraham, R. J.; Griffiths L. and Warne, M. A., *J. Chem. Soc. Perkin Trans. 2* 1997, 2151.
43. Abraham, R. J.; Mobli M. and Smith, R. J.; *Magn. Reson. Chem.* 2004, **42**, 436.
44. Kalinowski, H. O.; Berger S. and Braun, S.; *Carbon-13 NMR Spectroscopy.* J. Wiley & Sons, Inc.:New York, NY, 1984.
45. Beringer, F. M.; Chang, L. L.; Fenster A. N. and Rossi, R. R.; *Tetrahedron* 1969, **25**, 339.
46. Abraham, R. J.; Canton, M.; Reid M. and Griffiths, L., *J. Chem. Soc. Perkin Trans. 2* 2000, 803.
47. Foresman, J. B. and Frisch, Æ., *Exploring Chemistry with Electronic Structure Methods.* Gaussian, Inc.: Pittsburgh, PA, 1996.
48. Pouchert, C. J. and Behnke, J., *The Aldrich Library of 13C and 1H FT NMR Spectra.* Aldrich Chemical Company Inc.: Milwaukee, WI, 1993.
49. Abraham, R. J.; Fisher J. and Loftus, P., *Introduction to NMR Spectroscopy.* John Wiley & Sons, Ltd: Chichester, 1988.
50. Zanger, M.;*Org. Magn. Reson.* 1972, **4**, 1.
51. Silverstein, R. M. and Webster, F. X., *Spectrometeric Identification of Organic Compounds.* J. Wiley & Sons, Inc.: New York, NY, 1998.
52. Tiers, G. V. D., *High Resolution NMR Spectroscopy.* Pergamon Press: Oxford, 1966.
53. Altona, C.; Ippel, J. H., Hoekzema, A. J. A. W.; Erkelens, C.; Groesbeek M. and Donders, L. A., *Magn. Reson. Chem.* 1989, **27**, 564.

54. Gasteiger, J. and Marsili, M., *Org. Magn. Reson.* 1981, **15**, 353-360.
55. Neuman, Jr R. C. and Rahm, M. L., *J. Org. Chem.* 1966, **31**, 1857.
56. Pachler, K. G. R. and Wessels, P. L., *J. Mol. Spectrosc.* 1969, **3**, 207.
57. Stepanyants, A. U.; Lezina, V. P.; Zlokazova I. V. and Schvedchikov, A. P., *Zh. Anal. Khim.* 1976, **31**, 1770.
58. Fay, C. K.; Grutzner, J. B.; Johnson, L. F.; Sternhell S. and Westerman, P. W., *J. Org. Chem.* 1973, **38**, 3122.
59. Raynes, W. T.; Buckingham A. D. and Bernstein, H. J., *J. Chem. Phys.* 1962, **36**, 3481.
60. van Deursen, F. W. and Udding, A. C., *Recl. Trav. Chim. Pays-Bas* 1968, **87**, 1243.
61. Abraham, R. J. and Holker, J. S. E., *J. Chem. Soc.* 1963, 806.
62. Tribble, M. T., Miller M. A. and Allinger, N. L. *J. Am. Chem. Soc.* 1971, **93**, 16.
63. *Handbook of Chemistry and Physics.* 64th edn.; The Chemical Rubber Company, CRC Press, Boca Raton, Florida, 1984.
64. *Encyclopedia of Nuclear Magnetic Resonance.* John Wiley & Sons, Inc.: New York, NY, 1996.
65. Hall, L. D., *Tetrahedron Lett.* 1964, **23**, 1457.
66. Abraham, R. J.; Griffiths L. and Warne, M. A.; *J. Chem. Soc. Perkin Trans.2* 1998, 1751.
67. Abraham, R. J.; Byrne, J. J.; Griffiths L. and Koniotou, R., *Magn. Reson. Chem.* 2005, **43**, 611.
68. Rader, C. P., *J. Am. Chem. Soc.* 1969, **91**, 3248.
69. Farrar, T. C.; Ferris T. D. and Zeidler, M. D., *Mol. Phys.* 2000, **98**, 737.
70. Petterson, K. A.; Stein, R. S.; Drake M. D. and Roberts, J. D., *Magn. Reson. Chem.* 2005, **43**, 225.
71. Salman, S. R.; Farrant, R. D.; Sanderson P. N. and Lindon, J. C., *Magn. Reson. Chem. 2* 1993, 585.
72. Gallway, F. B.; Hawkes, J. E.; Haycock P. and Lewis, D., *J. Chem. Soc. Perkin Trans. 2* 1990, 1979.
73. Murcko, M. A. and Dipaola, R. A., *J. Am. Chem. Soc.* 1992, **114**, 10010.
74. Pearce, C. M. and Sanders, K. M., *J. Chem. Soc. Perkin Trans. 2* 1994, 1119.
75. Abraham, R. J.; Chambers E. J. and Thomas, W. A., *J. Chem. Soc. Perkin Trans.2*, 1993, 1061.
76. Kuhn, L. P., *J. Am. Chem. Soc.* 1951, **74**, 2492-2499.
77. Abraham, R. J., Sancassan F. and Koniutou, R. *J. Chem. Soc. Perkin Trans. 2* 2002, 2025.
78. Abraham R. J. and Mobli, M., *Magn. Reson. Chem.* 2007, **45**, 865.
79. Abraham, R. J.; Bardsley, B.; Mobli M. and Smith, R. J., *Magn. Reson. Chem.* 2005, **43**, 3.
80. Gottlieb, H. E.; Kotlyar V. and Nudelman, A., *J. Org. Chem.* 1997, **62**, 7512.
81. Byrne, J. J., *Ph.D. Thesis.* University of Liverpool, 2007.
82. Pople, J. A., *Disc. Faraday Soc.* 1962, **34**, 7.
83. Williamson M. P. and Asakura, T., *J. Magn. Reson.* 1993, **B101**, 63.
84. Pauling, L., *J. Chem. Phys.* 1936, **4**, 673.
85. Zumwalt L. R. and Badger, R. M., *J. Am. Chem. Soc.* 1940, **62**, 305.
86. Baker, A. W., *J. Am. Chem. Soc.* 1958, **80**, 3598.
87. Abildgaard, J.; Bolvig S. and Hansen, P. E., *J. Am. Chem. Soc.* 1998, **120**, 9063.
88. Tribble M. T. and Traynham, J. G., *J. Am. Chem. Soc.* 1961, **62**, 305.
89. Porte, A. L.; Gutowsky H. S. and Hunsberger, I. M., *J. Am. Chem.Soc.* 1960, **82**, 5057.
90. Griffiths V. S. and Socrates, G., *J. Mol. Struct.* 1966, **21**, 302.
91. Merril, J. R., *J. Am. Chem.Soc.* 1961, **65**, 2023.
92. Liddel U. and Becker, E. D., *Spectrochim. Acta* 1957, **10**, 70.
93. Kuhn, L. P., *J. Am. Chem. Soc* 1951, **74**, 2492.
94. Davies, M., *Trans. Faraday Soc.* 1938, **34**, 1427.
95. Davies, M., *Trans. Faraday Soc.* 1940, **36**, 1114.
96. Clausser W. F. and Wall, F. T., *J. Am. Chem. Soc*, 1939, **61**, 2679.

97. Vinogradov S. N. and Linnell, R. H., *Hydrogen Bonding.* Van Nostrand Reinhold Ltd: New York, NJ, 1971.
98. Hibbert F. and Emsley, J., *Adv. Phys. Org. Chem.* 1990, **26**, 255.
99. Liddle U. and Becker, E. D., *Spectrochim. Acta* 1957, **10**, 70.
100. Günther, H., *NMR Spectroscopy.* John Wiley & Sons, Ltd: Chichester, 1995.
101. Abraham, R. J., Personal Communication.
102. Abraham, R. J.; Mobli M. and Smith, R. J., *Magn. Reson. Chem.* 2003, **41**, 26.
103. Lampert, H.; Mikenda, W.; Karpfen A. and Kählig, H., *J. Phys. Chem. A* 1997, **101**, 9610.
104. Mobli, M., *Ph.D. Thesis.* University of Liverpool, 2004.
105. Pretsch, P. D.; Clerc, T.; Siebl J. and Simon, W., *Spectral Data for Structure Determination of Organic Compounds.* Springer-Verlag: New York, NY, 1989.
106. Alkorta I. and Elguero, J.; *Magn. Reson. Chem.* 2004, **42**, 955.
107. Abraham, R. J.; Byrne, J. J.; Griffiths L. and Perez, M., *Magn. Reson. Chem.* 2006, **44**, 491.
108. Basso, E. A.; Gauze G. F. and Abraham, R. J., *Magn. Reson. Chem.* 2007, **45**, 749.
109. McClellan, A. L., *Table of Experimental Dipole Moments*, Volume 3. Rahara Enterprises: El Cerrito, CA, 1989.
110. Abraham, R. J., *Prog. Nucl. Magn. Reson. Spectrosc.* 1999, **35**, 85.
111. Lampert, H., Mikenda, W.; Karpfen A. and Kählig, H., *J. Phys. Chem. A* 1997, **101**, 9610.
112. Eliel E. L. and Wilen, S. H., *Stereochemistry of Carbon Compounds.* J. Wiley & Sons, Inc.: New York, NY, 1994.
113. Patai S. and Rappaport, Z., *The Chemistry of Triple-Bonded Functional Groups.* J. Wiley & Sons, Ltd: Chichester, 1983.
114. Bothner-By A. A. and Pople, J. A., *Ann. Rev. Phys. Chem.* 1965, **16**, 43.
115. Reddy G. S. and Goldstein, J. H., *J. Chem. Phys.* 1963, **39**, 3509.
116. Cross A. D. and Harrison, I. T., *J. Am. Chem. Soc.* 1963, **85**, 3223.
117. ApSimon, J. W.; Beierbeck H. and Todd, D. K., *Can. J. Chem.* 1972, **50**, 2351.
118. McConnell, H. M., *J. Chem. Phys.* 1957, **27**, 226.
119. Abraham R. J. and Reid, M., *Magn. Reson. Chem.* 2000, **38**, 570.
120. Abraham R. J. and Fisher, J., *Magn. Reson. Chem.* 1986, **24**, 451.
121. Reid, M., *Ph.D. Thesis.* Unviersity of Liverpool, 2002.
122. Süess H. and Hesse, M., *Helv. Chim. Acta* 1979, **62**, 1040.
123. Burgess H. and Donnelly, J. A., *Tetrahedron* 1991, **47**, 111.
124. Castellano S. and Sun, C., *J. Am. Chem. Soc.* 1966, **88**, 4741.
125. Hayamizu K. and Yamamoto, O., *J. Mol. Spectrosc.* 1968, **28**, 89.
126. Hayamizu K. and Yamamoto, O., *J. Mol. Spectrosc.* 1968, **29**, 183.
127. Nir, M.; Hoffman, R. E.; Shapiro I. O. and Rabinovitz, M., *J. Chem. Soc. Perkin Trans. 2* 1995, 1433.
128. Buckingham, A. D., *Can. J. Chem.* 1960, **28**, 300.
129. Fraenkel, G.; Carter, R. E.; McLachlan A. and Richards, J. H., *J. Am. Chem. Soc.* 1963, **82**, 5846.
130. Huitric A. C. and Trager, W. F., *J. Org. Chem.* 1962, **27**, 1926.
131. Bullock, E., *Can. J. Chem.* 1963, **41**, 711.
132. Lauterbur, P. C., *J. Chem. Phys.* 1963, **38**, 1432.
133. Schneider H. and Hoppen, V., *J. Org. Chem.* 1978, **43**, 3866.
134. Craik, D. J.; Levy G. C. and Brownlee, R. T. C., *J. Org. Chem.* 1983, **48**, 1601.
135. Hiyama Y. and Brown, T. L., *J. Phys. Chem.* 1981, **85**, 1698.
136. Witanowski, M.; Biedrzycka, Z.; Grela K. and Wejroch, K., *Magn. Reson. Chem.* 1998, **36**, S85.
137. Arakawa, S.; Ueda, N.; Shigemasa Y. and Nakashima, R., *Chem. Expr.* 1987, **2**, 759.
138. Begunov, R. S.; Orlova, T. N.; Taranova O. V. and Orlov, V. Y., *J. Appl. Spectrosc.* 2002, **69**, 807.
139. Rasala D. and Gawinecki, R., *Magn. Reson. Chem.*, 1992, **30**, 740.

140. Martin, N. H.; Allen, N. W.; Minga, W. K.; Ingrassia S. T. and Brown, J. D., *J. Am. Chem. Soc.* 1998, **120**, 11510.
141. Mobli M. and Abraham, R. J., *J. Comput. Chem.* 2005, **26**, 389.
142. Li S. and Allinger, N. L., *Tetrahedron* 1988, **44**, 1339.
143. Vanelle, P.; Crozet, M. P.; Maldonado J. and Barreau, M., *Eur. J. Med. Chem.* 1991, **26**, 167.
144. Abraham R. J. and Reid, M., *Magn. Reson. Chem.* 2000, **38**, 570.
145. Fujita, T.; Koshimizu K. and Mitsui, T., *Tetrahedron* 1967, **23**, 2633.

7

Modelling ^{1}H Chemical Shifts, Divalent Substituents

7.1 Introduction

In this chapter the substituent effects of the more complex divalent substituents will be discussed. These include ketones, esters, amides, ethers and sulfones. These substituents are only complex from the conformational aspect, i.e. they can be part of the ring structure or alkyl chain of the molecule, whereas this is not possible with strictly monovalent substituents such as Cl. The theory of Chapter 3 still applies and there are no more interactions to consider, though of course the electronic (β and γ) effects may operate from both ends of the substituent. For example, for esters the γ effects of the both the CO group and O atom need to be considered. This is not the case for the simple CO group which is the first group to be considered.

7.2 Aldehydes and Ketones

7.2.1 Aliphatic Aldehydes and Ketones

Introduction. The influence of the carbonyl group on the chemical shifts of neighbouring protons was the subject of considerable debate and controversy in the beginning of organic NMR and the standard description of the C=O anisotropy shown in Chapter 1, Figure 1.5 is one of the most well known illustrations in NMR.[1]

The early investigations concentrated on the carbonyl anisotropy and Narasimhan and Rogers[2] concluded that the ^{1}H chemical shifts in formamide and DMF were entirely due to the C=O anisotropy. However even the C=O anisotropy was controversial as Jackman[3] suggested that there is a large diamagnetism in the direction normal to the nodal plane of the π orbitals whereas Pople's calculations[4] suggested a paramagnetism centred on the carbon atom, large in the x-direction and the largest diamagnetism on the O atom in the z-direction (i.e. along the C=O bond). An authoritative review of these early investigations has been given by Pople and Bothner-By.[5]

In his pioneering treatment of ${}^{1}H$ chemical shifts, Zurcher[6] was limited to observing only the methyl groups in steroids but concluded that both the C=O bond anisotropy and the electric field effect were needed to explain the observed SCSs. Zurcher used the full McConnell equation (Equation (3.15))[7] to calculate the C=O anisotropy and also used the carbonyl dipole to calculate the electric field effect. Due to lack of data Zurcher did not consider near (< 4 bonds) protons nor did he need to invoke any steric effects of the carbonyl group.

ApSimon and coworkers[8] again using only the methyl groups of steroids for their data, reformulated the McConnell equation in order to obtain the anisotropy effects on near nuclei (< 3 Å away from the substituent). They also found that both anisotropy and electric field effects were necessary to predict the SCSs of the carbonyl group. Subsequently Homer *et al.*[9] observed that the original McConnell equation was just as accurate in their investigations. Toyne[10] reviewed the literature calculations of the C=O anisotropy in which the position of the magnetic dipole varied from the carbon atom to the oxygen atom. He concluded that taking the dipole to be approximately mid-way along the C=O bond at 0.6 Å produced the best results. Subsequently Schneider *et al.*[11] obtained all the ${}^{1}H$ shifts in three keto steroids and analysed the SCSs in terms of both anisotropy and electric field effects. They obtained rather large values for the carbonyl anisotropy (see later) and also they were not able to calculate the chemical shifts of the protons vicinal to the carbonyl group. Williamson *et al.*[12-14] performed similar calculations for the α C—H protons in proteins. They used the known crystal structures of the proteins and included electric field and anisotropies, the latter from the C=O bonds and also from the aromatic residues present. They obtained good agreement with the observed data when both the electric field and anisotropy terms were included. As the ${}^{1}H$ shifts were measured in aqueous solution the electric field effect is considerably diminished compared to non-polar solvents. Again protons vicinal to the C=O bond were excluded from their treatment. Abraham and Ainger[15] assigned the ${}^{1}H$ spectra of a number of cyclic ketones and this plus previous literature results provided sufficient data for an analysis of carbonyl SCSs based on the CHARGE model.[1]

Theory, application to the carbonyl group. Abraham and Ainger[15] applied the CHARGE model to the carbonyl group as follows. The short range β (H.C=O) and γ (H.C.C=O, H.C.C.CO) effects needed to be determined together with the long-range steric, electric field and anisotropic effects. The steric effects of both the carbonyl carbon and oxygen atoms were unknown. The authors assumed the same value of the coefficient a_S in Equation (3.11) for the ketone carbon atom as for a saturated carbon. The value of a_S for the carbonyl oxygen atom needs to be obtained together with associated push–pull coefficient.

The electric field of the carbonyl group was calculated in an identical manner to that for any C—X bond (Chapter 3, Section 3.5.3). The partial atomic charges calculated in the CHARGE routine have been derived from the observed molecular dipole moments and the extent of the agreement provides one check of the electric field calculation. The calculated and observed (in parenthesis) dipole moments (in debye) of formaldehyde, acetaldehyde, acetone and cyclohexanone are 2.28 (2.34), 2.68 (2.68), 3.03 (2.86) and 3.03 (3.08) and the good agreement provides strong support for the use of these charges in the calculations. As the coefficient in Equation (3.12) is known together with the molecular geometries the electric field effect of the carbonyl group at any proton more than three bonds removed from the carbonyl oxygen atom is given immediately. These values will be discussed later.

The anisotropic effect of the carbonyl group also was calculated. The C=O group is not an axially symmetriic group and has different magnetic susceptibilities along principal axes (Chapter 3, Figure 3.5). There are two anisotropy terms required for a non-axially symmetric group and thus Equation (3.15) was used. The two anisotropy terms in Equation (3.15) are $\Delta\chi_1$ and $\Delta\chi_2$, the two anisotropies for the C=O bond which are usually termed the parallel and perpendicular anisotropy respectively. In order to apply the calculation to ketones the two anisotropies need to be determined and also it was necessary to determine the effect of the position of the dipole along the C=O bond.

The geometries of the compounds investigated were obtained by geometry optimizations using *ab initio* calculations (B3LYP/6-31G*); full details are given elsewhere.[16]

Observed versus calculated shifts. The assignments of compounds 2- and 4-t-butylcyclohexanones (**1**) and (**2**), fenchone (**3**), *trans* decalone (**4**), 5α-androstane-3-one (**5**), 5α-androstane-17-one (**6**), 5α-androstane-3,17-dione (**7**), 5α-androstane-3,11,17-trione (**8**), norbornanone (**9**) and camphor (**10**) are given in Tables 7.2 to 7.5. Full details of the assignments have been reported.[15,16] The ¹H NMR spectra were obtained at 400.14 MHz. Those for **2**, **4**, **5** and **7** were recorded at 600 MHz and **7** and **8** at 750 MHz.

The data in Tables 7.2–7.5 combined with the ¹H shifts of the parent compounds given previously[17] allows the carbonyl SCSs to be obtained in these compounds. The carbonyl SCSs for decalone (**4**) vs. *trans* decalin and norbornanone (**9**) vs. norbornane are given in Figure 7.1. Also the SCSs for the carbonyl group at the 3, 11 and 17 positions in the steroid nucleus obtained from data for compounds **6**, **7** and **8** together with the ¹H chemical shifts of androstane are given and compared with the results obtained by Schneider *et al.*[11] in Table 7.1. In Schneider *et al.*[11] only the SCSs were tabulated not the actual ¹H chemical shifts. Also the SCS for the 11-keto group was obtained in this investigation as δ(**8**)–δ (**7**) whereas Schneider *et al.*[11] obtained this SCS directly from the analysis of 11-keto androstane. The agreement of the two sets of results in Table 7.1 is very good and the additivity of the SCS values in the steroid nucleus is clearly shown by the agreement of the two sets of values for the SCS of the 11-keto group.

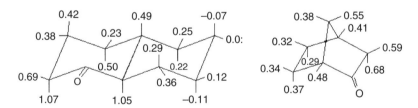

Figure 7.1   *Carbonyl SCSs in trans 1-decalone and norbornanone.*

The carbonyl SCSs in these well-defined systems are of some interest. In general the γ effect of the carbonyl oxygen atom (i.e. H.C.C=O) is strongly deshielding with however an orientational dependence. For example, in *trans* decalone the SCSs of the carbonyl group on H2ax (1.07) and H9 (1.05) are significantly greater than on H2eq (0.69) and this pattern is reproduced in the cyclohexanes and steroids. In contrast in norbornanone the SCS of the carbonyl on H3endo (0.68) is similar to that on H3exo (0.59) and again this is observed in camphor. The long-range (> 3 bonds) effects of the carbonyl group are also large and extend over both the bicycloheptene and decalin system. The effects are usually deshielding with only the 5ax and 6ax protons in *trans* decalone shielded. This pattern is also observed in the steroid nucleus (Table 7.1) where few of the protons show a negative SCS and these shifts are usually very small with the proton far removed from the keto group.

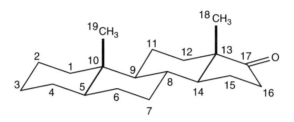

Figure 7.2   *Nomenclature used for 5α-androstan-17-one.*

The only marked exception to this is the SCS of the 11-keto group at the 1α proton (−0.15 ppm) and this is accompanied by a large positive SCS (0.74 ppm) at the 1β proton. The combined effect of these shifts is so large that these two methylene protons occur at the two extremes of the proton spectrum in **8** (apart from the methyl groups).

The data collected in Tables 7.2–7.5 provided a rigorous test of the application of both the CHARGE model and also present theories of carbonyl SCSs to these compounds. The compounds were all of fixed conformation with the possible exception of the five-membered rings of cyclopentanone and ring D of the steroid nucleus which exhibit conformational flexibility. The *ab initio* calculations gave the cyclopentanone geometry as the half-chair (Cs) conformation in agreement with both molecular mechanics calculations and the experimental gas phase geometries.[22] Similar calculations for the saturated ring D of androstan-3-one gave the same geometry as obtained for androstane,[17] i.e as a C13-envelope with C14, C15, C16 and C17 more or less in a plane with only a 9.5° twist. In the 17-keto compounds (**6,7** and **8**) the *ab initio* (and molecular mechanics) calculations gave the conformation of ring D as a C14 envelope with C13, C15, C16 and C17 almost coplanar and this is in agreement with the observed coupling constants for ring D.[16]

For the ketones studied the authors initially assumed that the electronic γ effects of the carbonyl carbon (H.C.C.C=O) were the same as for a saturated carbon atom which was already incorporated into the CHARGE scheme. Subsequently a small correction (0.1 ppm) was added. Inspection of the data of Figure 7.1 and the tables showed that there was an orientation dependence of the carbonyl γ SCS (H.C.C=O). In the similar analysis of saturated carbon (H.C.C.C) and oxygen (H.C.C.O) γ effects a simple angular function (Equation (3.5)) was found to be appropriate with values of the coefficients A and B determined by the observed data. Thus this approach was initially used here. However more detailed inspection of the observed data showed that the carbonyl γ SCS were also dependent on the bond angle (α) of the carbonyl group (C.C=O.C). In particular the five-membered ring ketones with carbonyl bond angles ca. 106–109° have quite different SCSs to the six-membered ketones

Table 7.1 ¹H SCSs for the 3-keto, 11-keto and 17-keto groups in 5α-androstane

Proton	3-keto[a]		11-keto[a]		17-keto[a]	
	b	c	b, d	c	b	c
1α	0.48	0.45	−0.15	−0.13	0.04	e
1β	1.67	1.66	0.74	0.74	−0.02	0.01
2α	0.81	0.77	0.04	e	e	e
2β	0.99	0.96	0.06	e	e	e
3α	—	—	—	−0.04	−0.03	e
3β	—	—	—	−0.02	e	e
4α	0.86	0.84	e	e	0.07	0.07
4β	1.05	1.02	0.02	e	0.03	0.07
5 (CH)	0.49	0.45	−0.05	−0.07	0.05	e
6α	0.10	0.11	0.03	e	0.03	0.03
6β	0.10	0.11	e	e	0.03	0.03
7α	0.05	0.03	0.19	0.19	0.06	0.04
7β	0.07	0.04	0.15	0.10	0.09	0.09
8 (CH)	0.05	0.05	0.34	0.36	0.28	0.26
9 (CH)	0.07	0.07	0.94	1.00	0.04	0.03
11α	0.03	0.02	—	—	0.14	0.12
11β	0.12	0.13	—	—	e	e
12α	0.03	0.02	1.05	1.15	0.14	0.12
12β	0.02	0.02	0.61	0.54	0.09	0.08
14(CH)	0.03	0.02	0.62	0.64	0.37	0.37
15α	0.02	e	0.14	0.12	0.30	0.27
15β	0.03	e	0.11	0.08	0.37	0.35
16α	0.02	0.03	0.18	0.16	0.48	0.49
16β	0.03	0.03	0.09	0.16	0.82	0.89
17α	0.03	e	—	0.22	—	—
17β	0.04	−0.02	—	0.03	—	—
18-Me	0.04	0.03	−0.07	−0.03	0.17	0.17
19-Me	0.24	0.23	0.18	0.22	0.02	0.02

[a] δ(ketone) − δ(androstane).
[b] δ(**8**) − δ(**7**).
[c] From Schneider et al.[11].
[d] From Abraham et al.[15].
[e] SCS <0.01 ppm.

with bond angles ca. 115–116°. This additional functionality was therefore incorporated into the carbonyl oxygen γ effect again as a simple cos α dependence. The coefficients in this equation were then determined from the observed SCSs by an iterative least mean squares calculation to give finally Equation (7.1) for the carbonyl gamma effect (GSEF).

$$\text{GSEF} = 0.09(2.0 - 3.0\cos\alpha)(2.0 - \cos\theta) \qquad (7.1)$$

This equation gave generally good agreement for all the vicinal protons in the data set (a total of 50 protons). These results will be discussed later.

Long-range effects. The interactions considered to be responsible for the long-range effects of the carbonyl group have been given earlier as steric, electric field and magnetic anisotropy effects. The electric field effect was determined with no additional parameters required to calculate the electric field effects of the carbonyl group from Equation (3.12). However both

Table 7.2   Observed versus calculated ^{1}H chemical shifts (δ) of acyclic and cyclic ketones

Compound		Obs.[a]	Calc.	Compound		Obs.[b]	Calc.
Acetaldehyde	Me	2.20	1.96	trans-1-Decalone	2a	2.33	2.27
	CHO	9.78	9.70		2e	2.36	2.34
Acetone	Me	2.17	1.83		3a	1.67	1.69
Cyclopentanone	Hα	2.17	2.22		3e	2.05	2.05
	Hβ	1.98	1.93		4a	1.43	1.34
Pinacolone	Me	2.14	1.88		4e	1.77	1.82
	tBu	1.13	1.26		5a	1.15	0.98
					5e	1.79	1.63
Cyclohexanone	H2,6	2.33	2.24		6a	1.18	1.27
	H3,5	1.88	1.82		6e	1.70	1.69
	H4	1.71	1.77		7a	1.14	1.20
					7e	1.79	1.67
					8a	1.25	1.34
					8e	1.91	1.77
					9	1.95	1.84
					10	1.37	1.31

[a] From Puchert and Behnke[18].
[b] From Abraham et al.[15].

Table 7.3   Observed versus calculated ^{1}H chemical shifts in substituted cyclohexanes

Proton	2-Methyl-cyclohexanone		3-Methyl-cyclohexanone		4-Methyl-cyclohexanone		2-t-Butyl-cyclohexanone		4-t-Butyl-cyclohexanone	
	Obs.[a]	Calc.	Obs.[a]	Calc.	Obs.[a]	Calc.	Obs.[b]	Calc.	Obs.[b]	Calc.
2a	2.43	2.30	2.01	1.84	2.32	2.22	2.15	1.95	2.27	2.23
2e	—	—	2.35	2.27	2.36	2.31	—	—	2.36	2.33
3a	1.38	1.44	1.89	1.77	1.41	1.31	1.47	1.47	1.45	1.32
3e	2.10	1.82	—	—	2.00	2.00	2.18	2.00	2.08	2.07
4a	1.67	1.65	1.34	1.31	1.89	1.77	1.64	1.64	1.45	1.45
4e	1.84	1.93	1.89	1.90	—	—	1.90	1.94	—	—
5a	1.67	1.64	1.66	1.66	1.41	1.31	1.66	1.59	1.45	1.32
5e	2.07	2.01	2.01	2.02	2.00	2.00	2.06	2.00	2.08	2.07
6a	2.30	2.21	2.25	2.21	2.32	2.22	2.32	2.23	2.27	2.23
6e	2.37	2.31	2.35	2.31	2.36	2.31	2.26	2.33	2.36	2.33
^{t}Bu	—	—	—	—	—	—	0.99	0.95	0.90	0.91

[a] Data from Griffiths[19].
[b] From Abraham et al.[15].

the steric coefficient a_s (Equation (3.11)) for the oxygen atom and the magnetic anisotropies $\Delta\chi_1$ and $\Delta\chi_2$ (Equation (3.15)) needed to be evaluated. In addition the push–pull coefficient for the steric effect and the position of the magnetic anisotropy along the carbonyl bond needed to be determined. The data set of the non-vicinal protons used comprised 112 proton shifts and the iterations were achieved using a non-linear least mean squares algorithm. All the iterations including the steric term plus the anisotropy terms gave no better results than the corresponding iterations without the steric term. Thus the steric term for the carbonyl oxygen atom was removed. Also the values of the parallel anisotropy ($\Delta\chi_1$) obtained from the iterations were always much larger than those for the perpendicular anisotropy $\Delta\chi_2$.

Table 7.4   *Calculated versus observed ¹H chemical shifts in bicycloheptane systems*

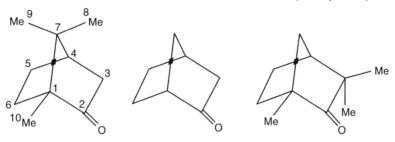

Proton	Camphor		Norbornanone		Fenchone	
	Obs.[a]	Calc.	Obs.[a]	Calc.	Obs.[b]	Calc.
1	—	—	2.60	2.62	1.15(Me)	1.00(Me)
3x	2.35	2.51	2.06	2.28	—	—
3n	1.84	1.78	1.84	1.89	—	—
4	2.09	2.18	2.67	2.61	2.14	2.16
5x	1.95	2.05	1.79	1.85	1.80	1.90
5n	1.34	1.37	1.45	1.46	1.70	1.52
6x	1.68	1.93	1.81	1.78	1.54	1.75
6n	1.40	1.64	1.53	1.58	1.37	1.66
7s	—	—	1.73	1.76	1.80	1.96
7a	—	—	1.56	1.63	1.54	1.36
8(Me)	0.84	0.98	—	—		
9(Me)	0.96	0.95	—	—		
10(Me)	0.92	1.05	—	—		
3x(Me)	—	—	—	—	1.04	1.07
3n(Me)	—	—	—	—	1.04	0.99

[a] From Ainger,[16] Abraham *et al.*[20] and Sanders *et al.*[21].
[b] From Abraham *et al.*[15].

These calculations were all performed with the carbonyl anisotropies placed at the mid-point of the C=O bond. It was found that the best iteration still gave significant errors for some protons in the bicycloheptanones (Table 7.4). In particular the observed 6exo and 6endo SCSs were much smaller than calculated. However placing the anisotropy at the carbonyl carbon atom gave much better agreement for these protons without any significant effect for the remaining protons in the data set. The final values of the anisotropies obtained were $\Delta\chi_1$ 17.1 and $\Delta\chi_2$ 3.2 (10^{-30} cm³ molecule⁻¹).

The general agreement of the observed versus calculated proton shifts for the ketones considered given in Tables 7.2–7.5 was good and most of the observed shifts were reproduced to <0.1 ppm. Also the general agreement for the steroid ketones was encouraging though in this quite sterically compressed system there were larger errors in the calculated shifts for some of the protons in the base hydrocarbon androstane. In particular the 7β and 15β protons are the only resolved protons in androstane with errors >0.2 ppm probably due to large steric interactions and this transfers to the steroid ketones. The good agreement for the C-1 protons in the 3,11,17-trione (Table 7.5) was noteworthy as the 1β proton in the 11-ketosteroids is very close to the 11-keto oxygen and the SCS for this proton provides a critical test of the model. Indeed Schneider *et al.*[11] noted that the 1β proton deviated

Table 7.5   *Observed[a] versus calculated ¹H chemical shifts in 5-α-androstanones*

Proton	3-one (5)		11-one		17-one (6)		3,17-dione (7)		3,11,17-trione (8)	
	Obs.	Calc.	Obs.[b]	Calc.	Obs.	Calc.	Obs.	Calc.	Obs.	Calc.
1α	1.35	1.44	0.76	0.80	0.91	0.97	1.35	1.44	1.22	1.26
1β	2.03	2.00	2.40	2.43	1.65	1.60	2.03	2.00	2.77	2.86
2α	2.29	2.21	1.50	1.51	1.49	1.53	2.31	2.21	2.27	2.19
2β	2.39	2.44	1.41	1.44	1.42	1.44	2.39	2.44	2.45	2.44
3α	—	—	1.19	1.19	1.18	1.25	—	—	—	—
3β	—	—	1.65	1.67	1.67	1.69	—	—	—	—
4α	2.08	2.03	1.23	1.08	1.29	1.38	2.11	2.05	2.12	2.02
4β	2.27	2.06	1.23	1.33	1.25	1.09	2.26	2.07	2.28	2.08
5	1.51	1.61	0.99	1.03	1.07	1.10	1.56	1.62	1.51	1.56
6α	1.32	1.48	1.23	1.42	1.25	1.43	1.38	1.53	1.41	1.56
6β	1.32	1.36	1.23	1.28	1.25	1.32	1.38	1.40	1.37	1.40
7α	0.96	1.19	1.12	1.25	0.97	1.26	1.01	1.31	1.20	2.16
7β	1.75	1.93	1.79	1.99	1.77	2.00	1.84	2.05	1.99	1.40
8	1.33	1.19	1.65	1.57	1.56	1.36	1.59	1.43	1.93	1.87
9	0.75	0.86	1.69	1.76	0.72	0.81	0.80	0.91	1.74	1.91
11α	1.56	1.56	—	—	1.67	1.56	1.70	1.60	—	—
11β	1.38	1.48	—	—	1.27	1.41	1.40	1.47	—	—
12α	1.12	1.01	2.25	2.01	1.23	1.18	1.27	1.20	2.32	2.21
12β	1.72	1.65	2.25	2.28	1.80	1.79	1.83	1.82	2.44	2.47
14	0.92	0.82	1.54	1.27	1.26	1.14	1.29	1.17	1.91	1.66
15α	1.65	1.65	1.77	1.75	1.93	2.18	1.95	2.20	2.09	2.31
15β	1.17	1.53	1.23	1.61	1.51	2.13	1.52	2.15	1.63	2.24
16α	1.60	1.56	1.72	1.62	2.06	2.26	2.08	2.27	2.26	2.35
16β	1.64	1.63	1.72	1.69	2.43	2.27	2.45	2.28	2.54	2.36
17α	1.15	1.05	1.35	1.14	—	—	—	—	—	—
17β	1.43	1.60	1.45	1.68	—	—	—	—	—	—
18(Me)	0.72	0.76	0.66	0.87	0.86	1.03	0.89	1.05	0.82	1.19
19(Me)	1.02	1.00	1.01	0.93	0.81	0.77	1.04	1.01	1.22	1.18

[a] From Abraham *et al.*[15].
[b] δ values from SCS (Table 7.1) and δ(5α-androstane)[17].

appreciably (by 0.6 ppm) from their calculated value, based on a dipole model of the electric field and Apsimon's anisotropy equation.

The calculated shifts in the bicycloheptanone systems were also in generally good agreement with the observed shifts (Table 7.4) though there were some significant errors. It may be significant that in the bicycloheptane system it was necessary to consider possible orbital interactions between the bridging C-7 carbon and the ring carbons in order to reproduce the observed shifts in these molecules using the CHARGE model (Chapter 4, Section 4.2.4). However the largest errors in Table 7.4 are for the H-6exo and H-6endo in camphor and fenchone in which both the calculated ¹H shifts and the SCSs are much less than the observed values (by ca. 0.2–0.3 ppm). This deviation does not appear to be a function of the bicyclic ring system as in norbornanone both the calculated shifts and the SCSs at the C-6 protons are in good agreement with the observed values. Why the introduction of methyl groups should affect the SCSs of the carbonyl group was not clear. The ¹H shifts of camphor were obtained in solvents of varying polarity (CCl₄, CDCl₃, acetone and methanol) in order to determine if any intramolecular hydrogen bonding between the carbonyl oxygen and the methyl protons was occurring but the shifts were as expected with no evidence of any such interaction.

The only acyclic compounds investigated are the simple compounds in Table 7.2 as all other acyclic ketones are conformationally mobile. The observed shifts for acetone and acetaldehyde are both slightly greater than calculated but the other calculated shifts in this table are in reasonable agreement with the observed.

The values of the carbonyl anisotropy determined were of interest. In all the iterations performed the value of the parallel anisotropy $\Delta\chi_1$ was reasonably constant at ca. 20 (10^{-30} cm^3 molecule^{-1}). In the final iteration the value obtained was 17.1. However the value of the perpendicular anisotropy $\Delta\chi_2$ varied considerably with both positive and negative values obtained during the iterations. The last iteration gave a value of 3.2. The variability is due to the small effect this parameter had on the ^{1}H chemical shifts. The only definitive method of determining this parameter would be to obtain SCSs from protons situated both at the sides and immediately above the carbonyl group, but these data were not available to the authors. These values will be considered with the results of previous investigations in the next section.

The actual magnitudes of the various contributions to the carbonyl SCSs are given in Table 7.6 for trans-1-decalone. This clearly shows that both electric field plus anisotropy contributions are important in determining carbonyl SCSs. This Table also shows that other contributions are present in determining the SCSs. For example, the sum of the electric field plus anisotropy contributions for the 8a and 8e protons are −0.04 and −0.01 ppm whereas the calculated SCSs are +0.40 and +0.30 ppm. The additional contribution in this case stems from the H..H steric interaction. H-8a and H-8e in *trans*-decalin are deshielded due to the proximity of H-1a and H-1e and these protons are to *low frequency* of the corresponding protons in cyclohexane (0.93 vs. 1.18δ for H-axial and 1.54 vs. 1.68δ for H-equatorial) as a result. This steric interaction is removed when these protons are replaced

Table 7.6   *Calculated versus observed SCSs for trans-1-decalone (4) with the electric field and anisotropy contributions*

Proton	Obs.	Calc.	Electric field	Anisotropy
2a	1.07	1.08	—	—
2e	0.69	0.73	—	—
3a	0.42	0.34	0.23	−0.07
3e	0.38	0.30	0.18	0.07
4a	0.51	0.47	0.12	0.30
4e	0.23	0.31	0.15	0.12
5a	0.22	0.11	0.04	0.07
5e	0.26	0.12	0.06	0.03
6a	−0.07	0.02	0.05	−0.04
6e	0.02	0.03	0.03	−0.01
7a	−0.11	−0.05	0.01	−0.05
7e	0.11	0.01	0.06	−0.05
8a	0.32	0.40	0.18	−0.22
8e	0.37	0.30	0.20	−0.21
9H	1.07	1.06	—	—
10H	0.49	0.35	0.24	−0.09

by the carbonyl group giving a deshielding effect. This effect is also observed in the SCS of H-10 in which there is a 1,3-diaxial H-H interaction with H-1ax in *trans*-decalin which is absent in 1-decalone. Apart from these special cases the anisotropic and electric field contributions determine the carbonyl SCSs though the relative size of these contributions varies considerably with the orientation of the proton from the carbonyl group.

7.2.2   Aromatic Aldehydes and Ketones

Introduction. When a carbonyl group is attached to an aromatic ring it will be conjugated and this is likely to affect the carbonyl anisotropy. This was noted by Jackman[23] who suggested that the inconsistency of Pople's model when applied to amides may be due to the assumption that the anisotropy of the carbonyl group in amides is similar to that in aldehydes and ketones.

The effect of conjugation on the carbonyl group anisotropy may also be determined from the chemical shielding tensor. Wasylishen *et al.*[24] used MASNMR to determine the chemical shift tensors for the carbonyl carbon of acetaldehyde, 3,4-dibenzyloxy-benzaldehyde and 3,4-dimethoxy benzaldehyde. There was a significant difference (>25 %) between the shift tensors for the carbonyl carbon of acetaldehyde and those for the carbonyl carbons of the benzaldehydes, which were identical (within the error margin of the observations). These results suggest strongly that the anisotropy of the carbonyl group should be treated separately for aromatic and aliphatic systems. Thus Abraham *et al.*[25] assigned the ¹H spectra of a number of aromatic aldehydes and ketones and these provided sufficient data for a complete analysis of the aromatic carbonyl substituent effects using the CHARGE model. The compounds studied are given in Scheme 7.1 and the ¹H chemical shifts in CDCl₃ and DMSO solvents in Tables 7.8 and 7.9. Only the shifts in CDCl₃ solvent were analysed here. ¹H shifts in DMSO are discussed in Chapter 8, Section 8.6.

Theory, application to aromatic carbonyl compounds. For the aromatic carbonyl compounds considered the only non-parameterized short-range effect in CHARGE was the C(Ar).CHO β effect. The electric field effect is calculated directly from the partial atomic charges thus the only long-range effects are the anisotropy of the carbonyl group ($\Delta\chi_{parl}$ and $\Delta\chi_{perp}$) and the CO steric effect. The steric effect of the aliphatic CO group was found to be due solely to the carbonyl oxygen.[26] Assuming the same for the aromatic carbonyl group, only the steric coefficient for the carbonyl oxygen (a_S in Equation (3.11)) needs to be determined. Thus only the above four parameters are required in the CHARGE routine to specify the ¹H shifts in the compounds considered.

¹H NMR were obtained at 400.13 MHz; the ¹H spectra of indanone (**7**), 9-acetylanthracene (**14**) and benzosuberone (**16**), were obtained at 700.13 MHz. The assignments of all the compounds are given in Tables 7.7–7.9. The geometries of the polycyclic molecules were obtained using the MMFF94 force field, but for the smaller molecules the geometries were further optimized using *ab initio* calculations (B3LYP/6-31G** or B3LYP/3-21G*). Full details of the spectral assignments and geometries are given elsewhere.[25,27]

The 2-substituted benzaldehydes (**2, 3** and **4**) and acetophenones (**6, 8**) can exist as *cis*- or *trans*-conformers (Figure 7.3). The *trans* conformer is usually the more stable form, due to steric effects, but where intramolecular hydrogen bonding occurs as in **3** and **8** the *cis* form

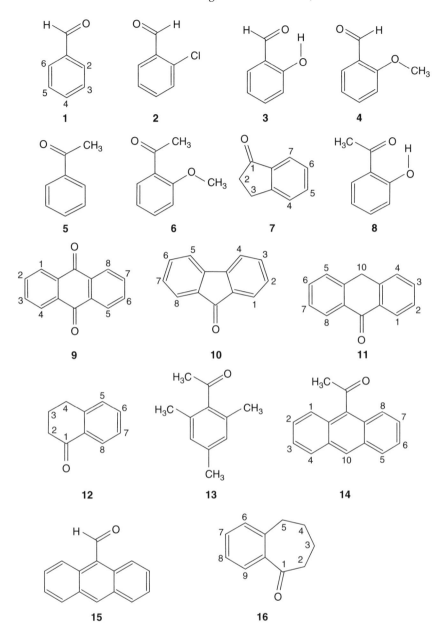

Scheme 7.1   *Molecules studied with their numbering.*

is the more stable. The geometries, energies and dipole moments of these molecules were calculated using PCMODEL and the results are given in Table 7.7. The conformer energy difference is so large for these compounds that they only exist in one conformation, **2, 4** and **6** in the *trans* form and **3** and **8** in the *cis* form. In all these compounds except **2** (see below)

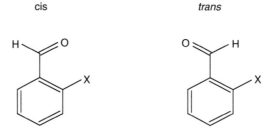

Figure 7.3   *cis- and trans-Conformers in 2-substituted benzaldehydes*

Table 7.7   *Energies (kcal mol⁻¹), dipole moments (D) and CO/ring dihedral angles of the trans- and cis-conformers of 2-substituted benzaldehydes and acetophenones*

Compound	Energy (trans)	(Energy cis)	ΔE $E_{cis} - E_{trans}$	Dipole (trans)	Dipole (cis)	CO/ring dihedral angle (trans)	(cis)
(**1**) Benzaldehyde	31.26			4.37		0	
(**2**) 2-Chlorobenzaldehyde	27.18	33.06	5.88	4.14	6.59	0	0
(**3**) 2-Hydroxybenzaldehyde	33.67	25.78	−7.89	2.90	4.87	11.2	0
(**4**) 2-Methoxybenzaldehyde	39.63	47.61	7.98	5.09	5.56	0	0
(**5**) Acetophenone	36.10			4.37		0	
(**6**) 2-Methoxyacetophenone	47.19			5.75	4.17	29.1	
(**8**) 2-Hydroxyacetophenone	36.36	30.57	−5.79	6.02	5.01	17.6	0

the carbonyl is coplanar with the aromatic ring. The remaining polycyclic compounds only exist in one conformation.

Compounds **7**, **9**, **10**, **11** and **15** are planar, **12** has an envelope cyclohexenone ring and **13** and **14** have the acetyl group orthogonal to the aromatic ring. LIS studies of the conformations of **7**,[28] **12**,[29] **13**,[30] **14**, **15**[34] and **16**[37] in chloroform solution agreed with these results.

All these compounds were used for the parameterization except for **2** and **16**. In **2** the PCMODEL geometry has a CO/ring dihedral angle of ca. 40⁰ but the *ab initio* geometry is planar. In **16** again the ring geometries for the molecular mechanics and various *ab initio* basis sets differ considerably. These molecules were therefore omitted from the calculations and will be considered subsequently.

Keto–enol tautomerism of anthrone.   The formation of 9-hydroxyanthracene from anthrone by the addition of NaOH has been known many years[32] but there was no mention in the literature of this tautomerism being observed by ¹H NMR and the ¹H NMR spectrum of the enol has not been described. Abraham *et al.*[25] observed that in chloroform there was no evidence of any enol form from the ¹H NMR spectrum, but in DMSO both conformers were observed as the slow exchange condition applied with the enol form being dominant (3:1). Presumably DMSO is stabilizing the enol form through hydrogen bonding. The OH proton occurs at 10.22δ in DMSO and this was evidence of strong hydrogen bonding with the solvent. However Novak *et al.*[33] observed that in pentane 1,3,5 triones the more polar

Table 7.8 *Observed ¹H chemical shifts of substituted benzaldehydes and acetophenones in CDCl₃ and DMSO versus calculated shifts*

Compound	Solvent	H2	H3	H4	H5	H6	CHO/ COCH₃	2- Substituent
(1) Benzaldehyde	CDCl₃	7.880	7.532	7.632	—	—	10.025	—
	Calc.	*7.876*	*7.555*	*7.640*	—	—	*10.026*	—
	DMSO	7.920	7.619	7.728	—	—	10.027	—
(2) 2-Chloro benzaldehyde	CDCl₃	—	7.457	7.530	7.389	7.928	10.492	—
	Calc.	—	*7.566*	*7.669*	*7.463*	*7.924*	*10.421*	—
	DMSO	—	7.632	7.709	7.543	7.879	10.349	—
(3) 2-Hydroxy benzaldehyde	CDCl₃	—	6.997	7.535	7.027	7.567	9.903	11.024
	Calc.	—	*7.071*	*7.533*	*7.167*	*7.662*	*9.893*	*10.298*
	DMSO	—	6.999	7.522	6.964	7.666	10.258	10.685
(4) 2-Methoxy benzaldehyde	CDCl₃	—	6.990	7.548	7.027	7.830	10.478	3.933
	Calc.	—	*7.115*	*7.520*	*7.140*	*7.900*	*10.574*	*3.870*
	DMSO	—	7.234	7.689	7.081	7.699	10.371	3.923
(5) Acetophenone	CDCl₃	7.960	7.459	7.562	—	—	2.604	—
	Calc.	*7.884*	*7.554*	*7.623*	—	—	*2.609*	—
	DMSO	7.958	7.528	7.639	—	—	2.582	—
(6) 2-Methoxy acetophenone	CDCl₃	—	6.967	7.459	6.995	7.728	2.611	3.913
	Calc.	—	*7.098*	*7.480*	*7.127*	*7.723*	*2.619*	*3.827*
	DMSO	—	7.167	7.535	7.017	7.569	2.523	3.888
(8) 2-Hydroxy acetophenone	CDCl₃	—	6.972	7.466	6.896	7.730	2.627	12.242
	Calc.	—	*7.088*	*7.524*	*7.171*	*7.608*	*2.627*	*12.833*
	DMSO	—	6.958	7.532	6.963	7.890	2.641	11.954
(13) 2,4,6-Trimethyl acetophenone[a]	CDCl₃	—	6.832	2.273[a]	—	—	2.449	2.215
	Calc.	—	*6.862*	*2.394[a]*	—	—	*2.509*	*2.336*
	DMSO	—	6.895	2.264[a]	—	—	2.449	2.217

[a] Methyl.

keto form was stabilized by DMSO compared to chloroform and they proposed that this was due to solvation of the more polar keto form.

The ¹H spectrum was obtained in a number of solvents and these results are given in Table 7.10. In pyridine, methanol and acetone the two forms could be observed and their proportions determined. In THF it was necessary to add a catalytic amount of base (NaOH) in order for the equilibrium to proceed. Even in this case there was still slow exchange between the keto and enol forms and thus the proportions could be readily determined. The proportion of enol varies from 77 % in DMSO to only 16 % in acetone. Table 7.10 gives the free energy difference $\Delta E = E(\text{keto}) - E(\text{enol})$ and also the relative permittivity and the Kamlett β parameter[34] of the solvents used. The latter is a measure of the hydrogen bonding ability of the solvent. It is clear from the results in Table 7.10 that the relative proportions of the keto and enol forms bear little relationship to the relative permittivity of the solvent, but there is an excellent correlation between the proportion of enol and the Kamlett β parameter. Analysis gives a linear equation with a correlation coefficient (r) of 0.92. If the uncertain data for MeOD is removed the r value is increased to 0.97. This is strong support for the proposal that the formation of the enol in this case is due mainly to hydrogen bonding with the solvent and not to polarity effects. This contrasts

Table 7.9 *Observed versus calculated ^{1}H chemical shifts (δ) of polycyclic aromatic carbonyl compounds and derivatives in $CDCl_3$ and DMSO*

Compound	Solvent	H1	H2	H3	H4	H5	H6	H7	H8	CHO/COCH$_3$
(7) Indanone	CDCl$_3$	—	2.695	3.152	7.480	7.586	7.371	7.766	—	—
	Calc.	—	*2.811*	*3.190*	*7.370*	*7.635*	*7.424*	*7.717*	—	—
	DMSO	—	2.629	3.110	7.587	7.665	7.418	7.639	—	—
(9) Anthraquinone	CDCl$_3$	8.325	7.805	—	—	—	—	—	—	—
	Calc.	*8.413*	*7.721*	—	—	—	—	—	—	—
	DMSO	8.231	7.948	—	—	—	—	—	—	—
(10) Flourene-9-one	CDCl$_3$	7.659	7.290	7.480	7.522	—	—	—	—	—
	Calc.	*7.745*	*7.307*	*7.410*	*7.603*	—	—	—	—	—
	DMSO	7.611	7.386	7.621	7.803	—	—	—	—	—
(11) Anthrone	CDCl$_3$	8.361	7.456	7.589	7.465	—	—	—	4.351[a]	—
	Calc.	*8.130*	*7.468*	*7.639*	*7.544*	—	—	—	*4.159[a]*	—
	DMSO	8.206	7.523	7.702	7.603	—	—	—	4.462[a]	—
(12) α-Tetralone	CDCl$_3$	—	2.656	2.141	2.967	7.248	7.461	7.300	8.034	—
	Calc.	—	*2.554*	*2.228*	*2.916*	*7.369*	*7.591*	*7.392*	*7.962*	—
	DMSO	—	2.598	2.042	2.945	7.350	7.540	7.341	7.863	—
(14) 9-Acetyl anthracene	CDCl$_3$	7.847	7.523	7.495	8.038	—	—	—	8.489[a]	2.820
	Calc.	*7.784*	*7.566*	*7.595*	*8.119*	—	—	—	*8.674[a]*	*2.793*
	DMSO	7.846	7.575	7.609	8.173	—	—	—	8.713[a]	2.804
(15) 9-Anthr aldehyde	CDCl$_3$	8.992	7.687	7.555	8.073	—	—	—	8.707[a]	11.541
	Calc.	*9.055*	*7.772*	*7.624*	*8.176*	—	—	—	*8.919[a]*	*11.077*
	DMSO	9.038	7.759	7.644	8.238	—	—	—	9.020[a]	11.493
9-Hydroxy anthracene	DMSO	8.430	7.429	7.469	7.978	—	—	—	8.033[a]	10.220[b]
	THF	8.423	7.267	7.329	7.840	—	—	—	7.799[a]	—
	Calc.	*8.581*	*7.427*	*7.559*	*7.893*				*7.933[a]*	*5.288[b]*
9-Methoxy anthracene	CDCl$_3$	8.300	7.470	7.470	7.996				8.224[a]	4.157
	Calc.	*8.535*	*7.611*	*7.607*	*8.042*				*8.545[a]*	*3.944*

[a] H10.
[b] OH.

Table 7.10 *The % enol form in the keto–enol tautomerization of anthrone in various solvents*

Solvent	n_{enol}	ΔE (kcal mol^{-1})	ε	β
DMSO	0.77	−0.720	46.7	0.76
Pyridine	0.66	−0.395	12.4	0.64
THF	0.37	0.317	7.6	0.55
MeOD	0.35	0.369	32.7	0.62[a]
Acetone	0.16	0.988	20.7	0.48

[a] Data are uncertain.

with a previous study on intra- vs. intermolecular hydrogen bonding in *cis*-cyclohexane-1,3-diol[35] in which the energy difference of the conformers involved (ax–ax vs. eq–eq) was shown to correlate with the polarity of the solvent but with different coefficients for H-bonding and non-H-bonding solvents.

The carbonyl anisotropy. The ^{1}H chemical shifts in CDCl$_3$ in Tables 7.8 and 7.9 are sufficient to parameterize the CHARGE routine for aromatic carbonyl groups. All the data in the tables were used except the chemical shifts for 2-chlorobenzaldehyde (**2**) and benzosuberone (**16**) (see later) and also the hydroxyl hydrogens of compounds **3** and **8**. This gave a total of 129 shifts ranging from 2.8 to 11.5δ. The four variables to be determined were the C(Ar).CHO β effect, the carbonyl anisotropy $\Delta\chi_{parl}$ and $\Delta\chi_{perp}$ and the oxygen steric effect. The values of the parameters were obtained by iteration which gave $\Delta\chi_{parl} = 6.36$, $\Delta\chi_{perp} = -11.88$ (10^{-30} cm³ molecule⁻¹) and the oxygen steric coefficient $a_s = 38.4$ ppm Å⁶. The rms error was 0.094 ppm for the whole dataset. The calculated and observed shifts are given in Tables 7.8 and 7.9. The agreement is good with the largest error for the ring protons ca. 0.15 ppm. This demonstrates the applicability of the CHARGE scheme to this class of compounds.

A small change was made to the π calculation in CHARGE (Equation (3.7)) for the 9-substituted anthracenes. In the unmodified routine H-10 of anthraldehyde was calculated at much too large a δ value (calc. 9.10 vs. obs. 8.71). Conversely H-10 of 9-hydroxy anthracene was calculated at too low a δ value (calc. 7.57 vs. obs. 7.80). The calculated SCSs have the correct signs (note H-10 in anthracene is 8.43δ)[36] but are much too large. This difference is not due to the carbonyl anisotropy or to steric or electric field effects as these effects decrease very rapidly with distance (see Table 7.11). This was thought to be due to Hückel theory exaggerating the π charges in compounds with very polarizable π systems such as the middle ring of anthracene. This effect did not happen with the 9-acetyl anthracene as the acetyl group is orthogonal to the ring, thus there is no conjugation with the π system.

In CHARGE the resonance integral coefficient (k_{rs} Equation (3.7)) is -1.0 for benzenoid aromatics. To account for the polarizability of the middle ring of anthracene the coefficient for the C9,10 bonds with the α-carbons was modified. Decreasing this integral to -1.25 gave reasonable agreement for both molecules and these calculated values are given in Table 7.9. Most interestingly the chemical shift of H-10 in 9-methoxyanthracene is also shielded with respect to anthracene. The SCS of the methoxy group at H-10 is -0.21 ppm which is comparable to that of the hydroxy group (-0.36 ppm) even though both the MM and *ab initio* calculations gave the methoxy group orthogonal to the anthracene ring and in consequence show no π effect.

7.2.3 Conformational Analysis

The CHARGE routine gave ¹H chemical shifts for the conformationally rigid molecules considered in good agreement with the observed shifts. The authors also considered whether the CHARGE routine could be used to obtain conformational information in conformationally mobile compounds and examined 2-chlorobenzaldehyde and benzosuberone. 2-Chlorobenzaldehyde is considered here, but the benzosuberone calculation is deferred to Chapter 8, Section 8.4.3.

2-Chlorobenzaldehyde (2). The theoretical calculations gave conflicting geometries for the stable *trans* conformer. Gaussian98 using the B3LYP density function theory with the 6-31G** basis set gave a planar molecule which was also the case with the MMF94 force field of PCMODEL. In contrast the MMX force field in PCMODEL gave a minimum energy for a 40° ring/aldehyde torsional angle. The ¹H chemical shifts were obtained from CHARGE for 10° rotations of the aldehyde from the plane and compared with the observed data. The best agreement was for a torsional angle of 25° with an rms error of 0.085 ppm. There is no experimental data to support this result but it would appear a reasonable value.

The carbonyl SCS. The components of the carbonyl SCS effect are given explicitly in CHARGE and some illustrative examples are given in Table 7.11 together with the C=O$\cdots$H distance. Comparison of the results for the near H-1/H-8 protons in 9-anthraldehyde (**15**) and 9-acetylanthracene (**14**) is of interest as in **15** the molecule is planar but in **14** the acetyl group is perpendicular to the anthracene ring plane. The effect of the carbonyl anisotropy is strongly deshielding in **15** but strongly shielding in **14**. In contrast the electric field and steric effects are the same sign in both molecules but much larger in **15** due to the closer proximity of the carbonyl group and H-1/8. Comparison of the anisotropy and electric field contributions is well illustrated by the results for anthrone (**11**). They are both long-range and all the protons of the compound except H-2 and H-3 have significant shifts but the electric field contribution is always larger and in the peri protons (H1/8) predominant.

Table 7.11   *Anisotropic, electric field and steric contributions of the carbonyl group SCSs*

Compound	Proton	CO$\cdots$H (Å)	$\delta_{C=O\text{-}AN}$	$\delta_{CO\text{-}EFLD}$	$\delta_{O\text{-}STERIC}$
(**15**) Anthraldehyde	H1/H8	2.32	0.246	0.337	0.241
(**14**) Acetylanthracene	H1/H8	2.93	−0.424	0.132	0.061
(**12**) α-Tetralone	H8	2.49	0.126	0.398	0.161
(**7**) Indanone	H7	2.83	0.086	0.311	0.074
(**11**) Anthrone	H1/H8	2.48	0.128	0.411	0.163
	H2/H7	4.84	0.019	0.064	0.003
	H3/H6	6.06	0.021	0.051	0.001
	H4/H5	5.73	0.048	0.073	0.001
	H10	4.90	0.094	0.161	0.003

The steric term only becomes significant in compounds where the oxygen–hydrogen distance is relatively short (2.5–3 Å). In some cases, e.g. H-8 in α-tetralone, it is larger than the anisotropy contribution but for the molecules studied here it is always less than the electric field term. At distances >4 Å the steric term is negligible.

It is of some interest to compare the values of the carbonyl anisotropy obtained here with those found in previous studies. The early investigations used different axes and nomenclature and these were converted by Abraham and Ainger[15] to the present nomenclature of Chapter 3, Figure 3.5 and Equation (3.15). The values of $\Delta\chi_{parl}$ and $\Delta\chi_{perp}$ obtained in this investigation were 6.4, −11.9, cf. Zurcher[6], 13.5, −12.2; ApSimon[8] 21, −6; Schneider[11] 24, −12; Williamson[12-14] 4, −9 and Abraham[26] 22.7, −14.8. There is a considerable difference between the present values and all the other investigations except those of Williamson. As noted earlier all the investigations except that of Williamson considered only aliphatic carbonyls. The values obtained by Williamson were based on the carbonyl anisotropy in peptides and proteins and it is interesting to see the close comparison between this value and the values for the aromatic carbonyl as this would be expected on chemical grounds. The π-electrons of the carbonyl group in amides are delocalized as also are the π-electrons in aromatic ketones and this delocalization cannot occur in saturated ketones. This delocalization appears to have a significant effect on the carbonyl anisotropy.

7.3 Esters

7.3.1 Introduction

The ester group is one of the commonest groups in chemistry and biology, being a constituent of all animal fats (glyceryl triesters) and the odour producing constituent in many fruits and berries. Two naturally occurring esters are the wine flavouring compound (**A**) and the hallucinogenic ester nepetalactone (**B**) from the Catnip plant (Scheme 7.2).

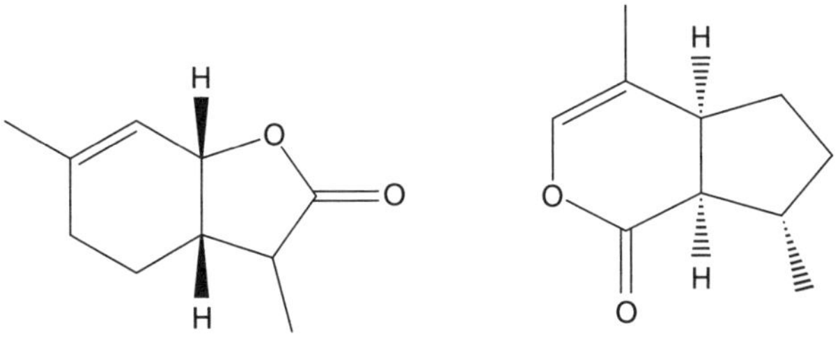

Scheme 7.2 *Important naturally occurring esters.*

Despite this common occurrence the effect of the ester group on the ¹H NMR chemical shifts of organic compounds was only known from additive tables of chemical shifts[1]. Esterification was initially used for assignment purposes,[38,39] e.g. acetylating an OH group as this gave a large change in the neighbouring protons chemical shift.

The first investigations of the magnetic anisotropy of the carbonyl group included esters with aldehydes and ketones.[3,23] Subsequently the carbonyl anisotropy in aldehydes, ketones and amides was obtained and shown to vary significantly in the different compounds[2,5,6,8,10,12,13,25,40–43]. Later studies[13,41,42] used these parameters for analysis of large systems such as proteins but certain parameters had to be obtained from *ab initio* calculations on small molecules.[42]

Conjugation with the aromatic system changes the anisotropy of the carbonyl group in aldehydes and ketones[25] to give values similar to those found for amides. It is possible that similar effects occur for esters and Abraham *et al.*[44] studied a range of esters using the CHARGE model to investigate the effect of the ester group. The compounds studied were in two categories. Those with rigid structures (Figure 7.5) were used to determine the parameters needed in the CHARGE model[45] and conformationally mobile compounds (Figure 7.6) were used for monitoring and modifying specific effects.

The ¹H chemical shifts of the compounds investigated were also calculated by the *ab initio* (GIAO) method (Chapter 3, Section 3.2). The recommended[46] B3LYP/6-31G** method was used. This was the first study of observed vs. calculated (GIAO and CHARGE) ¹H chemical shifts of esters. Comparison of ¹H chemical shift calculations by GIAO and CHARGE had been given[47] for a series of halo-compounds and general agreement with the observed data was found, though the GIAO calculations produced large errors for protons close to the halogens.

Molecular geometry. In order to calculate accurately the effect of an ester group on ¹H chemical shifts it is essential to obtain accurate molecular structures. The ester group is planar[48] with cis- and trans-conformers (Figure 7.4) with a barrier to interconversion of 10–15 kcal mol^{-1}.[49] Dipole moment studies of alkyl esters showed that the cis conformer is the more stable form[48] and a matrix IR study found ($E_{trans} - E_{cis}$) ca. 5 kcal mol^{-1} in methyl formate (R = H, R' = CH$_3$) and 8.5 kcal mol^{-1} for methyl acetate (R = R' = CH$_3$).[49] In t-butyl formate (R = H, R' = *t*-butyl) a ¹H NMR investigation at −90 °C found 15 % of the *trans* form[50] and a study[51] using ^{3}J(H.C.O.C) coupling constants showed an increase in the *trans* conformer from 12 % in toluene-d_8 to 30 % in DMSO-d_6 at room temperature.

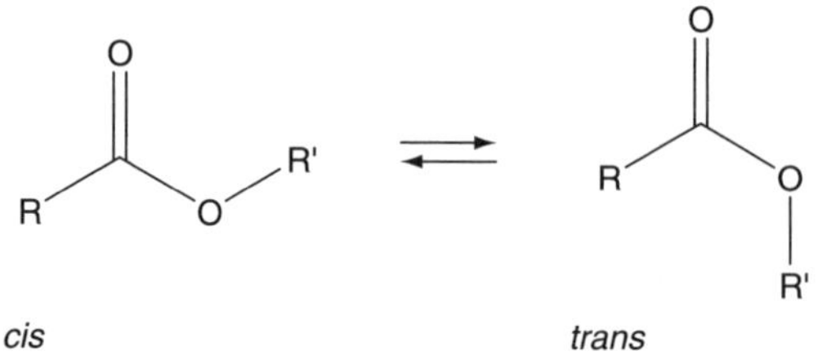

Figure 7.4   *The cis/trans isomerism of esters.*

The stability of the *cis* conformer has been explained as due to dipole–dipole interactions,[52] lone pair σ* interactions,[53] aromatic stabilization[54] and the electronegativity of the R' group.[55] Even for methyl formate the energy difference is such that only the planar *cis* conformer is observed by NMR. However the *trans* conformer is found in the cyclic compounds included in this study.

The geometries of all compounds (Figures 7.5 and 7.6) were obtained using the B3LYP theory with the 6-31G(d,p) basis set, except for **6** where the 6-311++G(d,p) basis set was used because of hydrogen bonding[56] and **14** (see below). The GIAO calculations were performed with both the geometry optimization and chemical shift calculations at the B3LYP/6-31G(d,p) level. Other calculations used the MMX and MMFF94 force fields.

The dihedral angles between the ester group and the aromatic ring in phenyl acetate (**5**) and methyl 9-carboxy anthracene (**14**) in solution were unknown. A Lanthanide Induced Shift (LIS) investigation of (**5**) gave a value of 45° for the dihedral angle[57–59] in good agreement with the crystal geometry. A series of *ab initio* (B3LYP) potential energy scans varying the ester dihedral angle were performed for (**14**) with different basis sets. The minimum energy dihedral angle varies from 25° to 58° which can be compared with the value in the crystal (70°)[60] and that from MM calculations with the MMFF94 force field (75°). The *ab initio* calculated value of 58° was used in the calculations. This case is discussed further in Chapter 8, Section 8.4.1.

7.3.2   Theory, Application to Esters

The separate contributions in the CHARGE model (Chapter 3, Section 3.5) must be determined for the ester group. These include the short-range γ effects from the carbonyl group and the oxygen, the steric effects of both oxygens in the ester group, the two magnetic anisotropies for the carbonyl group and the effect of the ester electric field. Also any aromatic ring currents in the hetero aromatic rings (e.g. pyrone) need to be determined.

π-Electron densities.   The π-electron densities were reproduced from those calculated from *ab initio* calculations. The results from *ab initio* calculations are dependent on the basis set used and it was found previously[61,62] that the 3-21G basis set at the B3LYP level gave the best values of the dipole moments for the compounds investigated. Thus the π-electron densities from this basis set were used to parameterize the Huckel calculations.

The π systems in the esters investigated range from the simple π system of methyl acrylate to the aromatic π systems of pyrone and coumarin and the complex systems of maleic and phthalic anhydride. Because of this diversity it was necessary for the CHARGE model to differentiate the various π systems. Thus the π systems of methyl acrylate, pyrone and coumarin were treated separately. This was achieved by determining the appropriate values of the Huckel integrals h_r and k_{rs} (Equation (3.7)) for the various π systems considered. Coumarin and isocoumarin have the same resonance integrals and the isocoumarin π densities were in good agreement with the 3-21G values.

The anhydrides were also defined separately and as in the pyrone/coumarin case maleic anhydride required different integrals from phthalic anhydride.

The accuracy of the π-electron densities calculated in the CHARGE scheme may be examined by comparing the calculated π-electron densities and dipole moments of these esters with those obtained by *ab initio* theory using the 3-21G basis set (Table 7.12). The

Table 7.12 π charges (me)[a] and dipole moments μ (D) for esters

Compound	Atom	3-21G (me)[a]	CHARGE	Exp.
Methyl acrylate (**15**) (*cis*)	C=O	95	81	
	CH	−19	−15	
	CH$_2$	77	77	
	μ	1.3	2.1	1.7[63]
Methyl acrylate (*trans*)	C=O	92	81	
	CH	−19	−15	
	CH$_2$	72	77	
	μ	1.9	2.8	1.7
Pyrone (**3**)	C$_2$	24	31	
	C$_3$	−91	−64	
	C$_4$	76	71	
	C$_5$	−55	−42	
	C$_6$	58	60	
	μ	4.42	4.58	5.18[64]
Coumarin (**12**)	C$_2$	−4	−26	
	C$_3$	−53	−20	
	C$_4$	69	72	
	C$_9$	−42	−50	
	C$_{10}$	68	21	
	μ	4.63	4.47	4.49[64]
Isocoumarin (**13**)	C$_3$	−11	−14	
	C$_4$	−86	−82	
	C$_9$	41	29	
	C$_{10}$	−57	−44	
	C$_1$	75	29	
	μ	4.22	3.52	4.24[64]
Phthalic anhydride (**11**)	C$_1$	99	98	
	C$_8$	−23	−24	
	μ	5.73	6.05	5.34[65,66]
Maleic anhydride (**17**)	C$_1$	94	97	
	C$_2$	43	48	
	μ	4.0	3.5	4.14[65,66]

[a] milli-electrons.

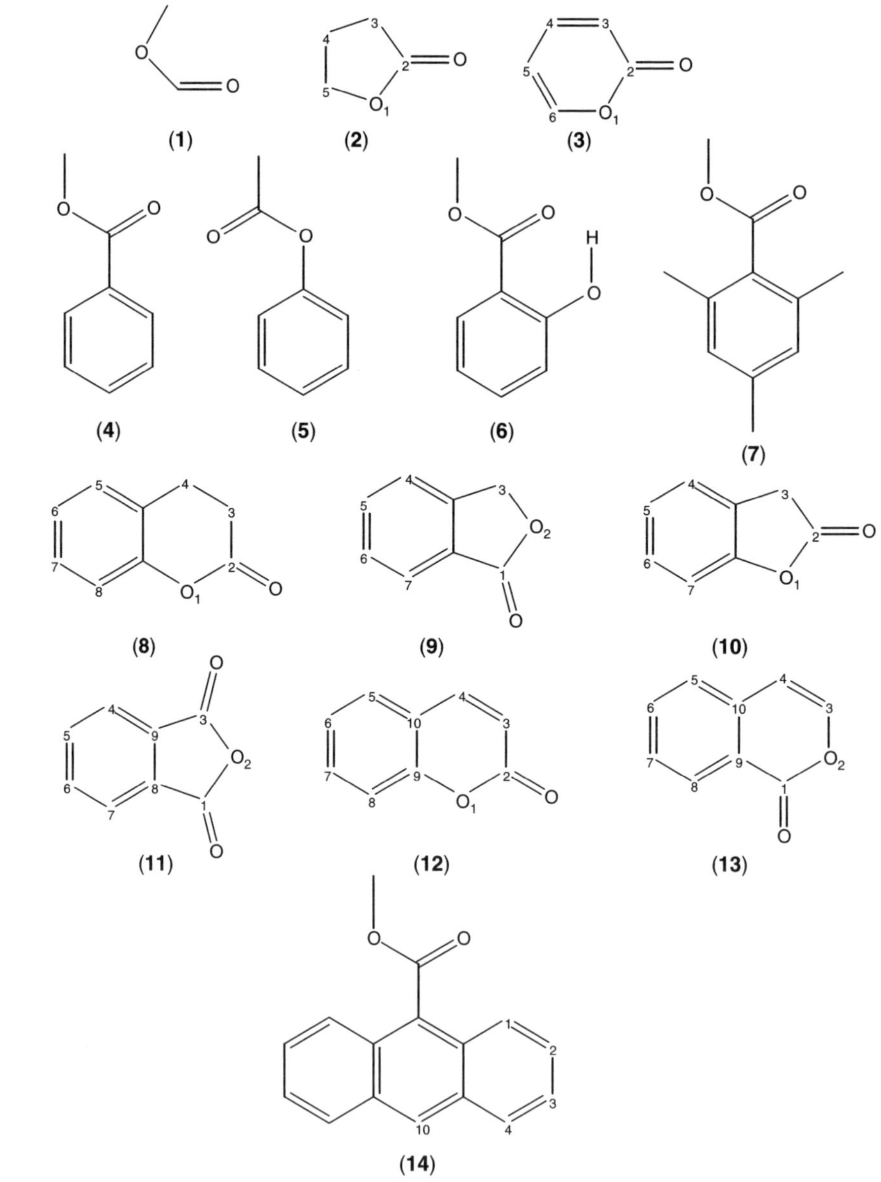

Figure 7.5   *Ester compounds used for parameterization of CHARGE.*

good agreement of the calculated vs. observed dipoles in this Table is strong support for the calculations.

7.3.3   Observed versus Calculated Shifts

¹H spectra were obtained at 400.13 MHz. The ¹H spectra of **4, 9, 10** and **14** were also obtained at 750 MHz. The assignments of all the compounds are given in Table 7.13 together

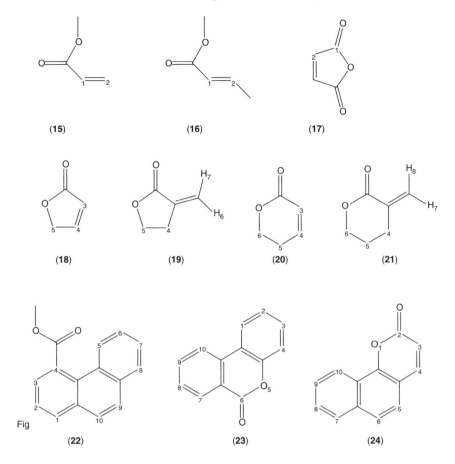

Figure 7.6   *Compounds used for refinement and monitoring of determined parameters.*

with the chemical shifts calculated from CHARGE and from the GIAO calculations. Full details of the assignments are given elsewhere.[27,44]

The short-range γ effects of both the ester oxygens to the aromatic, olefinic or aliphatic protons and the γ effect from the carbonyl oxygen atom were obtained directly from the ¹H shifts of compounds **1–16**, **18** and **20**, plus **15** and **16** for the O=C.CH fragment

The electric field in the CHARGE program is given directly from the partial atomic charges. For esters this resulted in a dipole acting along the C=O bond and two equal dipoles along each of the C—O—C bonds. As the charges must balance, the charge on the carbonyl oxygen produces an equal and opposite charge on the carbon atom. Similarly the charge on each of the carbons in the C—O—C fragment is half that of the divalent oxygen.

The electric field produced was too large due to treating each dipole separately as this produced a larger charge on the carbonyl carbon than is given by the CHARGE routine. In the CHARGE routine the sum of all the charges in a molecule equals zero, but the charge on the central carbon atom is less than the sum of the charges on the two oxygen atoms. To

Table 7.13   *Experimental and calculated ¹H chemical shifts of esters*

		H	Me				
(**1**) Methylformate	$CDCl_3$[18]	8.100	3.800				
	CHARGE	8.066	3.816				
	GIAO	8.024	3.628				
		H2	H3	H4			
(**2**) γ-Butyrolactone	$CDCl_3$[67]	2.490	2.260	4.320			
	CHARGE	2.503	2.028	4.296			
	GIAO	2.060	1.891	4.142			
		H3	H4	H5	H6		
(**3**) Pyrone	$CDCl_3$[68]	6.31	7.33	6.25	7.48		
	CHARGE	6.314	7.056	6.259	7.429		
	GIAO	6.045	7.000	5.835	7.650		
		H2	H3	H4	Me		
(**4**) Methyl benzoate	$CDCl_3$	8.042	7.434	7.552	3.918		
	CHARGE	8.239	7.424	7.511	3.764		
	GIAO	8.076	7.556	7.568	3.903		
		H2	H3	H4	Me		
(**5**) Phenylacetate	$CDCl_3$	7.084	7.372	7.221	2.294		
	CHARGE	7.199	7.292	7.140	2.226		
	GIAO	7.276	7.365	7.169	2.020		
		H3	H4	H5	H6	Me	OH
(**6**) Methyl salicylate	$CDCl_3$	6.978	7.449	6.873	7.833	3.949	10.727
	CHARGE	7.080	7.457	7.160	7.973	3.904	11.212
	GIAO	7.150	7.547	6.877	8.085	3.914	11.078
		H3,5	OMe	4Me	2,6Me		
(**7**) Methylmesitoate	$CDCl_3$[58]	6.840	3.880	2.280	2.270		
	CHARGE	6.911	3.834	2.397	2.335		
	GIAO	6.869	3.708	2.128	1.779		
		H3	H4	H5	H6	H7	H8
(**8**) 3,4-Dihydrocoumarin	$CDCl_3$[57]	2.785	2.993	7.198	7.127	7.250	7.036
	CHARGE	2.863	3.081	7.115	6.976	7.114	6.939
	GIAO	2.367	2.722	7.103	7.071	7.306	7.040
		H3	H4	H5	H6	H7	
(**9**) Phthalide	$CDCl_3$	5.332	7.512	7.694	7.552	7.935	
	CHARGE	5.376	7.388	7.576	7.433	7.627	
	GIAO	5.084	7.382	7.608	7.533	8.028	
		H3	H4	H5	H6	H7	
(**10**) 2-Cumaranone	$CDCl_3$	3.731	7.284	7.130	7.304	7.097	
	CHARGE	3.706	7.144	6.997	7.179	6.906	
	GIAO	3.287	7.262	7.107	7.318	7.029	
		H4	H5				
(**11**) Phthalic anhydride	$CDCl_3$	8.032	7.914				
	CHARGE	7.952	7.861				
	GIAO	8.013	7.785				
		H3	H4	H5	H6	H7	H8
(**12**) Coumarin	$CDCl_3$[57]	6.406	7.705	7.502	7.290	7.502	7.290
	CHARGE	6.446	7.695	7.624	7.233	7.513	7.270
	GIAO	6.142	7.310	7.296	7.203	7.515	7.274

		H3	H4	H5	H6	H7	H8
(**13**) Isocoumarin	$CDCl_3$[57]	7.276	6.505	7.432	7.717	7.520	8.286
	CHARGE	7.187	6.478	7.456	7.563	7.329	8.280
	GIAO	7.375	6.105	7.202	7.596	7.477	8.504

		H1	H2	H3	H4	H10	OMe
(**14**) Methyl-9-carboxy-anthracene	$CDCl_3$	8.006	7.533	7.477	8.028	8.516	4.172
	CHARGE	8.078	7.651	7.626	8.145	8.806	4.076
	GIAO	8.199	7.493	7.446	7.885	8.249	3.879

		H1	H2(c)	H2(t)	OME
(**15**) Methyl acrylate	$CDCl_3$[18]	6.12	6.41	5.83	3.77
	CHARGE	6.20	6.45	6.12	3.84
	GIAO	6.13	6.70	5.89	3.59

		Me	H1	H2	OMe
(**16**) Methyl crotonate	$CDCl_3$[69]	1.88	5.85	6.98	3.72
	CHARGE	1.81	5.80	7.05	3.85
	GIAO	2.06	5.85	6.62	3.59

		H
(**17**) Maleic anhydride	$CDCl_3$[18]	7.09
	CHARGE	6.96
	GIAO	6.65

		H3	H4	H5
(**18**) γ-Crotonolactone	$CDCl_3$[70]	6.18	7.59	4.91
	CHARGE	6.08	7.23	4.76
	GIAO	6.05	7.36	4.69

		H4	H5	H6	H7
(**19**) α-Methylene-γ-butyrolactone	$CDCl_3$[71]	3.00	4.37	5.67	6.26
	CHARGE	2.75	4.48	5.70	5.95
	GIAO	2.65	4.09	5.62	6.48

		H3	H4	H5	H6
(**20**) 5,6-Dihydropyran-2-one	$CDCl_3$[72]	6.05	7.10	2.50	4.44
	CHARGE	6.05	7.09	2.39	4.30
	GIAO	5.91	6.75	2.11	4.25

		H4	H5	H6	H7	H8
(**21**) 3-Methylene-tetrahydro-pyran-2-one	CCl_4[73]	2.60	2.00	4.31	5.46	6.29
	CHARGE	2.43	1.86	4.30	5.67	6.21
	GIAO	2.47	1.60	4.28	5.47	6.74

		H4
(**22**) Phenanthrene-4-carboxylic acid methyl ester	CCl_4[74]	8.05
	CHARGE	7.80
	GIAO	7.87

		H1	H2	H3	H4
(**23**) Benzo chromen-6-one	$CDCl_3$	8.064	7.359	7.499	7.329
	CHARGE	8.120	7.286	7.474	7.362
	GIAO	8.032	7.311	7.467	7.319

		H7	H8	H9	H10
	$CDCl_3$	8.408	7.584	7.825	8.125
	CHARGE	8.610	7.564	7.794	8.222
	GIAO	8.624	7.533	7.729	8.032

		H10
(**24**) Benzo chromen-2-one	$CDCl_3$[75]	8.53
	CHARGE	8.37
	GIAO	9.08

resolve this problem the ester group was modelled as one system (Figure 7.7). In this system the partial atomic charges on the O=C—O atoms produce the electric field. The charges on these atoms were then varied to give good agreement with the observed chemical shifts. A reduction of 25 % gave good agreement with the observed shifts.

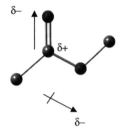

Figure 7.7   *The electric field model for esters in CHARGE.*

The parallel and perpendicular anisotropies of the carbonyl group were obtained from the chemical shifts of over 100 protons in compounds **1–14** in Table 7.13. The parameters were determined using a non-linear least-mean-squared algorithm. This gave a well-determined data set from which the values for the parallel and perpendicular anisotropies were 10.1 and -17.1 $(10^{-30}$ cm^3 molecule$^{-1})$ respectively. The anisotropy contribution to the ¹H chemical shift for compounds **7**, **13** and **14** is shown in Table 7.14. For the protons perpendicular to the carbonyl group in **7** and **14,** a large shielding is observed whilst in **13** for the proton parallel to the carbonyl group an equally large deshielding is found.

Table 7.14   *Effect of the carbonyl anisotropy on the proton chemical shift*

	$\delta_{(anis)}(\parallel)$	$\delta_{(anis)}(\perp)$	$\delta_{(anis)}$
Anthracene-9-carboxylic acid methyl ester (**14**)			
H 1/8	−0.10	−0.34	−0.44
H 4/5	0.00	0.02	0.02
COOCH$_3$	−0.15	0.19	0.04
Methylmesitoate (**7**)			
CH$_3$ (2,6)	−0.06	−0.21	−0.27
Isocoumarin (**13**)			
H 8	−0.20	0.35	0.15

The steric coefficients for both the oxygen atoms were determined. The carbonyl oxygen steric coefficient[15,25] was determined iteratively from the data set as 85 $(10^{-30}$ cm^3 molecule$^{-1})$. The steric term for the ether oxygen when included in the iteration (including compounds **1–14**) was very small. To better define this the chemical shift of H-10 in compound (**24**) was used as this proton is very near the oxygen atom. This gave the steric coefficient for the ether oxygen as 40 $(10^{-30}$ cm^3 molecule$^{-1})$.

Analysis of the ¹H chemical shifts in Table 7.13 suggested that pyrone had an aromatic ring current and when this was included in the CHARGE calculations a value of ca. 1/3 that of benzene was obtained. This was confirmed by GIAO calculations of the chemical shift of the methane hydrogen situated above the ring in pyrone and benzene (Figure 7.8).

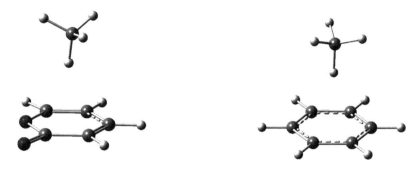

Figure 7.8   *Comparison of ring current shifts in pyrone and benzene (see color Plate 2).*

The chemical shift of the proton above the ring was -1.3 ppm for pyrone and -3.4 for benzene. There is excellent agreement between the GIAO result and the ring current found in CHARGE. This method was also used to confirm that there was no ring current in maleic anhydride.

The parameterization of the ester group has produced in CHARGE an accurate model for the calculation of the ^{1}H chemical shifts of esters. The calculated–observed shifts over the 100 protons used for the parameterization gave an average error of 0.09 ppm and a correlation coefficient (r^2) of 0.995 when all compounds are included. The largest error was found for the hydroxyl proton as observed previously.[25]

The model can now be used with confidence to predict the effect of the ester group on the ^{1}H chemical shifts in any chemical environment.

The GIAO calculations gave reasonable results (rms error of 0.2 ppm and correlation coefficient, r^2, of 0.992). However large discrepancies are present for protons close to the electronegative oxygens, the largest error being 0.5 ppm for H-10 of **24**. Similar errors were found for halogen compounds.[47] This may be related to the large steric hindrance produced by use of more restricted basis sets as seen for **14**. These calculations may be improved by using alternative basis sets or theories but this was beyond the scope of this study.

Conformation analysis of vinyl esters. The CHARGE program was then used to determine the conformational profile of methyl acrylate and methyl crotonate. These molecules rapidly interconvert on the NMR time scale between the s-trans and s-cis conformations (Scheme 7.3), thus the ^{1}H chemical shifts can be predicted from the calculated conformer shifts if the conformer populations are known.

Conversely the populations of the conformers can be found by comparing the calculated conformer shifts with the observed. The conformer shifts were calculated by CHARGE and by GIAO calculations and the averaged chemical shift, based on the populations of the *cis* and *trans* isomers was plotted against the *cis*-conformer population.[44] The best fit of the averaged chemical shifts from CHARGE vs. the observed shifts was at 90 % s-cis population for methyl acrylate (**15**), in good agreement with the 66 % found from an electron diffraction study.[76] For methyl crotonate (**16**) the best fit was found at 60 % s-*cis* conformation, again in good agreement with the 75 % *cis* from an LIS experiment.[69] The GIAO calculated average shift did not give a minimum when plotted against the observed shifts[44] though the rms error was only ca. 0.1 ppm. The calculated shifts were not sufficiently accurate for this purpose.

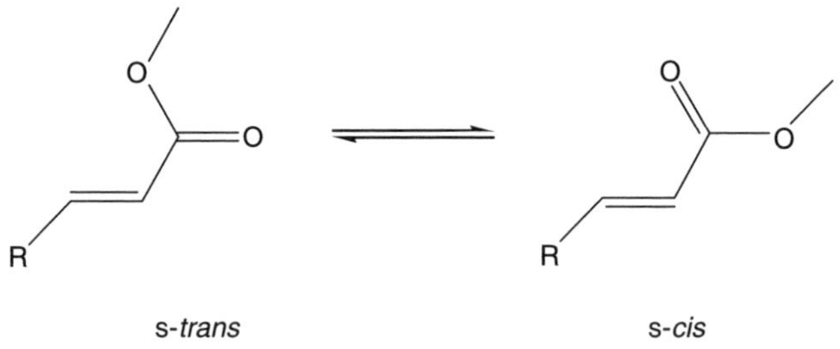

Scheme 7.3   *Conformers of methyl acrylate (R=H) and methyl crotonate (R=CH₃).*

The *cis/trans* populations were calculated from a range of *ab initio* and MM programs and the results are compared with the CHARGE and experimental values in Table 7.15. There is generally good agreement between these methods apart from the GIAO calculations, which is encouraging.

Table 7.15   *Conformer populations of methyl acrylate and crotonate*

Method	Methyl acrylate *cis* population (%)	Methyl crotonate *cis* population (%)
B3LYP/631G**	78	94
MMFF94	72	74
MMX	54	58
CHARGE	90	60
GIAO	0	0
Experimental	66	75

Carbonyl anisotropies. Table 7.16 gives the carbonyl anisotropies obtained in this study with those found for other carbonyl groups together with the oxygen steric coefficient, where known.

Table 7.16 $C=O$ anisotropies $\Delta\chi(\times10^{-30}\ cm^3\ molecule^{-1})$ and $=O$ steric coefficients, a_s ($\mathring{A}^6$)

Compound (Study)	Reference	$\Delta\chi_{parl}$	$\Delta\chi_{perp}$	a_s
Ketones (Zurcher,1967)	6	13.5	−12.2	
Ketones (Apsimon,1971)	8,43	21	−6	
Ketones (Schneider, 1985)	11	24	−12	
Aliphatic ketones (Abraham, 1999)	45	22.7	−14.8	67.9
Aromatic ketones (Abraham, 2003)	25	6.4	−11.9	38.4
Amides (Williamson, 1993)	13	4	−9	
Amides (Hunter, 2003)	46	12.6	−14.2	
Aliphatic amides (Perez, 2004)	47	13.5	−21.2	60.0
Aromatic amides (Perez, 2004)	47	10.5	−7.3	62.4
Esters (Abraham, 2005)	48	10.1	−17.1	85.0

There is general agreement in the data in Table 7.16 on the reduction in the anisotropy on conjugation of the carbonyl in aromatic compounds, amides or ethers. The anisotropy values given by Hunter are from *ab initio* calculations on formaldehyde[42], thus they should be compared with the data for aromatic ketones in this table.

7.4 Amides

7.4.1 Introduction

¹H NMR studies of the biologically ubiquitous amide group have been performed since the early days of NMR. Indeed the first study on hindered rotation was performed with *N,N*-dimethylformamide.[77] The partial double bond character of the amide group gives rise to an essentially planar amide group with a large barrier to rotation round the amide bond.[78,79] This leads to non-equivalence of the substituents and the observation of distinct ¹H NMR spectra for the *cis* and *trans* conformers[80] (Figure 7.9), which also gives the nomenclature used here. The early studies in this area have been well reviewed.[78]

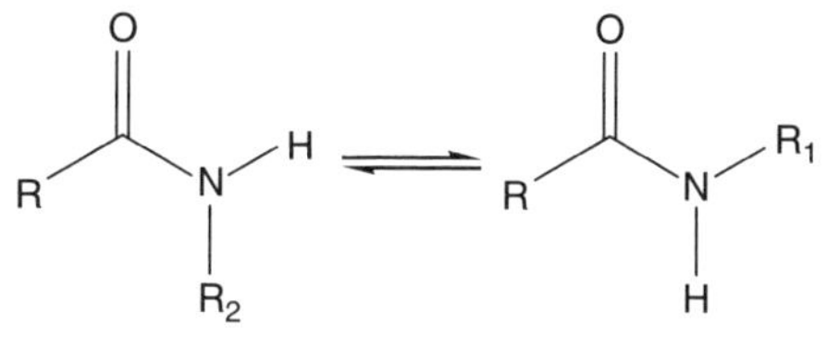

Figure 7.9 *Cis and trans conformers of amides.*

The planarity of the amide group has been a continuous source of contention for many years. For example, the structure of acetamide in the gas phase is essentially planar[81,82] but in the crystal it is almost pyramidal.[83] Again *ab initio* studies using HF/3-21G basis set gave a planar molecule[84,85] but when higher basis sets and more complex methods were used (e.g. HF/631G**+, MP2/631G** +)[86] the amide group was non-planar. However Perez[43] in the study considered here used B3LYP/631G** ++ and obtained a planar structure.

An extensive *ab initio* study of *N*-formyl l-alanyl l-alanine found that the peptide bond was not planar[87] and suggested that the origin of the non-planarity in peptides was due to steric effects.

Investigations of the effect of the amide group on ¹H chemical shifts in peptides and proteins were summarized by Williamson and Asakura.[13,14] The contributions they considered were the magnetic anisotropy and electric field of the amide group plus ring current shifts from aromatic residues. Two different models for the anisotropy were considered. In the first the anisotropies of the C=O, C—N and the Cα–N bonds were included, In the second the anisotropy was represented by one axially symmetric anisotropy running perpendicular to the peptide plane, as suggested by Osapay and Case.[88] Both models were tested over several proteins. The first model gave better answers and the results indicated that the Cα–N anisotropy was negligible.

Two models were also considered for the electric field effect. The first considered the field to arise from the charge dipoles of the C=O and (O)C—N bonds, the second calculated the electric field directly from point charges on the O,C and N atoms. The second was preferred.

7.4.2   Theory, Application to Amides

In order to apply the CHARGE theory of Chapter 3 to amides, both the short-range electronic effects of the amide group and the long-range electric field, anisotropy and steric effects all need to be considered. The short-range effects included an α effect for the NH proton and a number of γ effects, i.e. coefficients A, B_1 and B_2 (Equation (3.5)).

Electric field.   The electric field of the amide group must include the polar O=C.N atoms. The first model tried was of two independent dipole moments ($\overset{+}{N}=\overset{+}{C}, \overset{}{C}=\overset{-}{O}$) and gave much too large electric fields. The second model was as follows. The electric field was calculated first using the negative charge on the oxygen atom and an equal and opposite charge on the attached carbon atom as before and in addition using the positive π charge on the nitrogen atom ($\delta\pi^+$), and an equal and opposite negative π charge on the carbon atom. The σ charge on the nitrogen atom is compensated by opposite charges ($= 1/3\sigma$) on the three atoms attached to the nitrogen atom, and the resultant electric field averages to zero. This model gave much more reasonable values of the electric field contribution and was used for all the amides.

Anisotropy.   Different models for the anisotropy of the amide group were tried following the studies of Asakura and Williamson.[12–14] In all the models the carbonyl anisotropy was formulated as in ketones and taken from the centre of the CO bond. The C—N bond anisotropy proved more elusive and a number of models were attempted, including (a) a full C—N bond anisotropy, (b) only the parallel component of the C—N bond anisotropy (c) C—N anisotropy arising from the lone pair of electrons on the nitrogen atom and (d), as (c) but only considering shielding from the lone-pair. Of these four models only the last one provided good answers.

Steric effects.   The steric coefficients as given in Equation (3.11) for the O, N, N<u>H</u> and N<u>H</u>₂ atoms needed to be obtained.

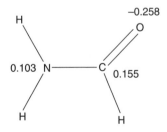

π *excess (electrons) in formamide*

π-effects. The Huckel integrals in Equation (3.7) were adjusted for the amide nitrogen to give the excess π density on the N atom and agreement with the observed dipole moments (cf. formamide μ obs. 3.71D, calc 3.92D) – see above.

Ring current. The amide group does not affect the aromatic ring current, thus the ring current model for condensed aromatics (Chapter 5, Section 5.1) was used unchanged.

There was a modification to the ring current for phenanthridone (see below). When the amide group is included into the aromatic system, it was necessary to include a small aromatic ring current for the amide ring ($f_c = 12.0$ ppm Å^3, cf. benzene $= 27.6$ ppm Å^3, Equation (3.17)). This is immediately rationalized in terms of the resonance contributions shown below.

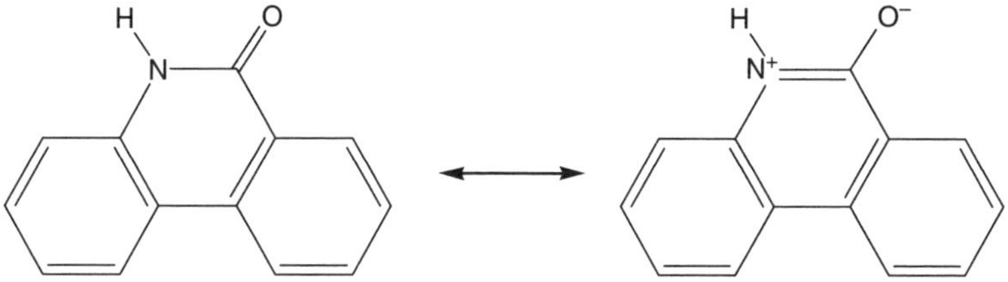

Resonance forms of phenanthridone

All models were tested using the data base of aliphatic, cyclic amides and aromatic amides by iteration. It was necessary to distinguish between the aliphatic and aromatic amides anisotropy in accord with previous results for the aromatic vs. aliphatic carbonyl compounds.[15,89]

7.4.3   Aliphatic and Cyclic Amides

Molecular geometries. The compounds investigated are shown in Figures 7.10 and 7.11. The molecular geometries were obtained from *ab initio* calculations at the B3LYP/6-31G++(d,p) level. For molecules too large to be handled conveniently by this program, the PCMODEL program was used. Both the *ab initio* and MM programs often generate different dihedral angles for the amide moiety. Table 7.17 shows some typical values of the O=C.N.H dihedral angle which can alter by ca. 10° depending on the force field used.

The geometries used for simple acyclic amides were mostly planar. In propionamide (**7**), the favoured conformer is with the O=C.C$_2$.C$_1$ angle $= 0°$ and in (**8**) one of the methyl groups eclipses the oxygen atom. The cyclic amides were all non-planar with usually a very slightly non-planar amide group. For example, (**9**) 8.3°, (**10**) and (**11**) 0.3°, (**12**) 3.4°, etc. (**11**) and (**13**) are interconverting between chair/half-chair and boat forms but (**14**) is solely in the chair conformation.[90,91] Full details of these geometries are given elsewhere.[43]

Observed versus calculated shifts. In Figure 7.10 compounds **1–16** and **22** were commercial samples, the remainder were synthesized using standard procedures.[43] ^{1}H NMR spectra were obtained in CDCl$_3$ solvent at 400 MHz. For **11** and **14** 500 MHz spectra were obtained.

Table 7.17   Calculated O=C−N−H dihedral angles (degrees) for amides

Method	1	2-cis	2-trans	4	17[a]	20[b]
MMFF94	0	5.7	172.2	4.3	−1.2	167.6
HF631**	0	0	180	0	−1.2	175.2
B3LYP	0	0	180	0	−1.2	175.3

[a] O=C−N−C₂.
[b] O=C−N−C₁.

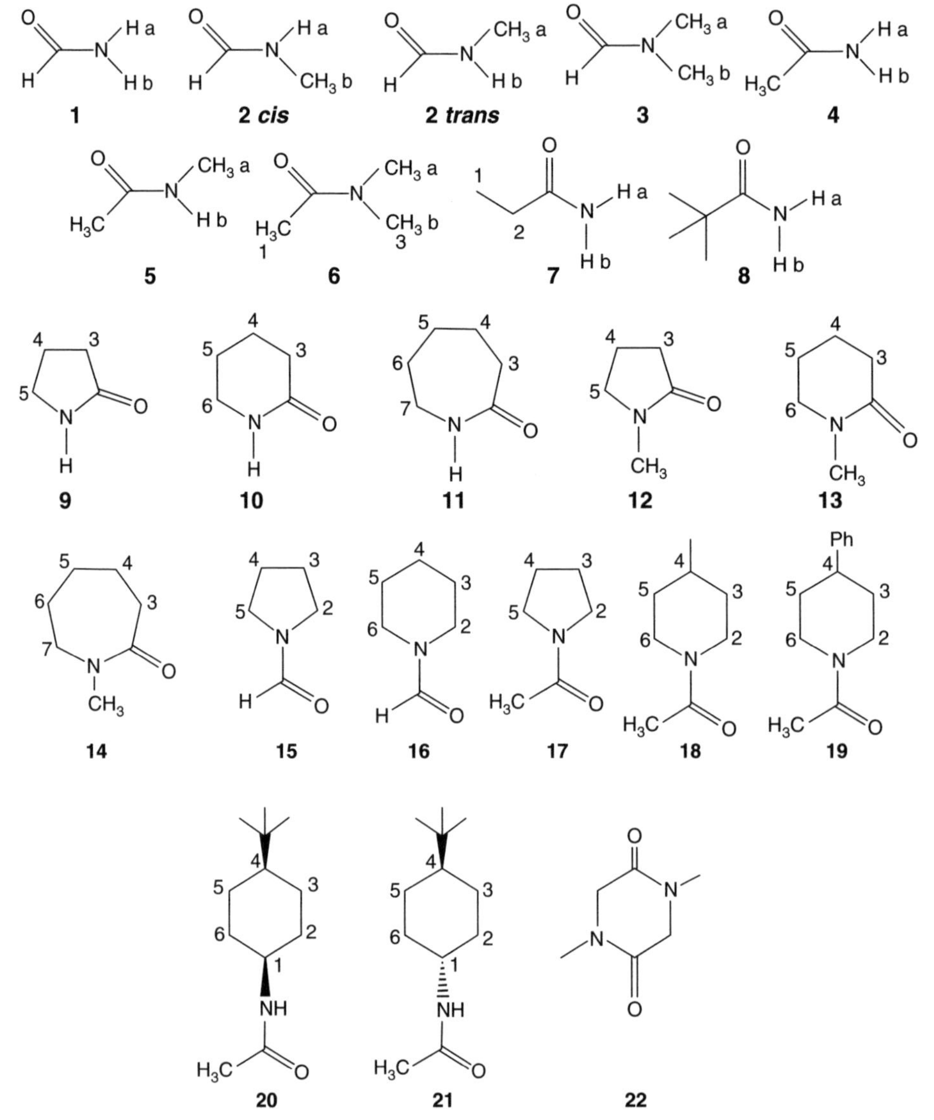

Figure 7.10   Aliphatic amides investigated.

The assignments of all the spectra are given in Tables 7.18 to 7.21, together with the shifts from CHARGE and GIAO calculations. Full details of all the spectra and assignments are given elsewhere.[43] One interesting assignment is that of the NH protons of formamide. An early unequivocal assignment of the ¹H spectrum of the pure liquid[92] gave NHa (Figure 7.10) *cis* to the oxygen as deshielded w.r.t. Hb. A COSY plot of the ¹H spectrum in CDCl₃ showed unambiguously that now Ha was shielded w.r.t. Hb.[43] The assignment of the two protons reverses in going from non-polar to polar solvents.

Table 7.18   *Observed and calculated ¹H chemical shifts for aliphatic amides*

Compound		CHO	NH$_a$	NH$_b$	NCH$_3^a$	NCH$_3^b$	–CH₃	CH₂	t-Bu
1	*Obs.*	8.228	5.799	5.478	—	—	—	—	—
	Calc.	8.245	5.795	5.479	—	—	—	—	—
	GIAO	8.407	4.853	4.174	—	—	—	—	—
2 *cis*	*Obs.*	8.057	5.986	—	—	2.944	—	—	—
	Calc.	8.127	5.876	—	—	2.848	—	—	—
	GIAO	8.466	4.549	—	—	2.763	—	—	—
2 *trans*	*Obs.*	8.194	—	5.550	2.855	—	—	—	—
	Calc.	8.134	—	5.546	2.703	—	—	—	—
	GIAO	8.411	—	4.575	2.702	—	—	—	—
3	*Obs.*	8.019	—	—	2.881	2.951	—	—	—
	Calc.	8.017	—	—	2.858	2.951	—	—	—
	GIAO	8.525	—	—	2.671	2.840	—	—	—
4	*Obs.*	—	5.415	5.415	—	—	2.027	—	—
	Calc.	—	5.610	5.610	—	—	2.032	—	—
	GIAO	—	4.668	4.801	—	—	1.835	—	—
5	*Obs.*	—	—	5.529	2.801	—	1.975	—	—
	Calc.	—	—	5.395	2.736	—	2.019	—	—
	GIAO	—	—	4.730	2.775	—	1.908	—	—
6	*Obs.*	—	—	—	3.010	2.929	2.078	—	—
	Calc.	—	—	—	2.920	2.860	2.014	—	—
	GIAO	—	—	—	3.586	3.280	1.958	—	—
7	*Obs.*	—	6.142	5.377	—	—	1.172	2.261	—
	Calc.	—	5.909	5.397	—	—	1.147	2.294	—
	GIAO	—	4.656	4.749	—	—	1.158	2.052	—
8	*Obs.*	—	5.562	5.223	—	—	—	—	1.228
	Calc.	—	5.970	5.453	—	—	—	—	1.200
	GIAO	—	4.565	5.266	—	—	—	—	1.395

^{a,b} see figure 7.10.

Compounds **20** and **21** were synthesized in order to determine the steric effects from the N atom and the −NH proton. This was required as there were no other compounds in the database in which these interactions were noticeable, and during the iterations of the different parameters, the steric coefficients changed randomly due to this lack of definition. Even with these molecules, the steric coefficient of the nitrogen was small.

*Sarcosine anhydride (**22**).* The ¹H chemical shifts for this compound were CH₂ obs.[93] 3.98, calc. 3.96; NMe obs.[93] 2.96, calc. 3.00. The agreement between observed and calculated chemical shifts is excellent.

Table 7.19   *Observed and calculated ¹H chemical shifts for cyclic amides 9–17*

Compound		CHO	1	2	3	4	5	6	7	CH₃—N	CH₃—CO
9	*Obs.*	—	6.062	—	2.301	2.142	3.403	—	—	—	—
	Calc.	—	6.078	—	2.379	2.130	3.435	—	—	—	—
	GIAO	—	4.438	—	2.100	2.064	3.390	—	—	—	—
10	*Obs.*	—	6.334	—	2.361	1.786	1.809	3.313	—	—	—
	Calc.	—	6.191	—	2.441	1.895	1.907	3.414	—	—	—
	GIAO	—	4.855	—	2.291	1.796	1.744	3.395	—	—	—
11[a]	*Obs.*	—	6.347	—	2.455	1.689	1.754	1.646	3.206	—	—
	Calc.	—	6.123	—	2.487	1.686	1.642	1.689	3.302	—	—
12	*Obs.*	—	—	—	2.370	2.033	3.391	—	—	2.854	—
	Calc.	—	—	—	2.344	2.122	3.361	—	—	2.928	—
	GIAO	—	—	—	2.180	1.893	3.327	—	—	2.893	—
13	*Obs.*	—	—	—	2.370	1.815	1.810	3.291	—	2.940	—
	Calc.	—	—	—	2.344	1.904	1.900	3.166	—	2.975	—
14	*Obs.*	—	—	—	2.519	1.664	1.700	1.646	3.358	2.980	—
	Calc.	—	—	—	2.481	1.660	1.625	1.687	3.212	3.010	—
	GIAO	—	—	—	2.204	1.879	1.855	1.856	3.426	2.855	—
15	*Obs.*	8.264	—	3.494	1.902	1.919	3.425	—	—	—	—
	Calc.	8.016	—	3.424	1.758	1.756	3.351	—	—	—	—
	GIAO	8.622	—	3.537	1.924	1.881	3.507	—	—	—	—
16	*Obs.*	8.005	—	3.302	1.579	1.541	1.688	3.483	—	—	—
	Calc.	8.018	—	3.323	1.568	1.582	1.680	3.605	—	—	—
	GIAO	8.188	—	3.622	1.544	1.713	1.548	3.228	—	—	—
17	*Obs.*	—	—	3.460	1.862	1.956	3.416	—	—	—	2.047
	Calc.	—	—	3.435	1.716	1.773	3.449	—	—	—	2.003
	GIAO	—	—	3.596	1.856	1.968	3.385	—	—	—	1.855

[a] GIAO calculation was not performed due to the fact that this compound is in two conformations in CDCl₃ solution.

Table 7.20   *Observed and calculated ¹H chemical shifts for compounds 18 and 19*

Proton	18			19		
	Obs.	*Calc.*	*GIAO*	*Obs.*	*Calc.*	*GIAO*
CO–CH₃	2.078	1.955	1.950	2.135	2.000	1.983
–CH₃	0.953	0.943	0.979	—	—	—
2a	2.543	2.910	2.363	2.625	2.998	2.484
2e	4.550	4.630	5.086	4.787	4.716	5.144
3a	1.075	1.057	1.165	1.613	1.644	1.710
3e	1.653	1.736	1.548	1.877	2.111	1.678
4	1.125	1.189	1.560	2.734	2.790	2.490
5a	1.601	1.437	1.226	1.645	1.784	1.797
5e	1.692	1.803	1.514	1.913	2.169	1.702
6a	3.022	2.847	3.018	3.164	2.896	3.111
6e	3.768	4.085	3.621	3.929	4.183	3.723
H$_o$	—	—	—	7.209	7.186	7.478
H$_m$	—	—	—	7.317	7.321	7.529
H$_p$	—	—	—	7.220	7.197	7.601

Table 7.21 *Observed and calculated ^{1}H chemical shifts for cyclic amides 20 and 21*

Proton	20		21	
	Obs.	Calc.	Obs.	Calc.
1	4.112	4.091	3.674	3.622
2a	1.507	1.575	1.073	1.172
2e	1.850	1.865	2.013	1.913
3a	1.026	1.053	1.107	1.078
3e	1.657	1.719	1.776	1.709
4	1.038	1.119	1.010	1.105
5a	1.026	1.053	1.107	1.078
5e	1.657	1.719	1.776	1.709
6a	1.507	1.575	1.073	1.172
6e	1.850	1.865	2.013	1.913
−NH	5.559	6.222	5.420	5.715
CO−CH$_3$	1.993	1.991	1.945	1.985
−(CH$_3$)$_3$	0.865	0.934	0.843	0.929

7.4.4 Aromatic Amides

The aromatic amides investigated are shown in Figure 7.11 and the observed ^{1}H shifts together with the calculated values, given in Tables 7.23–7.25.

Compounds **1–9** were commercial samples and the remainder synthesized using standard procedures. Full details are given by Perez.[43]

Molecular geometries. The major geometric problems were the planarity of the amide group and the dihedral angle of the amide and aromatic ring.

In **1**cis and **1** trans the amide group was planar but the angle of twist (τ_1 in Table 7.22) with the phenyl ring varied widely depending on the force field used (Table 7.22).

Table 7.22 *Dihedral angles for N-formyl aniline obtained with different force fields*

Compound		MMFF94	PM3MM	HF321	HF321+**	HF^d	B3LYP^d
1cis	$\tau_1{}^a$	0.0	0.0	33.4	31.9	44.5	31.1
	$\tau_2{}^b$	0.0	0.0	3.8	2.7	5.5	5.8
	$\tau_3{}^c$	180.0	180.0	−175.2	−176.2	−174.0	−173.5
1 trans	$\tau_1{}^a$	0.0	−0.1	−33.4	0.0	0.0	0.0
	$\tau_2{}^b$	−180	171.7	179.1	180.0	180.0	180
	$\tau_3{}^c$	0.0	−9.8	−1.9	0.0	0.0	0.0

^{a} C$_2$−C$_1$−N−C.
^{b} O−C−N−H.
^{c} H−N−C−H.
^{d} 631G++(d,p) level.

N-methyl formanilide is exclusively in the *exo* conformation shown in Figure 7.11 and again the amide group is twisted out of the ring plane. Conversely acetanilide (**3**) is exclusively *trans* and has a planar structure. Benzamide (**4**) is non-planar, but **5** is planar in the predominant *trans* form. In **6** and **7** the amide group is almost orthogonal with the ring plane. *N*-formyl indoline gives the separate ^{1}H spectra of the *endo* and *exo* forms at room

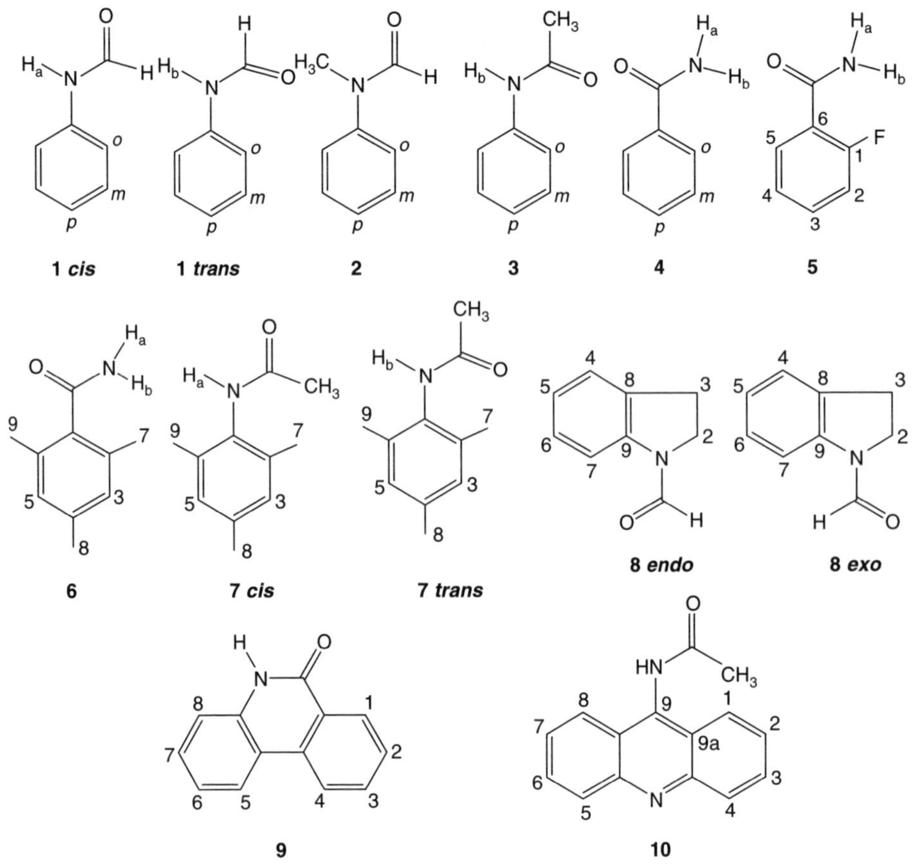

Figure 7.11   Aromatic amides investigated.

Table 7.23   Observed and calculated ¹H NMR chemical shifts for amides **1–4**

Compound		−CHO	−NHₐ	−NH_b	−NCH₃	−COCH₃	H_o	H_m	H_p
1 *cis*	Obs.	8.693	7.154	—	—	—	7.085	7.374	7.207
	Calc.	8.893	7.127	—	—	—	7.070	7.502	7.275
	GIAO	9.384	6.565	—	—	—	7.137	7.492	7.208
1 *trans*	Obs.	8.400	—	7.143	—	—	7.545	7.331	7.154
	Calc.	8.629	—	7.561	—	—	7.521	7.427	7.207
	GIAO	8.593	—	6.293	—	—	7.953	7.543	7.251
2	Obs.	8.483	—	—	3.322	—	7.175	7.414	7.285
	Calc.	8.675	—	—	3.393	—	7.072	7.594	7.459
	GIAO	8.738	—	—	3.371	—	7.314	7.644	7.410
3	Obs.	—	—	—	—	2.176	7.321	7.491	7.107
	Calc.	—	—	—	—	2.115	7.425	7.576	7.202
	GIAO	—	—	—	—	1.662	7.640	7.269	7.016
4	Obs.	—	6.075	6.075	—	—	7.821	7.448	7.532
	Calc.	—	6.041	6.041	—	—	7.798	7.567	7.597
	GIAO	—	4.809	5.375	—	—	8.126	7.643	7.710

Table 7.24   *Observed and calculated ^1H NMR chemical shifts for amides 5–7*

Compound		NH$_a$	NH$_b$	CO—CH$_3$	2	3	4	5	7	8	9
5	*Obs.*	6.088	6.715	—	7.151	7.513	7.288	8.137	—	—	—
	Calc.	6.167	6.567	—	7.348	7.617	7.348	8.309	—	—	—
	GIAO	5.263	6.753	—	7.253	7.679	7.503	8.780	—	—	—
6	*Obs.*	6.220	5.740	—	6.840	—	6.840		2.320	2.270	2.320
	Calc.	6.264	5.901	—	6.910	—	6.910		2.314	2.407	2.314
	GIAO	5.073	4.950	—	6.968	—	6.968		2.512	2.296	2.512
7 *cis*	*Obs.*	6.575	—	1.746	6.939	—	6.939		2.227	2.296	2.227
	Calc.	6.675	—	1.835	6.926	—	6.926		2.198	2.429	2.198
7 *trans*	*Obs.*	—	6.660	2.204	6.897	—	6.897		2.195	2.262	2.195
	Calc.	—	6.467	2.121	6.855	—	6.855		2.267	2.388	2.267

Table 7.25   *Observed and calculated ^1H NMR chemical shifts for amides 8–10*

Compound		—NH	—CHO	1	2	3	4	5	6	7	8	—CH$_3$
8*endo*	*Obs.*	—	8.530	—	4.124	3.201	7.240	7.073	7.209	8.079	—	—
	Calc.	—	8.503	—	3.987	3.093	7.187	7.191	7.301	7.915	—	—
	GIAO		8.840		3.975	3.173	7.369	7.286	7.440	8.650		
8*exo*	*Obs.*	—	8.945	—	4.071	3.163	7.255	7.058	7.212	7.177	—	—
	Calc.	—	8.711	—	3.925	3.033	7.253	7.258	7.284	7.129	—	—
	GIAO	—	9.254	—	4.051	3.121	7.360	7.205	7.331	7.256	—	—
9	*Obs.*	9.174	—	8.546	7.624	7.819	8.310	8.241	7.315	7.523	7.218	—
	Calc.	8.661	—	8.655	7.598	7.768	8.410	8.394	7.485	7.620	7.105	—
	GIAO	6.970	—	8.704	7.640	7.776	8.206	8.106	7.195	7.415	6.875	—
10	*Obs.*	—	—	7.857	7.625	7.812	8.311	7.857	7.625	7.812	8.311	2.291
	Calc.	—	—	7.831	7.677	7.836	8.466	7.831	7.677	7.836	8.466	2.172

temperature. Both forms are planar from *ab initio* calculations. Amide **9** is planar and in **10** the planar amide group is orthogonal to the anthracene ring. Full details of all these calculations are given by Perez.[43]

The complete data set of aliphatic and aromatic amide ^1H chemical shifts was used to obtain the best fit parameters for the amide calculations. There was a close relationship between the anisotropy and the steric terms in the iteration. For every different anisotropy term, the a_s coefficients were reparameterized. No change in the carbon steric term was necessary. The values used are those given previously.

The values of the anisotropy from the iterations were for aliphatic amides $\Delta\chi_1 = 13.5$ and $\Delta\chi_2 = -21.6$, and for aromatic amides $\Delta\chi_1 = 10.5$ and $\Delta\chi_2 = -7.3$. The major change is in the perpendicular component ($\Delta\chi_2$). The anisotropies obtained by Perez[43] are compared with other literature values for the carbonyl anisotropies in amides, ketones and aldehydes in Table 7.26. The values for aromatic amides are in good agreement with the values found for proteins. This is surprising as only aliphatic residues were considered in this study. Also note that the average of the values obtained in this study for aliphatic and aromatic amides is almost identical to the results of Zürcher and those previously determined for esters using both aliphatic and aromatic compounds (Section 7.2). Although the actual values differ, there is a general trend that the carbonyl anisotropy decreases with conjugation with an aromatic ring.

Another possible contribution to the anisotropy of the amide group is due to the nitrogen atom with the lone-pair. In the iterations performed in this study the parallel component of

Table 7.26 *Anisotropy components for different groups (see Table 7.16 for references)*

Author	Functional group	$\chi_1-\chi_2$	$\chi_3-\chi_2$	$\chi_1-\chi_3$
Pérez	Aliphatic amide	13.5	−21.6	35.1
Pérez	Aromatic amide	10.5	−7.3	17.8
Zürcher	Carbonyl	13.5	−12.2	25.7
ApSimon *et al.*	Carbonyl	21	−6	27
Schneider *et al.*	Carbonyl	24	−12	36
Williamson *et al.*	Amides	4	−9	13
Abraham	Aliphatic ketone	22.7	−14.8	37.5
Abraham and Mobli	Aromatic ketone	6.4	−11.9	18.3
Abraham and Mobli	Ester	10.1	−17.1	27.2

this anisotropy was very small and this component was removed. The nitrogen anisotropy was therefore cylindrically symmetrical along the lone pair, with a value for $\Delta\chi^N(\Delta\chi_2)$ of −9.8. This compares well with the value of $\Delta\chi_1 = 1.4$ and $\Delta\chi_2 = -12.4$ (these values have been converted to the nomenclature used here) found in proteins.[14]

Steric effect. The coefficient in Equation (3.11) (a_s) was obtained for the =O, N, N$\underline{H}$ and N$\underline{H}_2$ atoms in the CHARGE program. The values of a_s obtained are compared with those for other functional groups in Table 7.27.

Table 7.27 *The steric coefficient, a_s (ppm Å⁻⁶) for amides, ketones and esters (see Table 7.16 for references)*

Functional group	Al CO	Ar CO	N	NH₂	Al NH	Ar NH
Amides	62.4	60.0	37.4	−19.3	−31.2	−31.2
Ketones/aldehydes	67.8	38.4	—	—	—	—
Esters	85.0	85.0	—	—	—	—

The value of a_s for the amide oxygen is unchanged for aromatic and aliphatic amides and similar to the value for aliphatic ketones. The values for esters are larger, but the same trend as in amides is observed, that aliphatic and aromatic compounds have very similar values of a_s.

The steric effects from the NH protons were originally divided into NH to an aliphatic proton and NH to an aromatic proton but the value of a_s iterated to the same value for both cases (−31.23). For the NH₂ protons $a_s = -19.30$.

7.5   Steric and Electric Field Effects in Acyclic and Cyclic Ethers

7.5.1   Introduction

The effect of the electronegative oxygen atom on ¹H chemical shifts has been known since the beginning of NMR and a number of attempts at a definitive analysis of oxygen SCSs in ¹H NMR spectra were attempted. Zurcher[6] in his pioneering studies concluded that the C—O bond anisotropy was not important for the OH group SCSs. Schneider *et al.*[11] regarded the electric field term as the dominant term but Hall[94] suggested that the chemical shift difference between the anomeric protons of the C_2—O axial and C_2—O equatorial sugars could be accounted for by C—O anisotropy alone. Yang *et al.*[95] concluded that electric field, anisotropy and a constant term were necessary to reflect the observed ether SCSs in oxasteroids but did not consider any steric contributions.

One reason for the absence of any systematic investigation is simply the divalent nature of the oxygen atom which gives an additional degree of freedom when compared to mono-valent substituents. For example the SCSs of the hydroxyl group may well be dependent on the position of the hydroxyl proton. The value of the CH.OH coupling in alcohols shows that the OH proton is usually not in a single orientation but may have a preferred conform-ation.[96] However the precise populations of the different conformers cannot be estimated accurately. For these reasons Abraham *et al.*[97] in their analysis of oxygen SCSs restricted their investigation to ethers. Even in ethers the conformational mobility caused by the oxy-gen atom produced considerable problems. The only simple acyclic ethers that exist in one conformation are dimethyl ether and methyl-t-butyl ether. Methyl ethyl ether has two pop-ulated conformers, diethyl ether four and the number of conformations of more complex ethers (e.g. diglyme, 2-methoxy ethyl ether) is prohibitive for any quantitative calculation.

Thus apart from the two acyclic ethers mentioned above, only cyclic ethers of known conformation were selected. The ^{1}H chemical shifts in 1,3- and 1,4-dioxanes and their methyl derivatives had been obtained previously.[98–100] Abraham *et al.*[97] also assigned the ^{1}H spectrum of tetrahydropyran (**1**) at $-85\,°C$ where the ring inversion is slow on the NMR time scale and the spectra of 3-methyl (**2**), 2-hydroxymethyl (**3**), 2-methoxy (**4**), 4-hydroxy (**5**), 4-(4'-hydroxybenzyl) (**6**) THP, THP-4-carboxylic acid (**7**), 2-methyl-1,3-dioxolane (**8**) and two cyclic ethers of known conformation, 7-oxanorbornane (**9**) and 1,8-cineole (**10**) (see below).

9   10

7.5.2   Theory, Application to Ethers

These results plus previous literature data provided sufficient data for an analysis of oxygen SCSs. In their analysis the authors did not include any magnetic anisotropy contributions as these had not been required for any of the single bonded substituents considered previously, thus the only long-range interactions are the steric and electric field mechanisms. To apply the CHARGE model to ethers the only addition necessary was the calculation of the electric field of the ether oxygen. The electric field for a univalent atom (e.g. fluorine) is calculated as due to the charge on the fluorine atom and an equal and opposite charge on the attached carbon atom. The vector sum gives the total electric field at the proton concerned and the component of the electric field along the C—H bond considered is Ez in Equation (3.12).

Table 7.28 ^{1}H NMR chemical shifts (δ) in THP (**1**), 3-methyl-THP (**2**), 2-(hydroxymethyl)-THP (**3**), 2-methoxy-THP (**4**), 4-hydroxy-THP (**5**), 4-(4'-hydroxybenzyl)-THP (**6**), THP-4 carboxylic acid (**7**) and 2-methyl-1,3-dioxolane (**8**)

	1[a]	5[a]		2[b]	3[b]	4[b]	5[b]	6[b,g]	7[b]	8[b]
		eq-OH	ax-OH							
2ax	3.439	3.407	3.836	2.971	3.437	—	3.444	3.34	3.457	4.987
2eq	3.997	4.029	3.75[c]	3.801	—	4.517	3.962	3.954	3.99	—
3ax	1.682	1.579	d	1.686	1.323	1.6	1.566	1.315	1.803	—
3eq	1.566	1.915	d	—	1.53[e]	1.732	1.904	1.551	1.885	—
4ax	1.493	3.751	—	1.114	1.51[e]	1.816	3.86	1.692	2.593	3.857 (syn)
4eq	1.86	—	4.229	1.81	1.856	1.55	—	—	—	3.993 (anti)
5ax				1.618	1.56[e]	1.6				3.857 (syn)
5eq				1.563	1.56[e]	1.55				3.993 (anti)
6ax				3.316	3.463	3.856				
6eq				3.86	4.012	3.522				
CH3				0.805		3.405				1.384
CH2					3.569			2.47		
					3.507[f]					
Ph								6.75		
								7.006		

[a] In CDCl$_3$:CFCl$_3$ at −85 °C.
[b] In CDCl$_3$ at RT.
[c] Obscured by H4ax of eq-OH.
[d] Obscured by respective eq-OH protons.
[e] cf. 13C—1H correlations.
[f] Non-degenerate CH$_2$ group.
[g] Compound **6**, benzyl group δ (CH$_2$) 2.470, Ph 2,6H 7.006, 3,5H 6.750.

For the ether oxygen two possible models were attempted. In the first the electric fields due to the charge on the oxygen atom and of both the attached carbon atoms were calculated at the proton in question. The carbon atoms each have a positive charge equal to half the oxygen charge. An alternative procedure was to calculate the electric field due to the ether oxygen atom and a dummy atom placed midway between the attached carbon atoms with an equal and opposite charge. As the coefficient in Equation (3.12) is known the electric field effect is given immediately without any further parameterization. In practice the latter model gave better agreement with the observed data and this model is given here.

The steric effect of the oxygen atom is not known and therefore a value of the coefficient a_s in Equation (3.11) for oxygen must be determined. This and the associated push–pull coefficient are the only additional parameters required in order apply the theory to the more distant protons in ethers. The vicinal (H.C.C.X) effects are treated separately in CHARGE and for ethers an explicit H.C.C.O term was required.

The geometries of the compounds investigated were obtained by optimizations using *ab initio* calculations at the RHF/6-31G* level.

7.5.3   Oxygen SCSs in Ethers

Figure 7.12 shows the oxygen ring substitution effects of compound (**1**) relative to cyclohexane, (**2**) relative to methyl-cyclohexane (Chapter 4, Section 4.2.4) and also the SCS of the 2-methoxy group in (**4**).

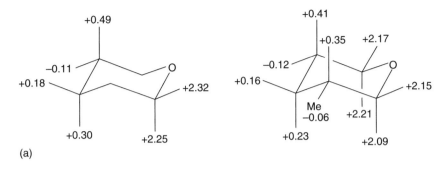

(a)

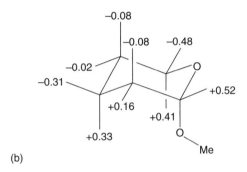

(b)

Figure 7.12   *The oxygen ring SCSs for compounds (1), (2) and (4).*

In THP (Figure 7.12a) the β effect of the oxygen atom on H-2ax and H-2eq is +2.1 to +2.3 ppm and non-orientational. The γ effect is strongly deshielding on H-3ax (+0.4 to +0.5 ppm) but shielding on H-3eq (ca. −0.1 ppm). This orientational effect is also seen in 1,3-dioxanes where the ring substitution shifts of 4-methyl-1,3-dioxane (Table 7.29) vs. methylcyclohexane (Table 4.3) on H-5ax is +0.83 ppm and for H-5eq −0.28 ppm. Here the effect of two oxygen atoms would appear roughly additive. Finally, the long-range oxygen effect on H-4 is deshielding at +0.18 to +0.3 ppm, with the greatest effect on the axial proton. The oxygen ring substitution effect of 7-oxa-norbornane versus norbornane (Table 4.2) may be compared with the results for THP (1). For the methine proton the oxygen shift of +2.37 ppm is in accord with the β THP protons. The oxygen SCS of +0.22 ppm (*Hexo*) and −0.01 ppm (*Hendo*) again demonstrate the orientational dependence, though less pronounced than for THP.

Even more remarkable is the SCS of the 2-methoxy group in 2-methoxy-THP (Figure 7.12(b)). The 2-methoxy group is predominantly axial (ΔG⁰ (eq–ax) is ca. −0.8 kcal mol⁻¹)[101] as shown in Figure 7.12. The values of the methoxy SCSs are confirmed by the similar data for the *trans*-2-methoxy-4-methyl THP (Table 7.29) in which due to both the substituent groups being in their favourable conformation the molecule is entirely in the conformation with eq-methyl and axial-methoxy. The corresponding values for the 2-methoxy SCSs at the 6ax and 6eq protons are +0.38 and −0.34 ppm.[101]

These data together with the other data in Table 7.29 provided a rigorous test of the application of the theoretical model to these compounds. The six-membered ring compounds in Table 7.29 all exist predominantly in one chair conformation but the five membered rings of THF and 1,3 dioxolane exhibit pseudorotation. For THF both the O-envelope (C_2) and half-chair (Cs) conformations were considered (at the 6-31G* level of theory the Cs conformer was favoured by 0.39 kcal mol^{-1}). The *ab initio* calculations for dioxolane iterated preferentially to the Cs conformation which was used in the calculations. However the calculations for the 2-methyl-1,3-dioxolane gave an envelope conformation with C_2 out of the plane of the other ring atoms. The calculated shifts for the 2-methoxy-THP derivatives are given for the axial conformation of the methoxy group. In this conformation there are only two populated conformations of the methoxy methyl group as the conformer with the methyl over the THP ring (i.e. with the H.C$_2$.O.Me dihedral angle ca. 180°) is very unfavoured. The calculations for the two *gauche* conformers (H.C$_2$.O.Me dihedral) are given in Table 7.29, although the rotamer with ∠O.C.O.Me of −64° is preferred over ∠O.C.O.Me of 145° at the RHF/6-31G*level by 4.0 kcal mol^{-1}.

The electric field effect is given directly by Equation (3.12), thus only the steric coefficient a_s and the push–pull coefficient need to be determined. The data in Figure 7.12 suggest that the push–pull coefficient for the methoxy group should be ca. −1.0 and this was confirmed in the calculations together with an appropriate value of a_s for oxygen as 100.0.

7.5.4   Observed versus Calculated Shifts

The observed vs. calculated shifts for a range of ethers is given in Table 7.29. The general agreement of the observed vs. calculated shifts is good and the majority of the observed shifts are reproduced to better than 0.1 ppm. The agreement is particularly striking for the chair conformations of THP and 1,3-dioxane with an overall rms error (observed vs. calculated shifts) of < 0.05 ppm. In both cases the calculations reproduce all the oxygen SCSs very well and in particular the deshielding of the axial proton with respect to the corresponding equatorial proton in H-3a vs. H-3e (THP) and H-5a vs. H-5e (3-methyl THP and 1,3-dioxanes) is strikingly reproduced. The only significant discrepancy for these compounds is H-4ax for which the calculated shift is ca. 0.2 ppm too low.

The agreement between the observed and calculated shifts is not as good in the case of the boat structures of 7-oxanorbornane and 1,8-cineole. This may well be due to the fact that the proton shifts in the parent hydrocarbons of norbornane and bicyclo-2,2,2-octane are not as well reproduced as those of cyclohexanes in this theoretical model (Table 4.2) and these discrepancies may well carry over to the oxygen analogues. However even in these cases the model does provide a basis for estimating the ¹H shifts in these compounds and indeed the correct order of the ¹H shifts is given in every case.

As a further check on the accuracy of the calculations Table 7.30 gives the calculated δ values for 4-oxa-5-α-androstane and also the comparable values for the parent 5-α-androstane from which the calculated SCSs can be obtained to compare with the observed SCSs.[95] The agreement is good and again the vicinal equatorial proton in the tetrahydropyran ring (2α in Table 7.30) shows a negative SCS. The only disagreement between the observed and calculated SCSs occurs for the H-1α and the C-6 protons. The H-1α is exactly analogous to H-4a of THP and the C-6 protons in 5-α-androstane are part of a complex unresolved multiplet with the C-4 protons and thus the observed SCSs may not be very accurate.[97]

Table 7.29 *Observed versus calculated ¹H chemical shifts (δ) of acyclic and cyclic ethers*

		Obs.[a]		Calc.	
Dimethyl ether	Me	3.24^{102}		3.31	
Methyl-t-butyl ether	Me	3.22^{18}		3.43	
	t-Bu	1.19		1.27	
Tetrahydrofuran	a-CH_2	3.83^{18}		3.80(C2), 3.84(Cs)	
	b-CH_2	1.85		1.58(C2), 1.54(Cs)	
Tetrahydropyran	2e	4.00		3.99	
	2a	3.44		3.44	
	3e	1.57		1.54	
	3a	1.68		1.65	
	4e	1.86		1.8	
	4a	1.49		1.25	
3-Methyl-THP	2e	3.8		3.92	
	2a	2.97		3.1	
	3e (Me)	0.81		0.86	
	3a (CH)	1.69		1.85	
	4e	1.81		1.77	
	4a	1.11		0.87	
	5e	1.56		1.56	
	5a	1.62		1.64	
	6e	3.86		3.99	
	6a	3.32		3.47	
2-Methoxy-THP	2e (CH)	4.52		4.79^{b}	4.68^{c}
	2a (OMe)	3.41		3.37	3.32
	3e	1.73		1.77	1.72
	3a	1.6		1.53	1.41
	4e	1.55		1.53	1.53
	4a	1.82		1.74	1.75
	5e	1.55		1.57	1.57
	5a	1.6		1.64	1.62
	6e	3.52		3.64	3.61
	6a	3.86		4.19	4.21
trans-2-Methoxy-4-methyl-THP	2e (CH)	4.64^{99}		4.76^{b}	4.70^{c}
	2a (OMe)	3.3		3.38	3.33
	3e	1.67		1.73	1.69
	3a	1.24		1.17	1.05
	4e (Me)	0.87		0.87	0.87
	4a (CH)	1.88		1.95	1.97
	5e	1.52		1.54	1.53
	5a	1.23		1.28	1.26
	6e	3.53		3.66	3.63
	6a	3.7		4.18	4.2
1,4-Dioxane	CH_2	3.70^{18}		3.85	
1,3,5-Trioxane	CH_2	5.00^{102}	5.15^{18}	4.93	
2,4,6-Trimethyl-1,3,5-trioxane	CH	5.05^{103}		4.59	
	Me	1.4		1.37	
1,3-Dioxane	2-H	4.84^{18}		4.85	
	4-H	3.9		3.73	
	5-H	1.78		1.75	

Table 7.29   (Continued)

		Obs.[a]		Calc.
2-Methyl-1,3-dioxane	2e (Me)	1.15[95]		1.29
	2a (CH)	4.5		4.59
	4e	3.95		4.17
	4a	3.61		3.52
	5e	1.23		1.44
	5a	1.95		2.12
4-Methyl-1,3-dioxane	2e	5.05	4.89[96]	5.19
	2a	4.71	4.55	4.53
	4e (Me)	1.24	1.17	1.27
	4a (CH)	3.73	3.61	3.59
	5e	1.48	1.38	1.42
	5a	1.76	1.69	1.67
	6e	4.09	4.02	4.15
	6a	3.71	3.58	3.53
2-Methyl-1,3-dioxolane	2(Me)	1.38		1.3
	2(CH)	4.99		4.75
	4,5 (anti)	3.99		3.99
	4,5 (syn)	3.86		3.79
7-Oxa-norbornane	H1 (CH)	4.56		4.17
	Hex	1.69		1.86
	Hen	1.44		1.3
1,8-Cineole	2ex	1.66		2
	2en	1.5		1.28
	3ex	2.02		1.9
	3en	1.5		1.32
	H4 (CH)	1.41		1.38
	9-Me	1.05		1.18
	10/11-Me	1.24		1.19
1,3-Dioxolane	a-CH$_2$	4.90[18]		5.02
	b-CH$_2$	3.88		3.89

[a] Abraham *et al.*[97] unless otherwise specified.
[b] O.C.O.Me dihedral −64°.
[c] O.C.O.Me dihedral 145°.

The good agreement between the observed and calculated shifts over the diverse range of compounds studied was strong support for the original assumption to neglect any C—O bond anisotropy contributions and this is also in accord with Zurcher's earlier conclusions.[6]

The theoretical interpretation of the SCSs is of some interest. The oxygen γ effect is deshielding for a *gauche* O.C.C.H orientation and shielding for a *trans* orientation, in contrast to the carbon γ effect which is shielding for a *gauche* C.C.C.H orientation and deshielding for the *trans* orientation. Thus the large differences in the oxygen γ effects shown in H-3ax vs. H-3eq in THP etc. are due to the replacement of the carbon at C-1 by the oxygen atom. The consequent replacement of the carbon γ effect by the oxygen γ effect gives rise to the observed oxygen SCS.

The oxygen SCS at the more distant H-4 proton (and H-4 and H-6 in 1,3-dioxane) are also due to the replacement of the CH$_2$ group interactions by the oxygen atom. In this case

Table 7.30 Calculated ¹H chemical shifts (δ) for 5α-androstane and 4-oxa-5α-androstane and calculated versus observed SCSs[a]

	Androstane[b] Calc.	4-Oxa- Calc.	SCS Calc.	SCS Obs.[c],[95]
1α	1.00	1.09	0.09	0.23
1β	1.53	1.69	0.16	0.08
2α	1.57	1.45	−0.12	−0.10
2β	1.44	1.90	0.46	0.44
3α	1.19	3.42	2.23	2.21
3β	1.74	4.08	2.34	2.32
5 (CH)	1.15	3.27	2.12	1.88
6α	1.35	1.95	0.60	0.37
6β	1.40	1.92	0.52	0.21
7α	0.97	0.95	−0.02	0.06
7β	1.94	1.96	0.02	0.05
11α	1.51	1.54	0.03	−0.03
11β	1.34	1.47	0.13	0.05
12α	1.05	1.06	0.01	−0.01
12β	1.60	1.63	0.03	0.00
14 (CH)	0.83	0.83	0.00	−0.03
19-Me	0.80	0.94	0.14	0.15

[a] Protons with SCSs < 0.01 ppm excluded.
[b] See Chapter 3, Table 3.6.
[c] Observed SCSs cf. 4-oxa-androstan-17-one vs. androstan-17-one.

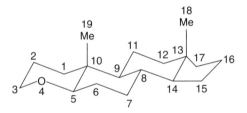

Figure 7.13 Nomenclature used for 4-oxa-5α-androstane.

the interactions concerned are the steric and electric field terms and these have much less orientation dependence in these systems. Both terms are quite small at these protons, e.g. in THP the steric interactions with the oxygen atom are 0.03 and 0.08 ppm respectively for H-4e and H-4a and the corresponding electric field contributions 0.13 and 0.12 ppm. The alternative model in which positive charges were placed on the carbon atoms attached to the oxygen atom gave generally smaller effects and poorer agreement with the observed shifts.

In contrast the steric term and the associated push–pull contribution dominate the SCS of the axial 2-methoxy group. For example, for H-4a the calculated contributions are 0.32 ppm (steric) and 0.17 ppm (electric field) and these well reproduce the observed shifts. There are similar calculated values for H-6a but here the calculated shift is somewhat larger than the observed shift.

7.6   Sulfoxides and Sulfones

7.6.1   Introduction

Sulfoxides and sulfones are common in pharmaceuticals, are versatile intermediates in organic synthesis and in the preparation of biologically and medically important products.[104,105] For example, the β-lactamase inhibitor sulbactam (penicillanic acid sulfone) is a penicillinase inhibitor used to improve the efficacy of penicillin.[106] However there are few investigations in the literature of their effects on the chemical shifts of neighbouring protons. They were included in early studies of chemical shift substituent parameters.[107] Beauchamp *et al.*[108] produced an algorithm for aliphatic sulfoxides and sulfones, using the base values for CH_3, CH_2 and CH protons as 0.9, 1.2 and 1.5 ppm respectively. The substituent constants for α (HCS), β and γ groups on these base values are 1.6, 0.5 and 0.3 for sulfoxides and 1.8, 0.5 and 0.3 for sulfones. This dealt only with short range (≤ 3 bonds) effects. A similar study was conducted for the sulfinyl chloride and sulfinate ester functional groups,[109] again including only short-range effects.

Byrne[110] and Abraham *et al.*[111] presented a ¹H NMR investigation of a series of aliphatic and aromatic sulfoxides and sulfones, with the aim of including the sulfoxide and sulfone functional groups into the CHARGE program.

The aliphatic compounds investigated (Figure 7.14) were: dialkylsulfoxides, R_2SO, R = Me (**1**), Et (**2**), Pr (**3**), Bu(**4**), t-Bu (**5**), tetramethylenesulfoxide (**6**), pentamethylenesulfoxide (axial) (**7a**) and (equatorial) (**7e**), 4-t-butyl-pentamethylenesulfoxide (axial) (**8**) and (equatorial) (**9**); dialkylsulfones, R_2SO_2, R = Me (**10**), Et (**11**), Pr (**12**), Bu (**13**), tetramethylenesulfone (**14**), pentamethylenesulfone (**15**), 4-t-butyl-pentamethylenesulfone(**16**) and 2-adamantyl methyl sulfone (**17**).

The aromatic compounds (Figure 7.15) were methylphenylsulfoxide (**19**), p-tolyl methyl sulfoxide (**20**), dibenzothiopheneoxide (**21**), phenylmethylsulfone (**22**), diphenylsulfone (**23**), dibenzothiophenedioxide (**24**), *E*-9-phenanthrylmethylsulfoxide (**25**), *Z*-1-methylsulfinyl-2-methylnaphthalene (**26**) and *E*-1-methylsulfinyl-2-methylnaphthalene (**27**). It was necessary to include dibenzothiophene (**18**) in the study to accurately model the aromatic ring currents for compounds **21** and **24**. The observed chemical shifts from the studied molecules served as the dataset for parameterization of CHARGE which was performed by iterative calculation using non-linear least mean squares.

7.6.2   Theory, Application to Sulfoxides and Sulfones

The CHARGE program was adapted to include the effect of the SO and SO_2 groups as follows. The short range effects are treated by CHARGE as due to electronic effects, and so are modelled on an empirical basis. The long-range effects include magnetic anisotropy, steric and electric fields. The magnetic anisotropy of the SO bond was modelled using the full McConnell equation[7] (Equation (3.15)) and the relevant geometrical parameters for the SO bond are shown in Figure 7.16. This model was used for sulfoxides and sulfones.

7.6.3   Molecular Geometries

The molecular geometries were obtained using the MMFF94 force field. The conformational profiles of five-membered rings are often complex,[112] thus the geometries of **6** and **14**

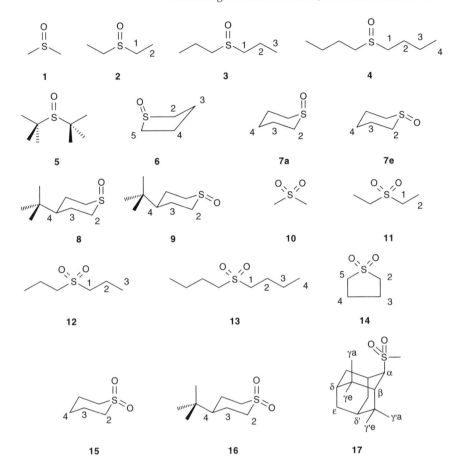

Figure 7.14   *Aliphatic compounds investigated.*

were obtained from *ab initio* calculations using RHF/3-21G and B3LYP/6-311G+(3df,2p) (shown by Denis[113] to give good agreement with experimental geometries for sulfoxides and sulfones).

For tetramethylene sulfoxide (**6**) the starting geometry was the envelope conformation (Figure 7.17), which always minimized to the corner/*endo* form (Figure 7.17(b)).

The DFT(B3LYP(6-311G+(3df,2p)) *ab initio* optimization of tetramethylenesulfone (**14**) minimized to the C₂ conformation (Figure 7.18), which was the geometry derived from experimental coupling constants.[114]

7.6.4   Observed versus Calculated Shifts

The ¹H shifts for compounds **2**,[115] **5**,[115] **8**,[116] **9**,[116] **16**,[116] **17**,[117] **18**,[118] **19**,[119] **21**,[118] **23**,[120] **24**,[118] **25**,[121] **26**,[122] and **27**[122] were from the literature. All other compounds were commercial samples. The ¹H spectra were obtained at 400 MHz. The spectra of **7** and **15** were obtained at 200 K to resolve the separate conformations of the six-membered ring. The axial

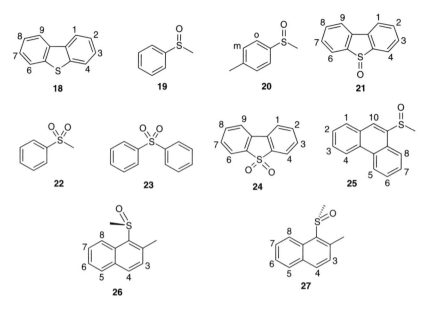

Figure 7.15 *Aromatic compounds investigated.*

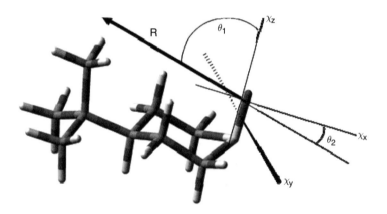

Figure 7.16 *The anisotropy of the SO bond (See color Plate 3).*

(**7a**) and equatorial (**7e**) conformers were assigned and for (**15**) the frozen chair form. The assignments of all the spectra are given in Tables 7.31–7.33. Full details are given elsewhere.[110]

The above data combined with the ^{1}H chemical shifts of the parent compounds allow the SCS for the SO/SO$_2$ groups to be obtained in these compounds. Figure 7.19 shows the SO/SO$_2$ SCS for **7a, 7e** and **15**, obtained by comparison with cyclohexane.[17] Both the short- and long-range effects of the SO and SO$_2$ functional groups are deshielding, with the γ effect of the oxygen atom (i.e. H.C.S=O) exhibiting an orientational dependence.

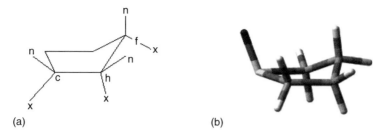

(a)   (b)

Figure 7.17   *(a) Geometry of tetramethylene sulfoxide (**6**): c, corner; h, hinge; f, flap; n, endo; x, exo. (b) DFT, B3LYP/6-311+G(3df,2p) optimized geometry (See color Plate 4).*

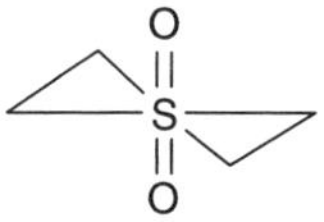

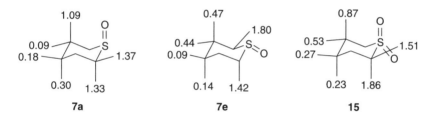

Figure 7.18   *B3LYP/6-311G+(3df,2p) geometry of tetramethylenesulfone (**14**).*

	1.09				0.47				0.87		

0.09 ─   S═O   0.44 ─   S═O   0.53 ─   O
0.18 ─   ─ 1.37   0.09 ─   ─ 1.80   0.27 ─   S ─ 1.51

        0.30   1.33   0.14   1.42   0.23   1.86

         7a   **7e**   **15**

Figure 7.19   *Observed SCSs (ppm) for cyclohexane-type compounds.*

The short-range effects for the cyclic and acyclic compounds were dealt with together. The conformational profile (*gauche/trans*) of the straight-chain molecules was not known but the short-range effects were well reproduced using the dihedral angle dependent functions defined from the cyclic molecules.

For the cyclic compounds short-range effects were included which differed for the two functional groups. The two-bond β effect for the sulfoxide functional group (which is a three-bond effect to the γ oxygen and carbon atoms) consists of two overlapping functions of the form $A + B \cos \theta + C \cos^2 \theta$, where A, B and C are empirically derived constants and θ is the H.C.S.O dihedral angle for one of the functions and the H.C.S.C dihedral angle for the other. Two functions were required to match the observed data for the 2,6 protons of the six-membered ring structures **7a**, **7e**, **8** and **9**.

For the sulfone functional group, a single function of the same form was used, taking θ as the H.C.S.C dihedral angle which, due to the symmetry of the SO_2 group, is effectively a combined function for the two H.C.SO dihedral angles.

The three-bond γ effects (H.C.C.S) for both sulfoxides and sulfones were again of the form $A + B \cos \theta + C \cos^2 \theta$, with the parameters optimized separately for the two functional

Table 7.31   *Observed versus calculated ¹H chemical shifts (ppm) of the aliphatic compounds*

Compound	Proton	Obs.	Calc.	Error
DMSO (**1**)	CH$_3$	2.620	2.620	0.000
Diethylsulfoxide (**2**)	1	2.770	2.693	0.077
	2	1.450	1.408	0.042
Dipropyl sulfoxide (**3**)	1	2.620	2.645	−0.025
	2	1.850	1.860	−0.010
	3	1.080	0.894	0.186
Dibutylsulfoxide (**4**)	1	2.658	2.665	−0.007
	2	1.760	1.801	−0.041
	3	1.430	1.279	0.151
	4	0.971	0.891	0.080
t-Butylsulfoxide (**5**)	CH$_3$	1.370	1.351	0.019
Tetramethylenesulfoxide (**6**)	3 *cis*	2.466	2.397	0.069
	3 *trans*	2.036	2.111	−0.075
	4 *cis*	2.466	2.397	0.069
	4 *trans*	2.036	2.111	−0.075
	5 *trans*	2.874	2.862	0.012
	5 *cis*	2.890	2.827	0.063
	2 *trans*	2.874	2.862	0.012
	2 *cis*	2.890	2.827	0.063
Pentamethylenesulfoxide (ax) (**7a**)	2a/6a	2.519	2.454	0.065
	2e/6e	3.047	3.268	−0.221
	3a/5a	2.281	2.323	−0.042
	3e/5e	1.770	1.834	−0.064
	4a	1.488	1.157	0.331
	4e	1.862	1.844	0.018
Pentamethylenesulfoxide (eq) (**7e**)	2a/6a	2.611	2.609	0.002
	2e/6e	3.483	3.278	0.205
	3a/5a	1.660	1.697	−0.037
	3e/5e	2.119	1.952	0.167
	4a	1.325	1.276	0.049
	4e	1.770	1.795	−0.025
4-t-Butylpentamethylenesulfoxide (ax) (**8**)	2a/6a	2.420	2.539	−0.119
	2e/6e	3.080	3.223	−0.143
	3a/5a	2.120	2.051	0.069
	3e/5e	1.790	1.894	−0.104
	4a	1.180	0.862	0.318
	t-Bu	0.900	0.883	0.017
4-t-Butylpentamethylenesulfoxide (eq) (**9**)	2a/6a	2.590	2.644	−0.054
	2e/6e	3.410	3.290	0.120
	3a/5a	1.450	1.408	0.042
	3e/5e	2.090	2.004	0.086
	4a	1.200	0.977	0.223
	t-Bu	0.880	0.863	0.017
Dimethylsulfone (**10**)	CH$_3$–SO$_2$	3.140	3.047	0.093
Diethylsulfone (**11**)	1	3.000	2.990	0.010
	2	1.400	1.373	0.027
Dipropylsulfone (**12**)	1	2.880	2.944	−0.064
	2	1.530	1.788	−0.258
	3	1.060	0.897	0.163
Dibutylsulfone (**13**)	1	2.970	2.964	0.006
	2	1.850	1.735	0.115

	3	1.530	1.280	0.250
	4	1.010	0.892	0.118
Tetramethylenesulfone (**14**)	2/5	3.030	3.071	−0.041
	3/4	2.220	2.191	0.029
Pentamethylenesulfone (**15**)	2a/6a	3.050	2.897	0.153
	2e/6e	3.192	3.110	0.082
	3a/5a	2.061	2.084	−0.023
	3e/5e	2.214	2.221	−0.007
	4a	1.420	1.325	0.095
	4e	1.949	1.919	0.030
4-t-Butyl pentamethylenesulfone (**16**)	2a/6a	2.930	2.932	−0.002
	2e/6e	3.080	3.127	−0.047
	3a/5a	1.890	1.806	0.084
	3e/5e	2.160	2.268	−0.108
	4a	1.220	1.026	0.194
	t-Bu	0.920	0.894	0.026
2-Adamantane methyl sulfone (**17**)	α	3.120	3.130	−0.010
	β	2.570	2.492	0.078
	γe	1.620	1.484	0.136
	γa	2.480	2.560	−0.080
	γ'e	2.000	1.707	0.293
	γ'a	1.760	1.658	0.102
	δ	1.800	2.060	−0.260
	δ'	1.800	2.059	−0.259
	ε	1.700	1.644	0.056
	CH₃	2.870	3.055	−0.185

Table 7.32 *Observed versus calculated chemical shifts (ppm) of the aromatic compounds*

Compound	Proton	Obs.	Calc.	Error
Dibenzothiophene (**18**)	1/9	8.150	8.138	0.012
	2/8	7.440	7.402	0.038
	3/7	7.440	7.487	−0.047
	4/6	7.850	7.866	−0.016
Methylphenylsulfoxide (**19**)	o	7.654	7.528	0.126
	m	7.525	7.459	0.066
	p	7.504	7.455	0.049
	CH₃	2.874	2.798	0.076
p-Tolylmethylsulfoxide (**20**)	o	7.580	7.474	0.106
	m	7.310	7.258	0.052
	CH₃—Ph	2.420	2.402	0.018
	CH₃—SO	2.710	2.803	−0.093
Dibenzothiopheneoxide (**21**)	1/9	7.790	7.915	−0.125
	2/8	7.580	7.478	0.102
	3/7	7.480	7.433	0.047
	4/6	7.970	7.959	0.011
Phenylmethylsulfone (**22**)	o	7.930	7.910	0.020
	m	7.580	7.527	0.053
	p	7.670	7.563	0.107
	CH₃	3.060	3.060	0.000

Table 7.32   *(Continued)*

Compound	Proton	Obs.	Calc.	Error
Diphenylsulfone (**23**)	o	7.950	8.000	−0.050
	m	7.520	7.525	−0.005
	p	7.530	7.549	−0.019
Dibenzothiophenedioxide (**24**)	1/9	7.770	7.988	−0.218
	2/8	7.610	7.573	0.037
	3/7	7.500	7.468	0.032
	4/6	7.800	7.788	0.012
E-9-Phenanthrylmethylsulfoxide (**25**)	1	7.900	8.021	−0.121
	2	7.600	7.616	−0.016
	3	7.700	7.676	0.024
	4	8.700	8.338	0.362
	5	8.770	8.388	0.382
	6	7.600	7.674	−0.074
	7	7.700	7.642	0.058
	8	8.050	7.917	0.133
	10	8.460	8.628	−0.168
	CH$_3$—SO	2.900	2.974	−0.074
Z-1-Methylsulfinyl-2-methylnaphthalene (**26**)	3	7.500	7.501	−0.001
	4	7.800	7.897	−0.097
	5	7.800	7.803	−0.003
	6	7.500	7.434	0.066
	7	7.500	7.524	−0.024
	8	9.300	9.029	0.271
	CH$_3$-2	2.660	2.578	0.082
	CH$_3$—SO	3.000	2.947	0.053
E-1-Methylsulfinyl-2-methylnaphthalene (**27**)	3	7.500	7.548	−0.048
	4	7.800	7.884	−0.084
	5	7.800	7.822	−0.022
	6	7.500	7.465	0.035
	7	7.500	7.536	−0.036
	8	8.250	7.789	0.461
	CH$_3$-2	3.000	2.593	0.407
	CH$_3$—SO	3.000	2.951	0.049

groups. This γ effect was parameterized as part of the same CHAP8 iteration as the long-range effects, because both long- and short-range effects are applied to protons three bonds from the SO$_n$ sulfur atom during the CHARGE calculation.

In order to make the program more generally applicable to sulfoxide and sulfone compounds, the long-range effects were parameterized together for the entire data set. These effects are the magnetic anisotropy of the SO bond, the electric field effect from the SO bond and steric effects from the sulfur and the oxygen atoms.

The steric effects from the sulfur atom of the two groups were set to those already in CHARGE for thioethers, as there were no protons in the data set which were significantly affected by this contribution. The steric effect from the oxygen atom of the groups was parameterized for all the compounds together to give a coefficient of 78.73 ppm Å^6, larger than the value of 60.56 ppm Å^6 for the oxygen atom of the carbonyl group and more similar to that of the ester group (see Table 7.27).

The electric field effect of the SO bond was calculated as discussed previously (see Chapter 3 and previous sections of this chapter). The magnetic anisotropy of the SO bond in sulfoxides and sulfones was considered separately and as one value. Both iterations gave the same rms error and therefore the SO bond in both series has the same anisotropy. The values of $\Delta\chi_1$ and $\Delta\chi_2$ (Figure 7.16) were optimized as -1.23 and 6.77 (10^{-30} cm^3 molecule^{-1}) respectively. These values are smaller than those determined previously (see Table 7.26) for the carbonyl anisotropy ($\Delta\chi_1 = 17.1$ and $\Delta\chi_2 = 3.2$) and these smaller values for the anisotropy are evidence for the single bond character of the SO bond in these functional groups, with the bond configuration S^+-O^-.

The overall rms error for the 354 proton included in this study was 0.11 ppm. The compounds methylparatoluenesulfonate (**28**), paratoluenemethylsulfonamide (**29**) and ethylenesulfite (**30**), were not included in the parameterization process, but were used as a test for the applicability of the SO bond model to related compounds containing this group. The observed versus calculated chemical shifts (ppm) for these compounds are shown in Table 7.33 and the agreement is good, apart from the NH of paratoluene methylsulfonamide (**29**), which may be due to concentration effects.

Table 7.33   *Observed versus calculated ^{1}H chemical shifts (ppm) of related compounds*

Compound	Proton	Obs.	Calc.	Error
Methylparatoluenesulfonate (**28**)	o	7.790	7.827	−0.037
	m	7.360	7.361	−0.001
	CH$_3$—Ph	2.450	2.441	0.009
	CH$_3$—O	3.730	3.698	0.032
Paratoluenemethylsulfonamide (**29**)	o	7.750	7.871	−0.121
	m	7.310	7.348	−0.038
	CH$_3$—Ph	2.420	2.436	−0.016
	NH	4.990	4.410	0.580
	CH$_3$—N	2.610	2.572	0.038
Ethylenesulfite (**30**)	*cis*	4.680	4.757	−0.077
	trans	4.350	4.457	−0.107

7.7   Summary

In this chapter the ^{1}H chemical shifts of compounds containing the 'divalent' functional groups CO, CO.O, CO.NH, OR, SO and SO$_2$ are presented together with their SCSs and the methods used to predict these SCS using the CHARGE routine.

A variety of aliphatic and aromatic aldehydes and ketones are reported including cyclo-hexanones, decalone and androstanones in the aliphatic series and indanone, anthraquinone, fluorenone, etc. in the aromatic series. Their shifts were predicted by the CHARGE routine by including the carbonyl anisotropy and electric field plus the oxygen steric contribution. The carbonyl anisotropy in the aromatic ketones was much less than the value in aliphatic ketones.

The keto–enol tautomerism in anthrone/9-hydroxy anthracene was observed by NMR and the percentage of enol was shown to be proportional to the Kamlett β hydrogen bonding effect of the solvent and not to the solvent polarity.

In the ester series 24 esters containing 150 protons including aliphatic, cyclic and aromatic esters were examined. The derivation of a realistic electric field, implementation of the π system to the Huckel model, together with the parameterization of the existing ring current, anisotropy, steric and charge effects produced a model which accurately describes the ester SCSs. The experimental data were reproduced with an rms error of 0.1 ppm, compared with an rms of 0.2 ppm for the GIAO calculated values. In the latter calculations large discrepancies were found for protons close to the electronegative atom.

The CHARGE model was used to derive the conformer populations of methyl acrylate and methyl crotonate from the observed ¹H chemical shifts and the results compared well with other theoretical and experimental values.

The ¹H shifts of 22 aliphatic amides including two types of cyclic amides, e.g. piperidone, *N*-acetyl pyrrolidine and 13 aromatic amides (e.g. benzamide, acetanilide, phenanthridone, etc.) were given. The geometries were discussed and a conformational analysis was carried out on the aromatic amides. New models for the anisotropy and the electric field effects in amides were presented. Combining the data of the aliphatic and aromatic amides refined the values for the anisotropy, oxygen and nitrogen steric coefficients and electric field effects needed to predict the shifts in the CHARGE routine. The value found for the carbonyl anisotropy was similar to that for aromatic ketones and much less than that for aliphatic ketones.

GIAO calculations were also performed. The results were generally good, but the observed shifts for the NH and most of the CHO protons were not well reproduced.

The data set for ethers comprised 78 proton shifts in 17 compounds of known geometry, including tetrahydropyran at 188 K, where the ring inversion is slow on the NMR time scale. The observed data were well reproduced by the CHARGE routine with the ether oxygen SCSs given by the electric field and oxygen steric effect. No C—O anisotropy was required.

A number of alphatic and aromatic sulfoxide and sulfones are reported including acyclic and cyclic compounds. Pentamethylene sufoxide and sulfone were measured at 200 K to slow the ring inversion. Models for the SO and SO_2 anisotropy are presented. The predicted ¹H shifts were in good agreement with the observed, the major contributions to the SO SCS is the electric field and oxygen steric contribution. The SO anisotropy is very small, supporting the S^+—O^- bond character.

References

1. Abraham, R. J.; Fisher, J.; Loftus, P., *Introduction to NMR Spectroscopy.* John Wiley & Sons, Ltd.: Chcichester, 1988.
2. Narasimhan, P. T. and Rogers, M. T., *J. Phys.Chem.* 1959, **63**, 1388.
3. Jackman, L. M. *Applications of NMR. in Organic Chemistry.* Pergamon Press: London, 1959.
4. Pople, J. A. *J. Chem. Phys.* 1962, **37**, 53.

5. Bothner-By A. A. and Pople, J. A. *Ann. Rev. Phys. Chem.* 1965, **16**, 43.
6. Zürcher, R. F. *Prog. Nucl. Magn. Reson. Spectrosc.* 1967, **2**, 205.
7. McConnell, H. M. *J. Chem. Phys.* 1957, **27**, 226.
8. ApSimon, J. W., Demarco P. V. and D. W. Mathieson, *Tetrahedron* 1970, **26**, 119.
9. Homer J. and Callagham, D. *J. Chem. Soc.*, 1968, 439.
10. Toyne, K. J. *Tetrahedron* 1973, **29**, 3889.
11. Schneider, H. J., Buchheit, U., Becker, N., Schmidt G. and Siehl, U. *J. Am. Chem. Soc.* 1985, **107**, 7027.
12. Williamson, M. P. and Asakura, T. *J. Magn. Reson.* 1991, **94**, 557.
13. Williamson, M. P. and Asakura, T. *J. Magn. Reson.* 1993, **B101**, 63.
14. Williamson, M. P., Asakura, T., Nakamura E. and Demura, M. *J. Biomol. NMR* 1992, **2**, 83.
15. Abraham, R. J. and Ainger, N. J. *J. Chem. Soc. Perkin Trans. 2* 1999, 441.
16. Ainger, N. J. *Ph.D. Thesis.* University of Liverpool, 1999.
17. Abraham, R. J., Griffiths L. and Warne, M. A. *J. Chem. Soc. Perkin Trans. 2* 1997, 31.
18. Puchert, C. J. and Behnke, J. *Aldrich Library of 13C and 1H FT NMR Spectra.* Aldrich Chemical Company: Milwaukee, WI, 1993.
19. Griffiths, L. *Ph.D. Thesis.* University of Liverpool, 1979.
20. Abraham, R. J.; Barlow A. P. and Rowan, A. E. *Magn. Reson. Chem.* 1989, **27**, 1024.
21. Sanders J. K. M. and Hunter, B. K. *Modern NMR Spectroscopy – A Guide for Chemists.* Oxford Univsersity Press: Oxford, 1993.
22. Landholt-Bornstein, *Structure Data of Free Polyatomic Molecules.* Springer-Verlag: New York, NY, 1976.
23. Jackman, L. M. and Sternhell, S. *Applications of Nuclear Magnetic Resonance Spectroscopy in Organic Chemistry*, 2nd edn. Pergamon Press: New York, NY, 1969.
24. Wu, G.; Lumsden, M. D.; Ossenkamp, G. C.; Eichele K. and Wasylishen, R. E. *J. Phys. Chem.* 1995, **99**, 15806.
25. Abraham, R. J.; Mobli, M.; and Smith, R. J. *Magn. Reson. Chem.* 2003, **41**, 26.
26. Abraham, R. J. Unpublished Results.
27. Mobli, M. *Ph.D. Thesis.* University of Liverpool, 2004.
28. Abraham, R. J.; Chadwick D. J. and Sancassan, F. *J. Chem. Soc. Perkin Trans.2* 1989, 1377.
29. Abraham, R. J. and Lucas, M. S. *J. Chem. Soc. Perkin Trans. 2* 1988, 1269.
30. Abraham, R. J.; Bergen H. A. and Chadwick, D. J. *J. Chem. Soc. Perkin Trans.2* 1983, 1161.
31. Abraham, R. J.; Chadwick D. J. and Sancassan, F. *J. Chem. Soc. Perkin Trans. 2* 1988, 169.
32. Meyer, K. H. *Justus Liebigs Ann. Chem.* 1911, **379**, 70.
33. Novak, P.; Skare, D.; Sekusak S. and Vikic-Topic, D. *Croat. Chim. Acta* 2000, **73**, 1153.
34. Kamlet, M. J.; Abboud, J. M.; Abraham M. H. and Taft, R. W. *J. Org. Chem.* 1983, **48**, 2877.
35. Abraham, R. J.; Chambers E. J. and Thomas, W. A. *J. Chem. Soc. Perkin Trans. 2* 1993, 1061.
36. Abraham, R. J.; Canton, M.; Reid M. and Griffiths, L. *J. Chem. Soc. Perkin Trans. 2* 2000, 803.
37. Epsztajn, J.; Bieniek, A.; Brzezinski J. Z. and Kalinowski, H. *Tetrahedron* 1986, **42**, 3559.
38. Narayanan, C. R. and Iyer, K. N. *Tetrahedron* 1965, **42**, 3741.
39. Narayanan, C. R. and Parkar, M. S. *Indian J. Chem.* 1971, **9**, 1019.
40. ApSimon, J. W. and Beierbeck, H. *Can. J. Chem.* 1971, **49**, 1328.
41. Hunter, C. A. and Packer, M. J. *Chem. Eur. J.* 1999, **5**, 1891.
42. Packer, M. J.; Zonta C. and Hunter, C. A. *J. Magn. Reson.* 2003, **162**, 102.
43. Perez, M. *Ph.D. Thesis.* University of Liverpool, 2004.
44. Abraham, R. J.; Bardsley, B.; Mobli, M. and Smith, R. J. *Magn. Reson. Chem.* 2005, **43**, 3.
45. Abraham, R. J.; Canton, M.; Edgar, M.; Grant, G. H.; Haworth, I. S.; Hudson, B. D.; Mobli, M.; Perez, M.; Smith, P. E.; Reid M. and Warne, M. A. *CHARGE7.* University of Liverpool, 2004.
46. Foresman, J. B. and Frisch, A. *Exploring Chemistry with Electronic Structure Methods.* Gaussian Inc.: Pittsburgh, PA, 1996.

47. Abraham, R. J.; Mobli M. and Smith, R. J. *Magn. Reson. Chem.* 2004, **42**, 436.
48. Jones, G. I. L. and Owen, N. L. *J. Mol.Struct.* 1973, **18**, 1.
49. Blom, C. E. and Gunthard, H. *Chem. Phys. Lett.* 1981, **84**, 267.
50. Öki, M. and Nakanishi, H. *Bull. Chem. Soc. Jpn* 1970, **43**, 2558.
51. Jung, M. E. and Gervay, J. *Tetrahedron Lett.* 1990, **31**, 4685.
52. Wiberg, K. B. and Laidig, K. E. *J. Am. Chem. Soc.* 1987, **109**, 5935.
53. Larson, J. R.; Epiotis, N. D. and Bernardi, F. *J. Am. Chem. Soc.* 1978, **100**, 5713.
54. Mark, H.; Baker, T. and Noe, E. *J. Am. Chem. Soc.* 1989, **111**, 6551.
55. Pawar, D. M.; Khalil, A. A.; Hooks, D. R.; Collins, K.; Elliott, T.; Stafford, J.; Smith, L. and Noe, E. A. *J. Am. Chem. Soc.* 1998, **120**, 2108.
56. Lampert, H.; Mikenda, W.; Karpfen, A. and Kählig, H. *J. Phys. Chem. A* 1997, **101**, 9610.
57. Abraham, R. J.; Ghersi, A.; Pertillo, G. and Sancassan, F. *J. Chem. Soc. Perkin Trans. 2* 1997, 1279.
58. Abraham, R. J.; Angiolini, S.; Edgar, M. and Sancassan, F. *J. Chem. Soc. Perkin Trans. 2* 1995, 1973.
59. Abraham, R. J.; Mobli, M.; Ratti, J.; Sancassan, F. and Smith, T. A. D. *J. Phys. Org. Chem.* 2006, **19**, 384.
60. Heller, E. and Schmidt, G. M. J. *Isr. J. Chem.* 1971, **9**, 449.
61. Abraham, R. J. and Reid, M. *J. Chem. Soc. Perkin Trans. 2* 2002, 1081.
62. Abraham, R. J. and Reid, M. *Magn. Reson. Chem.* 2000, **38**, 570.
63. McClellan, A. L. *Table of Experimental Dipole Moments*, Volume 3. Rahara Enterprises, El Cerrito, CA, 1989.
64. Chong, Y. S. and Huang, H. H. *J. Chem. Soc. Perkin Trans. 2* 1986, 1875.
65. Caswell, L. R.; Soo, L. Y.; Lee, D. H.; Fowler, R. G. and Campbell, J. B. *J. Org. Chem.* 1974, **39**, 15427.
66. Alonso, J. L.; Pastrana, M. R.; Pelaez, J. and Arauzo, A. *Spectrochim. Acta* 1983, **39**, 215.
67. Shi, M. *J. Chem. Res. (S)* 1998, 592.
68. Imagawa, T.; Haneda A. and Kawanisi, M. *Org. Magn. Reson.* 1980, **13**, 244.
69. Sancassan, F. Personal Communication.
70. Minami, I. and Tsuji, J. *Tetrahedron* 1987, **43**, 3903.
71. Choudhury, P. K.; Foubelo, F. and Yus, M. *Tetrahedron* 1999, **55**, 10779.
72. Moriarty, R. M.; Vaid, R. K.; Hokins, T. E.; Vaid, B. K. and Prakash, O. *Tetrahedron* 1990, **31**, 197.
73. Mori, M.; Washioka, Y.; Urayama, T.; Yoshiura, K.; Chiba, K. and Ban, Y. *J. Org. Chem.* 1983, **48**, 4058.
74. Bartle, K. D. and Smith, J. A. S. *Spectrochim. Acta* 1967, **23A**, 1715.
75. Harvey, R. G.; Cortez, C.; Ananthanarayan, T. P. and Schmolka, S. *J. Org. Chem.* 1988, **53**, 3936.
76. Egawa, T.; Maekawa, S.; Fujiwara, H.; Takeuchi H. and Konaka, S. *J. Mol. Struct.* 1995, **352/353**, 193.
77. Gutowsky, H. S. *J. Chem. Phys.* 1956, **25**, 1228.
78. Stewart, W. E.; Siddall III, T.H., *Chem. Rev.* 1970, **70**, 517.
79. Phillips, W. D. *J. Chem. Phys.* 1955, **23**, 1363.
80. LaPlanche, L. A. and Rogers, M. T. *J. Am. Chem. Soc.* 1964, **86**, 337.
81. Kitano, M.; Fukuyama, T. and Kuchitsu, K. *Bull. Chem. Soc. Jpn* 1973, **46**, 384.
82. Hansen, E. L.; Larsen, N. W. and Nicolaisen, M. *Chem. Phys. Lett.* 1980, **69**, 327.
83. Jeffrey, G. A.; Ruble, J. R.; McMullan, R. K.; DeFrees, D. J. and Pople, J. A. *Acta Crystallogr. Sect. B* 1980, **36**, 2292.
84. Lim, K. T. and F. M.M., *J. Chem. Phys.* 1987, **91**, 2716.
85. Jasien, P. G.; Stevens, W. J. and Krauss, M. *J. Mol. Struct.* 1986, **139**, 197.
86. Wong, M. W.; Wiberg, K.B., *J.Phys.Chem.* 1992, **96**, 668.

87. Ramek, M.; Yu, C. H.; Sakon, J. and Schäfer, L. *J. Phys. Chem. A* 2000, **104**, 9636.
88. Ösapay, K. and Case, D. A. *J. Am. Chem. Soc.* 1991, **113**, 9436.
89. Abraham, R. J.; Mobli, M. and Smith, R. J. *J. Chem. Soc. Perkin Trans. 2* 2002, **41**, 26.
90. Barfield, M.; Babaqi, A.S., *Magn. Reson. Chem.* 1987, **25**, 443.
91. Piaggio, P. University of Genoa, Personal Communication, 1996.
92. Sunners, B.; Piette, L.H., Sheneider, W.G., *Can. J. Chem.* 1960, **38**, 681.
93. Yoshimura, J.; Sugiyama, Y. and Nakamura, H. *Bull. Chem. Soc. Jpn* 1973, **46**, 2850.
94. Hall, L. D. *Tetrahedron Lett.* 1964, **23**, 1457.
95. Yang, Y.; Haino, T.; Usui, S. and Fukazawa, Y. *Tetrahedron* 1996, **52**, 2325.
96. Rader, C. P. *J. Am. Chem. Soc.* 1969, **91**, 3248.
97. Abraham, R. J.; Griffiths, L. and Warne, M. A. *J. Chem. Soc. Perkin Trans. 2* 1998, 1751.
98. Buys, H. R. and Eliel, E. L. *Tetrahedron Lett.* 1970, 2779.
99. Eliel, E. L. and Wilen, S. H. *Stereochemistry of Carbon Compounds.* J. Wiley & Sons, Inc.: New York, NY, 1994.
100. Pihlaja, K. and Ayras, A. *Acta Chem. Scand.* 1979, **24**, 531.
101. Booth, H.; Khedhair, K. A. and Readshaw, S. A. *Tetrahedron* 1987, **43**, 4699.
102. Tiers, G. V. D. *High Resolution NMR Spectroscopy.* Pergamon Press: Oxford, 1966.
103. Bhacha, N. F. *Varian NMR Spectra Catalogue.* Varian Associates: Palo Alto, CA, 1962.
104. Firouzabadi, H.; Iranpoor, N.; Jafari, A. A. and Riazymontazer, E. *Adv. Synth. Catal.* 2006, **348**, 434.
105. Patai, S. *The Chemistry of Sulfones and Sulfoxides.* John Wiley & Sons, Ltd: Chichester, 1988.
106. Sutherland, R. *Infection* 1995, **23**, 191.
107. Shoolery, J. N. *J. Chem. Phys.* 1953, **21**, 1899.
108. Beauchamp, P. S. and Marquez, R. *J. Chem. Educ.* 1997, **74**, 1483.
109. O'Donnell, J. S.; Faragher, R. J.; Motto, J. M. and Schwan, A. L. *J. Sulfur Chem.* 2004, **25**, 29.
110. Byrne, J. J. *Ph.D. Thesis.* University of Liverpool, 2007.
111. Abraham, R. J.; Byrne; J. J. and Griffiths, L. *Magn. Reson. Chem.* 2008, **46**, 667.
112. Pople, J. A. *J. Am. Chem. Soc.* 1975, **97**, 1358.
113. Denis, P. A. *J. Chem. Theory. Comput.* 2005, **1**, 900.
114. Barbarella, G.; Rossini, S.; Bongini, A. and Tugnoli, V. *Tetrahedron*, 1985, **41**, 4691.
115. Alnaimi, I. S. and Weber, W. P. *J .Organomet. Chem.* 1983, **241**, 171.
116. Renaud, P. *Helv. Chim. Acta* 1991, **74**, 1305.
117. Greidanus, J. W. *Can .J. Chem.* 1970, **48**, 3593.
118. Balkau, F. and Heffernan, M. L. *Aust .J .Chem.* 1971, **24**, 2305.
119. Abraham, R. J.; Pollock, L. and Sancassan, F. *J. Chem. Soc. Perkin Trans .2* 1994, 2329.
120. Abraham, R. J. and Haworth, I. S. *Magn. Reson. Chem.* 1988, **26**, 252.
121. Donnoli, M. I. *Org. Biol. Chem.* 2003, **1**, 3444.
122. Casarini, D. *J. Org. Chem.* 1993, **58**, 5674.

8

^{1}H Chemical Shifts and Structural Chemistry

8.1 Introduction

In previous chapters the various factors that influence the ^{1}H chemical shift have been described and it is clear that the molecular geometry can substantially perturb the calculated values. The interactions most sensitive to the molecular geometry are through space effects such as the electric field, steric and anisotropic shifts. The 3-bond γ effect is also sensitive to the intervening dihedral angle. In ^{1}H chemical shift calculations through space effects can be dominant as (i) hydrogen atoms are usually on the periphery of the molecule which can bring them close to the functional groups and (ii) their chemical shift range is very small hence even small effects can have a noticeable effect upon the chemical shift. Although this sensitivity to the molecular geometry is a nuisance for prediction purposes it does allow the chemist to detect small changes in the molecular configuration, and is one reason for the popularity of ^{1}H NMR in organic chemistry.

It is apparent that an accurate molecular geometry is essential for accurate ^{1}H shift calculations. Since most molecules contain single bonds with a low barrier to rotation and can therefore form many conformations it is difficult to arrive at the correct conformation without a method of generating optimized 3D structures. Thus there are two steps in NMR chemical shift calculations that will be considered in the beginning of this chapter, where we also consider a few examples to highlight the importance of these steps. Once an optimized structure has been obtained any low energy conformation will contribute to the experimental spectrum. It is therefore necessary to include conformational averaging which in NMR may be considered as a special case of chemical exchange.

Another important factor on ^{1}H chemical shifts is the effect of solvent. This may be either an indirect effect due to changes in the populations of the conformers contributing

Modelling ^{1}H NMR Spectra of Organic Compounds: Theory, Applications and NMR Prediction Software
Raymond Abraham and Mehdi Mobli © 2008 John Wiley & Sons, Ltd

to the averaged NMR spectrum <u>or</u> due to a direct solvent effect through interactions with the solute molecules. Often it is a combination of both factors. The indirect solvent effects will not be discussed here as these affect the conformational energies but do not directly affect the solute chemical shift. Such contributions are included in conformational energy calculations and can then be accounted for in the subsequent chemical shift calculations. The direct solvent effects, however, do contribute to the chemical shift, and in the final section of this chapter a semi-empirical model is described which can reproduce these effects using extensive experimental data.

8.2   Electronic Structure Calculations

The general aim of electronic structure calculations is to determine the energy and electron-density distribution in molecules, using approximations of the Schrödinger equation. The full Schrödinger equation (Equation (8.1)) for a molecule involves a Hamiltonian containing the kinetic energies of each of the N electrons and M nuclei as well as coulombic interactions between these particles. The solution of the Schrödinger equation gives the energy and other properties

$$\mathbf{H}\psi = \mathbf{E}\psi \qquad (8.1)$$

of a molecule, but in this form it has only been solved for a two particle system.[1,2]

The Schrödinger equation has to be simplified in order to solve chemical problems. One fundamental assumption made is the Born–Oppenheimer approximation. This states that since nuclei are much heavier than electrons they move relatively slowly and can therefore be regarded as stationary while the electrons move relative to them.[3] Thus only the electronic form of the Schrödinger equation needs to be solved, where the nuclear translational, vibrational and rotational kinetic energy factors are ignored.[3,4] Also all relativistic effects are usually ignored. These effects only become important when dealing with large nuclei where the substantial positive charge in the nucleus can cause electrons in the innermost orbitals to exhibit significant relativistic effects.[5]

There are a number of strategies proposed to solve the electronic Schrödinger equation and some common strategies and the associated approximations are as follows. The molecular orbitals are described by a linear combination of atomic orbitals, the (LCAO) method. The atomic orbitals are given by a set of basis functions (Basis Sets) see Section 8.2.1 on to determine the coefficients in the molecular orbitals. The iterative manner in which the molecular orbitals are obtained is known as the self consistent field (SCF) method, developed by Hartree and later improved by Fock and Slater.[4] The SCF method calculates the spin orbital for an electron taking into account the spin orbitals of all other electrons in the system, starting from an initial guess. Once the spin orbital for one electron is calculated the same procedure is repeated for all other electrons in the molecule. The resulting spin orbitals will be more accurate than the initial guess and the procedure is repeated with the new spin orbitals until a convergence criterion is satisfied.[1,4,6] The quality of the result will be dependent on the number of basis functions used and their flexibility to describe the molecular orbitals.[2] The Hartree–Fock (HF) SCF method however has a serious limitation in the way it calculates the probability of two electrons of opposite spins occupying the same orbital, such as the two electrons in the sigma bond of H_2. According to the HF method each electron has the same probability distribution in the orbital. In reality the electrons prefer to be separated, due to their charges, and this is known as electron correlation.[4,6] The way in which the electron correlation is handled differentiates the common *ab initio* methods. These are the configuration interaction, perturbation and density functional methods and are collectively known as post SCF or post HF methods.[4,6]

The HF SCF method builds the electronic structure of a molecule by adding two electrons of opposing spins to each orbital until all the electrons are accounted for. From quantum theory there is a probability of finding an electron in an unoccupied or virtual orbital. Therefore including such orbitals to account for excited states of electrons will improve the representation of the electronic structure. By mixing all the possible electronic states of a molecule, each of which has some probability of being occupied, this should give a complete representation of the molecule within the limitations of the SCF method. This is what is known as full configuration interaction (CI). However this is only possible for the smallest systems and therefore the methods in use add single excitations in the HF calculation (CIS), or double excitations (CID). The CI methods are sometimes grouped with multi configuration SCF (MCSCF) and complete active space SCF (ASSCF) and are then collectively known as variational methods. The methods are similar in that they use the variational principle that states that the expectation value or average value of the energy for an approximate wave function always lies above or equal to the exact solution of the Schrödinger equation for the same Hamiltonian operator. One therefore seeks to minimize the energy for the system through the SCF calculation.[2,4,6]

Perturbation theory states if the exact solution to a problem is known the answer to a closely related problem is a perturbation of the known solution. The Hamiltonian is then written:

$$\mathbf{H} = \mathbf{H}_0 + \lambda \mathbf{V} \tag{8.2}$$

where $\mathbf{H}_0$ has an exact solution and $\lambda \mathbf{V}$ is a small perturbation applied to $\mathbf{H}_0$. The choice of $\mathbf{H}_0$ was made in 1934 by C. Møller and M. S. Plesset, and is now the basis for the MP_n methods. $\mathbf{H}_0$ is the zero order Hamiltonian from the HF-SCF method, and is in effect the sum of the orbital energies. The first order energy correction (MP1) uses the two-electron integrals that correct the sum of orbital energies to give the normal HF energy. The next correction is the second order correction and this leads to MP2, which is the most commonly used perturbation theory in *ab initio* calculations. MP3 and MP4 calculations are sometimes also used but are computationally very expensive.[3,4,6]

The density functional theories (DFT) were originally established in the 1920s and further developed in the 1960s but have only gained popularity in the last 15 years. The method is completely different from any of the above mentioned in the way it constructs the wave functions. The basic idea behind DFT is that the energy of an electronic system can be written in terms of the electron probability density ρ. The energy is then a functional (a function of a function) of the electron density. The theory does however not tell us the form of the functional's dependence of energy on the density, it only confirms that such a functional exists. The choice of the functional is the only limitation of the DFT method. At the present time, there is no systematic way of choosing the functional and the most popular ones in the literature have been derived by careful comparison with experimental data.[3,4,6]

8.2.1   Basis Sets

Once a theory has been chosen for electronic structure calculations it is necessary to define a basis set. As mentioned above all *ab initio* methods assume that molecular orbitals can be expressed as linear combinations of atomic orbitals, LCAOs. The general procedure is to use combinations of atomic orbitals on the constituent atoms in the molecule. The coefficient of each AO in the MO is varied to find the combination giving the lowest energy. Each atomic orbital is approximated using several mathematical functions called primitives which are selected prior to calculation. Gaussian functions ($\exp(-r^2)$) are now most often used and if a number of these are used together, they can approximate Slater ($\exp(-r)$) orbitals.[2] Slater orbitals are better approximations to the electron wavefunctions but are computationally more cumbersome than the combination of Gaussians. The selection of how many of these primitive Gaussian functions are to be used for each atomic orbital depends on the element and the polarization or not of these orbitals (by including higher energy unoccupied orbitals) and whether the valence orbitals are to be composed of two or more distinct concentric groups of functions (split valence). This is referred to as the basis set. In general, the more sophisticated the basis set chosen, the more accurate, computationally expensive and time-consuming the calculation will be.[2] The basis sets used in this work are 3-21G, 6-31G+ and 6-311G++(2d,p). 3-21G demands that the core (non valence) orbitals are made up of three primitive Gaussian functions, with the three primitives describing the valence orbitals split into groups of two and one. This is a relatively simple basis set and leads therefore to fast but less accurate calculations. 6-31G+ is a more sophisticated basis set, having six primitives to describe the core orbitals, four primitives, grouped as three and one, describing the valence orbitals, and a diffuse function (large s and p type functions) added to the heavy atoms. 6-311G++(2d,p) is a more elaborate basis set and is expected to give the most accurate results. This basis set uses six primitive Gaussians for the core orbitals, triply split (three/one/one) valence orbitals for the heavy atoms ('311G') and diffuse functions for both heavy and hydrogen atoms ('++') to deal with electrons relatively far from the nucleus. The 2d,p in brackets indicates that polarization functions are added to both heavy atoms and hydrogen atoms, with 2d character added to the p orbitals of heavy atoms and p character added to the s orbitals of hydrogen.[7] There are many books and reviews available detailing all of the above topics.[2,3,5,7–10]

8.3   Molecular Mechanics Calculations

We have seen in the case of chemical shift calculations how the observed chemical shifts can be used to determine unknown parameters in the physical model describing the chemical shift. Similarly experimental structure data can be used to parameterize models for molecular geometry calculations. The molecular properties are now described by potential energy calculations of bonded and non-bonded interactions. In some cases however simplified molecular orbital calculations may be used to describe conjugated π-electron systems.[9]

Molecular mechanics (MM) algorithms treat the atoms and bonds in molecules in a classical balls and springs analogy. The electrons are ignored during the calculations, but the electronic characteristics are implicit due to the parameterization of the programs. The energy of the molecule is the sum of a number of potential energy functions. A typical example is Equation (8.3) from the MMFF94 program in which each function describes certain features of the molecule.[9]

$$V_{tot} = V_{bnd} + V_{ang} + V_{bnd/ang} + V_{oop} + V_{tor} + V_{vdW} + V_{elec} \qquad (8.3)$$

Each of the functions is constructed so that the minimum energy is at the equilibrium state of that function. The equilibrium states are commonly taken from experimental data such as X-ray diffraction, electron diffraction or NMR data, however *ab initio* calculations have also been used as a basis for parameterization. The total potential function given by Equation (8.3) is termed the molecular force field. The terms included in force fields are generally very similar (e.g. bond lengths, bond angles, etc.) however the basis for the parameterization can vary appreciably. The terms in Equation (8.3) defining the MMFF94 force field from Merck, which is one of the more popular force fields available have been determined from *ab initio* calculations. Other force fields commonly used for small molecules are the MM2 and MM3 force fields developed by the Allinger group and a modification of the MM2 force field called MMX is used in the PCModel program available from Serena Software.[14] Other well known force fields are the AMBER and CHARMM force fields[11,12] but these are not covered here because they have been developed for specific systems such as proteins. The MMFF94 and MMX force fields will be used here and are complementary because they are parameterized using *ab initio* and experimental data respectively.[9,13,14]

The different terms in Equation (8.3) describing the total potential energy V_{tot} are:

V_{bnd}:   is the term describing the equilibrium bond length between two nuclei.
V_{ang}:   is the equilibrium angle formed in a three atom sequence.
$V_{bnd/ang}$: is a cross term present to describe the effect of non bonded repulsion from an atom I in an I–J–K sequence to the atom K, such that an increase in bond length will decrease steric repulsion and can decrease the I–J–K angle.
V_{oop}:   is the out of plane angle bending component of an atom bonded to three other atoms.
V_{tor}:   describes the dihedral angle dependence of the potential energy in a four atom sequence.
V_{vdw}:   is the van der Waals non bonded steric repulsion term between atoms separated by three or more bonds.
V_{elec}:   is the electrostatic interaction between polar atoms separated by three or more bonds.

The main advantage of MM force fields is that they are based on classical physical concepts and are therefore computationally inexpensive. The disadvantage is that they can only be used for atom types and chemical environments on which they have been trained.

8.3.1   Conformer Generation

One of the fastest growing branches of MM is the development of algorithms for conformational searching either by statistical methods or by using dynamics data. These methods produce multiple variations of a chemical structure in order to retrieve the most energetically stable conformers. There are two main ways of altering molecular geometries to evaluate minimum energy conformations of molecules, molecular dynamics (MD) and conformational searching. Molecular dynamics simulations work by assigning a trajectory to each atom in a molecule. To calculate a trajectory, one only needs the initial positions of the atoms, an initial distribution of velocities and the acceleration, which is determined by the gradient of the potential energy function (from the force field used). The distribution of velocities is usually chosen from the Boltzmann distribution at a given temperature. The molecular structure is then altered over a time t and energy minimized by the force field used. These calculations have found their greatest use in studying the interaction of two molecules, such as the interaction of a substrate with a protein.[15,16]

The second approach involves the stepwise alteration of conformational parameters of a molecule such as a dihedral angle followed by energy minimization using a force field. This can be done either systematically (deterministic) or by including an element of random variation of the molecular structure (stochastic or Monte Carlo). The deterministic methods will guarantee a complete coverage of all regions of space but will of course demand more computational time. The Monte Carlo (MC) methods will however not be as consistent but by using sophisticated algorithms for conformational searching they have been shown to find more minimum energy structures given the same computational time as the deterministic methods.[17]

If one wishes to retrieve minimum energy structures from MD calculations it is necessary to use statistical methods to find the most often reoccurring geometries.[15] Although generally these would represent energy minima they can also include local energy maxima, which would obviously be a probable conformation (e.g. a broad local maxima with a small energy difference compared to two nearby local minima). These calculations would then become computationally expensive and produce a large number of structures to be evaluated. Monte Carlo conformational searches will however find distinct energy minima and rank these according to their energies and require no further processing.

For use in chemical shift calculations obviously very accurate molecular geometries are needed. The preferred method would be to exclusively use *ab initio* methods with as few restrictions as possible (i.e. larger basis sets). The time required for these calculations however sometimes makes their use impractical, in particular for large and flexible molecules. It is therefore very tempting to use molecular mechanics calculations to generate molecular geometries. In the next section we shall investigate the validity of MM geometries for chemical shift calculations. Indeed the approach most often taken is to couple MM force field calculations with MC conformational searching. This approach is very attractive from a user point of view (and therefore also a commercial one) as it requires minimal user interactions. Once the energies of a number of minimized structures are calculated, the relative populations of these is given by the Boltzmann distribution (Equation (8.4)) where f is the fractional population of the higher energy state and ΔE is the energy difference $(E_a - E_{min})$.

$$f = \exp\left(-\Delta E / RT\right) \tag{8.4}$$

If chemical shifts are calculated for the nuclei in all of these structures, then a weighted average can be obtained which should be a better fit of the experiment data obtained at any given temperature (see also Section 8.5). This procedure is used in the accompanying NMRPredict software package.

8.4   Molecular Geometries and ¹H Chemical Shift Calculations

As mentioned in the previous section, a common approach to ¹HNMR predic-
tion/calculations is to use MM geometries for chemical shift calculations. It is therefore
surprising that no scientific evaluation of this approach is available in the literature. This
leaves the potential users without any appreciation of the limitations and quality of the
simulated data. It is possible to evaluate the validity and accuracy of this approach by
using a data set of a wide variety of organic compounds and comparing the values of the
chemical shifts obtained against those from *ab initio* calculations. We use the MMX and
the MMFF94 force fields as examples as both are available in the accompanying PCModel
software. They are also incorporated in the NMRPredict software.[18]

The data set consists of conformationally rigid compounds used in the proceeding
chapters for the parameterization of the CHARGE program. The result will therefore be
primarily dependent on the accuracy of the structures produced by the molecular mechanics
force fields. Since the original calculations have mainly been based on quantum chemical
calculations, the results will also indicate the accuracy and validity of this approach in
general. The molecules used cover a range of functionalities and sizes from simple alkanes
to multi functional aromatics and steroids. The data were broken down into functional
groups so that the performance of the different computational methods could be more easily observed. All protons attached to hetero-atoms were removed to avoid complications
due to chemical exchange and intermolecular hydrogen bonding. The classification of the
compounds is given in Table 8.1 together with the rms errors from the two force fields
with the published rms errors from *ab initio* calculations. In Figure 8.1 the shifts calculated
using geometries from the MMFF94 force field are plotted against the observed chemical
shifts (correlation coefficient, 0.998). The data cover the entire ¹H chemical shift spectrum

Table 8.1   *Evaluation of ¹H chemical shifts calculation based on MM generated structures*

Compounds (Hs)		rms MMX	rms MMFF 94	rms Published	Compounds (Hs)
Alkanes	8 (130)	0.06	0.06	0.09	4 (90)
Alkenes	7 (78)	0.17	0.17	0.14	6 (68)
Alkynes	2 (16)	0.10	0.09	0.08	2 (16)
Aromatics	6 (48)	0.14	0.13	0.10	6 (48)
Heteroaromatics	14 (77)	0.10	0.09	0.09	14 (77)
Halogens	11 (125)	0.16	0.17	0.15	8 (67)
Nitriles	4 (29)	0.10	0.10	0.08	3 (18)
Alcohols	6 (66)	0.15	0.15	0.13	4 (44)
Ethers	3 (36)	0.13	0.15	0.13	3 (36)
Amines	8 (120)	0.11	0.11	0.09	8 (120)
Ketones	10 (132)	0.11	0.11	0.09	8 (106)
Esters	8 (54)	0.16	0.13	0.11	8 (54)
Amides	10 (93)	0.16	0.14	0.10	10 (93)
Nitros	4 (28)	0.30	0.09	0.05	4 (28)
Bi-functional	10 (88)	0.14	0.14	0.08	7 (51)
Sulfides	1 (8)	0.11	0.09	0.09	1 (8)
Total	**113 (1128)**	0.138	0.127	0.103	**96(924)**

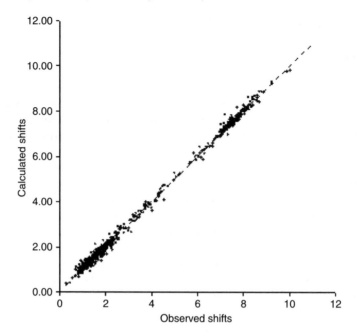

Figure 8.1   *Observed versus calculated ^{1}H chemical shifts from MMFF94 geometries.*

(0–10 ppm) for general organic compounds. The corresponding correlation coefficient for the MMX geometries and from the publications are 0.997 and 0.999 respectively. These are all excellent correlations and the order agrees with the rms errors shown in Table 8.1.

This simple exercise clearly shows that structures generated by MM force fields can be used for ^{1}H chemical shift calculations. It can also be seen from the results in Table 8.1 that there is a gain in using *ab initio* calculated structures especially in cases where conjugated π systems are involved such as alkenes, esters, amides, nitro and bi-functional compounds.

In every case halogen compounds proved to be the most difficult to predict, in particular iodo compounds were poorly predicted by both geometry optimization and chemical shift calculations.[19]

The MMFF94 force field is seen to perform slightly better than the MMX force field. However the MMX force field is still being developed[9] which may affect these findings. All the chemical shifts used above are available in PDF spreadsheets in the accompanying CD.

The above study gives the results of a statistical analysis. Although these results are statistically valid, it is important to realize that these conclusions may not be valid for any particular case and we give some examples to illustrate this point.

8.4.1   Methyl Anthracene-9-carboxylate

One of the most difficult geometrical parameters to determine is the angle of twist of substituents with an aromatic ring, yet this will have a large impact on any calculation of ^{1}H chemical shifts in the molecule.

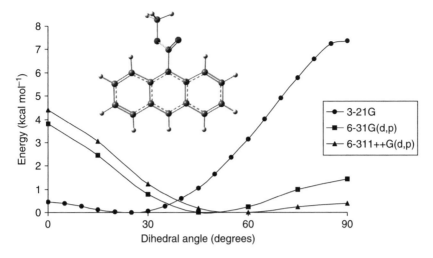

Figure 8.2 *Potential energy scans of methyl anthracene–9-carboxylate about the O=C—C=C dihedral angle, using various basis sets.*

In *methyl anthracene–9-carboxylate* (Figure 8.2) the dihedral angle from the ester group to the anthracene ring system in solution is unknown and it was necessary to obtain this angle to determine the ¹H spectrum[65] (see Chapter 7, Section 7.3). The angle in the crystal structure is 70 degrees[20] but this may not be the conformation in solution. Thus the *ab initio* (B3LYP) method was used with different basis sets to run a series of potential energy scans varying the ester dihedral angle (Figure 8.2) where the minimum energy is set to 0 kcal mol^{-1} for all calculations to simplify the comparison.

The minimum energy dihedral angle varies considerably depending on the basis set used, from 23° for 3-21G, 48° for 6-31G(d,p) to 58° for 6-311++G(d,p). Larger basis sets more accurately approximate the orbitals by imposing fewer restrictions on the locations of the electrons in space.[2] Thus larger basis sets will increase the steric repulsion of the ester group and increase the minimum energy dihedral as observed. However there is another intriguing observation from Figure 8.2. The global energy maximum when using the more restricted 3-21G basis set is at 90 degrees but with the larger basis sets the trend is reversed and the energy maximum is for the planar conformation. This is very significant as in the former case the molecule interconverts via the planar form and in the other case it is via the 90° conformation.

It is of interest that even at the level where the optimized dihedral angle is at its largest (58°) it is still smaller than that found in the crystal structure (70°) in which crystal packing forces may have been expected to flatten the structure. A molecular mechanics calculation using the MMFF94 force field gave the dihedral angle to be 75°, a very reasonable result. in view of this analysis. The results also illustrate the dangers of a very common procedure for calculating conformational energies. Often the energy profile is obtained at a minimum basis set (e.g. 3-21G) and then the minimum energy conformations are refined using higher basis sets. Because of the large variation in energy profile using different basis sets one could fail to find global energy minima.

8.4.2 *N*-Formyl Aniline (1)

The angle of twist of the formamide (NH.CHO) group bonded to an aromatic ring is unknown in many cases. The example shown here is *N*-formyl aniline studied by Perez

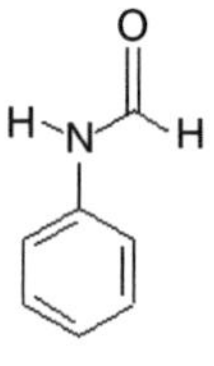

1 *cis* 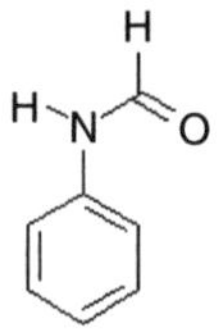   1 *trans*

(Chapter 7, Section 7.4.2 and Perez[22]) which exists in two forms (**1** *cis*) and (**1** *trans*).

Separate ¹H spectra of the *cis* and *trans* forms are observed due to hindered rotation about the N.CO bond (see next section), but in both conformers both the *ortho* protons resonate at the same frequency, showing that there are no large barriers to rotation about the C_1–N bond. The *cis/trans* ratio for these two isomers is 1/1 in dilute $CDCl_3$ solution and the ratio varies with the concentration of the amide in solution.[21] Geometry optimizations were performed with different modeling packages and the results are shown in Table 8.2. The two MM force fields predict a planar structure for both compounds, but the *ab initio* calculations diverge. For the *cis* compound the *ab initio* calculations predict an angle of twist varying from 31 to 44° and for the *trans* conformer the *ab initio* calculations differ even more, with twist angles of 33° for the simplest basis set (HF/321) and 0° for the more complex basis sets. A lanthanide induced shift (LIS) investigation on the conformations of these compounds was performed[22] and these results indicated dihedral angles of ca. 50° for the *cis* conformer and 20° for the *trans*.

Table 8.2 *Dihedral angles (degrees) for N-formyl aniline obtained with different force fields*

		MMFF94	MMX	HF/321	HF/321+**	HF/631G++(d,p)	B3LYP/ 631G++(d,p)
1 *cis*	τ_1^a	0.0	0.0	33.4	31.9	44.5	31.1
	τ_2^b	0.0	0.0	3.8	2.7	5.5	5.8
	τ_3^c	180.0	180.0	−175.2	−176.2	−174.0	−173.5
1 *trans*	τ_1^a	0.0	0.0	−33.4	0.0	0.0	0.0
	τ_2^b	−180.0	180.0	179.1	180.0	180.0	180.0
	τ_3^c	0.0	0.0	−1.9	0.0	0.0	0.0

^{a} C_2–C_1–N–C.
^{b} O–C–N–H.
^{c} H–N–C–H.

It is very clear from these results that even the complex *ab initio* calculations do not necessarily give the fine details of the molecular conformations required for accurate ¹H chemical shift predictions. The above calculations did not include solvent effects and it is possible that in these very polar compounds solvation effects could play a significant part.

8.4.3   Benzosuberone (2)

Another aspect related to the molecular geometry is how sensitive are the calculated ¹H chemical shifts to the variations in the molecular geometry? This of course will depend critically on the position of the proton concerned with respect to the functional groups. Thus in any given molecule one proton may be very dependent on the molecular geometry yet another neighbouring proton may be completely unaffected. A typical example is the ¹H spectrum of benzosuberone (2) (see Chapter 7, Section 7.2.2 and Abraham et al.[53]).

2

The seven-membered ring is in a chair conformation and is interconverting rapidly with its mirror image at room temperature, thus the two protons in each ring CH₂ group are equivalent. The molecule was minimized using the MMFF94 force field but the calculated ¹H chemical shift for H-9 peri to the carbonyl was in error by ca. 0.4 ppm (Table 8.3). This could be due to an incorrect geometry as the torsional strain in such a molecule is not easy to reproduce by molecular mechanics calculations. Thus the calculations were repeated with different optimized geometries including the MMX force field and *ab initio* calculations with both 3-21G* and 6-31++G(d,p) basis sets. The calculated shifts from CHARGE using these geometries are shown in Table 8.3 with the observed shifts. The calculated shifts for protons other than H-9 do not change appreciably but the values for H-9 range from 7.175 to 7.836δ (experimental 7.717). Interestingly the best agreement for H-9 was with the 3-21G basis set. However the rms error decreases as the level of theory increases with the best result for the larger basis set. The dihedral angle of the carbonyl with respect to the benzene ring changes significantly for the different geometries, from 67° (MMX), 53° (MMF94)

Table 8.3   Observed versus calculated ¹H chemical shifts of benzosuberone using different geometries

Proton	Exp.	MMX	MMFF94	B3LYP (3-21G*)	B3LYP (6-31++G(d,p))
2	2.733	2.850	2.796	2.677	2.755
3	1.813	1.808	1.832	1.862	1.859
4	1.882	1.835	1.875	1.903	1.906
5	2.931	2.684	2.715	2.740	2.741
6	7.196	7.249	7.317	7.377	7.354
7	7.415	7.389	7.473	7.533	7.495
8	7.297	7.262	7.317	7.348	7.323
9	7.717	7.175	7.349	7.836	7.557
rms		0.134	0.109	0.098	0.088

to 32° (3-21G) and 42° (6-31G). Thus the results in Table 8.3 suggest that an appropriate value for this dihedral angle is 37° ± 5°. A LIS experiment[23] suggested that the dihedral angle of the carbonyl with respect to the benzene ring is 56 degrees, which agrees with the value obtained with the MMX force field, but differs considerably from the *ab initio* values.

This result emphasizes the necessity of using the correct geometry as input to CHARGE to obtain accurate ^{1}H chemical shifts. However the values of the calculated shifts from CHARGE for all the geometries used are reasonable with all rms errors ≈ 0.1 ppm.

8.5   Rate Processes and NMR Spectra

There are innumerable examples of rate processes affecting NMR spectra and a number of books and reviews[24-28] have dealt with aspects of this large field of research. We present here a revision of the basic principles of dynamic NMR in order to better understand the ^{1}H NMR spectra which will be discussed later. The treatment given follows that of Abraham *et al.*[29]

We consider here only ^{1}H NMR but the basic principles are identical for all nuclei.

8.5.1   Theory

Consider a molecule interconverting between two states A and B, or a nucleus exchanging between two molecules A and B. What NMR spectrum will be observed?

It is first necessary to define the molecular parameters which govern the equilibrium considered. The equilibrium equation (Equation (8.5)) where n_A and n_B are the mole fractions of A and B is characterized by two parameters.

$$A \leftrightarrow B \tag{8.5}$$

$$n_A \qquad n_B$$

The position of the equilibrium is determined by ΔG, the free energy of the process, namely

$$n_A/n_B = \exp(-\Delta G/RT) \tag{8.6}$$

and

$$n_A + n_B = 1$$

The rate of interconversion is determined by the free energy of activation (ΔG^*), i.e. the rate of the reaction A → B is given by

$$k = RT/Nh \exp(-\Delta G^*/RT) \tag{8.7}$$

It is very important to distinguish these parameters. This may seem obvious, but, for example, the commonly used phrase 'a freely rotating fragment' may mean a molecular fragment which has equal rotamer populations (i.e. $\Delta G = 0$) *or* a fragment undergoing fast rotation (i.e. ΔG^* is small). The two cases are fundamentally different.

Consider a hydrogen nucleus in the molecule. In state A it has chemical shift v_A and coupling J_A and in state B shift v_B and coupling J_B. Note that all these quantities are measured in Hz.

There are three cases to consider:

(i) *Slow exchange.* If the *rate* of interconversion of A and B is slow (on the NMR time scale), then the NMR spectra of the two separate species A and B will be observed. That is we observe signals at ν_A and ν_B with couplings J_A and J_B and the relative intensity of the signals gives directly n_A and n_B and therefore ΔG.

(ii) *Fast exchange.* If the *rate* of interconversion is fast, the NMR spectrum observed is *one* 'averaged' spectrum in which the chemical shifts and couplings are the weighted averages of the values in A and B. Thus the nucleus will give rise to *one* signal with a position (ν_{Av}) given by

$$\nu_{av} = n_A \nu_A + n_B \nu_B$$

and coupling (J_{AV}) by

$$J_{AV} = J_A \nu_A + J_B \nu_B \tag{8.8}$$

(iii) *Intermediate rates of exchange.* In this case broad lines are observed in the NMR spectrum, and indeed this is one of the few cases in which the resolution of the spectrum is not due solely to the spectrometer. The other common examples are the presence of quadrupolar nuclei (in particular ^{14}N) and paramagnetic species, the latter often as an impurity; and poor sample preparation, resulting in solid impurities being present. Apart from these easily recognized features, the presence of broad lines, therefore, is indicative of a rate process.

Consider the simplest system of two equal-intensity peaks collapsing to one, with no coupling present. This is the case for N,N-dimethylacetamide, which at room temperature shows three single peaks in the proton spectrum (Figure 8.3). As the temperature is raised, rotation about the central bond becomes appreciable and the two N–Me groups exchange positions.

As the exchange rate increases, the N–Me resonances broaden, coalesce into one broad resonance and finally give one sharp single 'average' peak characteristic of fast exchange. During this process the $CH_3.CO$ peak remains sharp, as the interconversion of the N–Me groups does not affect the chemical shift of this peak. The position where the two separate peaks just merge into one is called the coalescence point. At this point the lifetime of nucleus A (or B) in a discrete state is given by

$$\tau = \sqrt{2}/\pi.\delta\nu \text{ (in seconds)} \tag{8.9}$$

where $\delta\nu = \nu_a - \nu_b$ (Hz.).

For ^1H NMR $\delta\nu$ is of the order of 0–100 Hz and therefore τ is approximately 10^{-2} s at coalescence; for ^{13}C the chemical shifts are larger and therefore the lifetimes are less at coalescence; a 10 ppm separation (1000 Hz at 100.8 MHz) requires a lifetime of ca. 2.5×10^{-2} s for coalescence.

The rate constant (k) for the reaction A $\rightarrow$ B is given by

$$k = 1/\tau \tag{8.10}$$

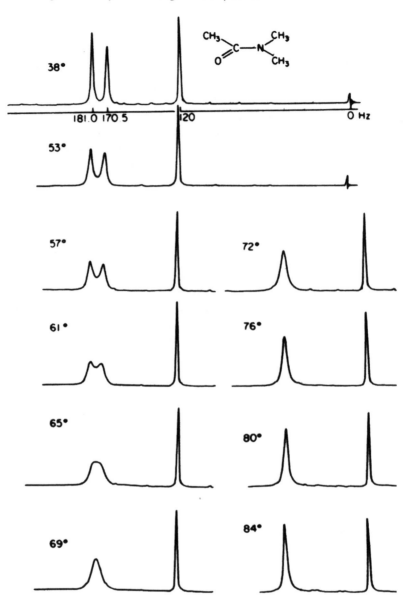

Figure 8.3 *The ¹H NMR spectrum of N,N-dimethylacetamide at the temperatures shown (from Abraham et al.[29]).*

and therefore combining Equations (8.7), (8.8) and (8.9) gives the free energy of activation ΔG^* from the coalescence point and temperature (T_c):

$$\Delta G^*/RT_c = \log_e(\sqrt{2R}/\pi Nh) + \log_e(T_c/\delta v)$$
$$= 22.96 + \log_e(T_c/\delta v) \qquad (8.11)$$

In NMR the normal range of values of ΔG^* are from ca. 5 to 25 kcal mol^{-1}, i.e. T_c values from -100 to $+200\,°$C. Equation (8.11) is *only* valid for the case given, i.e. $n_A = n_B$ and no coupling, but it is often used for more complex cases to obtain approximate values of ΔG^*.

It is, however, possible to solve the NMR rate equations (the Bloch equations) for exchanging systems and thus calculate the spectrum for any given exchange rate. This requires a computer analysis for complex systems, but for the simple case considered here two useful equations can be derived from the rate equations.

The linewidth (the total width at half height) of the single peak above coalescence (h) can be related to the exchange rate by Equation (8.12). This is valid to the coalescence point.

$$k = \pi/2.\delta v \, \{(\delta v/h)^2 - (h/\delta v)^2 + 2\}^{1/2} \tag{8.12}$$

At coalescence $h \approx \delta v$ and Equation (8.12) condenses to Equation (8.9).

When the signals are separate, the broadening of these separate signals under slow exchange is related to the exchange rate by Equation (8.13) where h_0 is the linewidth in the absence of exchange.

$$k = \pi(h - h_0) \tag{8.13}$$

These equations and the computer analysis method enable exchange rates to be determined over a range of temperatures and, therefore, in principle, allow the determination of ΔH^* and ΔS^* from the plot of $\ln k$ against $1/T$. This method has been used many times to obtain such parameters, but the values of ΔH^* and particularly ΔS^* obtained in this manner are often inaccurate. The major problems are the relatively small range in temperature over which meaningful values of the rate can be measured, the difficulty in measuring the sample temperature accurately in the probe and the errors involved in the estimation of the NMR parameters needed (e.g. δv and h_0) for the calculations. Undoubtedly, the most accurate molecular parameter obtained from such experiments is the value of the free energy of activation (ΔG^*) at the coalescence temperature. However, the combined use of ^{1}H and ^{13}C NMR can, in favourable cases, considerably extend the temperature range investigated, and this allows more accurate values of ΔH^* and ΔS^* to be obtained.

It is of interest to consider some simple examples of ^{1}H NMR and rate processes, as follows.

8.5.2   Amide Rotation

The room temperature ^{1}H spectrum of *N,N*-dimethylacetamide (**3**) shows two N–Me signals (Figure 8.3). This arises from slow rotation about the amide bond due to the partial double bond character, which also makes the molecule essentially planar.

On warming, the N–Me peaks broaden, coalesce (T_c, ca. 65 °C) and finally give one sharp peak. From the coalescence temperature and the separation of the N–Me signals, Equation (8.11) immediately gives $\Delta G^*(338\,\text{K}) = 17.8\,\text{kcal mol}^{-1}$. Here, therefore, the activation parameter is obtained even though chemically the rate process is between identical molecules.

The ^{13}C spectrum of this molecule is very similar and confirms the presence of a rate process. Again, all the signals at room temperature are sharp and now the separation of the N–Me signals is 3.0 ppm compared with 0.17 ppm in the proton spectrum. Exactly the same phenomenon is observed on increasing the temperature and, as before, the $CH_3.CO$ peak remains sharp at all times. Use of equation (8.11) with T_c equal to 80 °C and $\Delta\nu$ 75.6 Hz gives $\Delta G^* = 17.2$ kcal mol^{-1} in complete agreement with the proton data. Here the difference in the chemical shifts of the exchanging species in going from ^{1}H to ^{13}C is not sufficient to affect the coalescence temperature appreciably and thus two independent values of ΔG^* are obtained.

8.5.3   Proton Exchange Equilibria

There are many examples of such rate processes in NMR. The example given (Figure 8.4) is of a rate process involving coupling. The first spectrum (a) shows the ^{1}H spectrum of clean, dry ethanol in which there is coupling between the methyl and methylene protons as

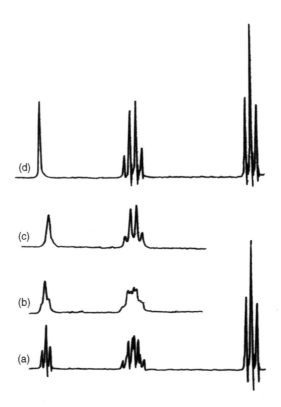

Figure 8.4   *The ¹H NMR spectra of ethanol: (a) pure; (b), (c) and (d) with increasing amounts of acid added (from McFarlane et al.[30]).*

well as between the methylene and the hydroxyl proton. The OH signal is therefore a triplet and the methylene signal a quartet of triplets. On the addition of traces of acid the rate of the exchange equilibrium below increases. Again no chemical reaction has occurred, but H_A and H_B have interchanged.

$$CH_3CH_2OH_A + H_B{}^+ \leftrightarrow CH_3CH_2OH_B + H_A{}^+$$

The rate of exchange increases dramatically with each addition of acid and it is only necessary to add ca. 10^{-5} mol of acid to remove the OH coupling completely (Figure 8.4(d)).

In this case the lifetime of the OH proton on *one* molecule is too short to relay coupling information. There is thus no coupling and the familiar single sharp peak for the hydroxyl proton and quartet pattern for the methylene signals is observed. Also the lifetime at coalescence, $\tau_c \approx 1/J$ (CH.OH). It is the coupling, not the chemical shift separation, which determines the appearance of the spectrum. Thus in Figure 8.4(b) the lifetime of each OH proton on a molecule is ca. 0.2 s.

If ethanol is dissolved in a strong hydrogen bonding solvent such as acetone or DMSO, the exchange rate is reduced considerably, owing to the competing hydrogen bonding with the solvent, and often the CH.OH coupling appears. Indeed, this has been suggested as a method for detecting primary, secondary and tertiary alcohols, i.e. from the fine structure of the OH proton in dilute DMSO solution.

8.5.4   Rotation about Single Bonds, Ring Inversion Processes

In simple ethanes, the barrier height to rotation about the C.C bonds is so low (ca. 3–5 kcal mol^{-1}) that the NMR spectra of these compounds will always be in the 'fast exchange' limit and one averaged spectrum over the different conformers will be observed. Thus, changes in the proportions of the conformers merely alter the observed averaged values of the chemical shifts and couplings.

However, if sufficient steric interactions are present, the barrier height to rotation in ethanes can be increased so that it is directly observed by NMR. A good example is the ¹H spectrum of 2,2,3,3-tetrachlorobutane (**4**) (Figure 8.5), which gives a sharp line at room temperature, but on cooling the signal broadens (T_c, ca. $-30\,°C$) and finally separates into two unequally intense peaks. The chemical shifts of the two methyl groups within each conformer are the same, by symmetry, but the resonances of the two conformers may differ. Thus, the low-temperature spectrum consists of separate signals from the two conformers, which are now interconverting slowly on the NMR time scale.

The analysis of these spectra provides a complete picture of the energy profile involved in this rotation. The relative intensity of the signals in the low-temperature spectrum gives directly the populations of the conformers and, therefore, their energy difference; and the analysis of the coalescence process gives the barrier height to rotation. It is necessary to assign the signals to the individual conformers, and in this example, both solvent effects and chemical shift considerations *(gauche* halogens tend to deshield methyl groups) assigned the deshielded resonance to the *trans* isomer. This gave $\Delta G_{t-g} = 0.17$ kcal mol^{-1} and an analysis of the coalescence curves by simulated computed curves (shown in Figure 8.5) gave $\Delta G^*(243\,K) = 13.5$ kcal mol^{-1}.

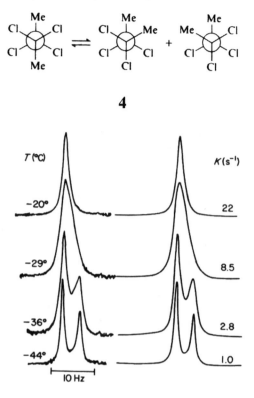

4

Figure 8.5   *Observed and calculated ¹H spectra of 2,2,3,3-tetrachlorobutane (4) in acetone-d₆ at the temperatures shown (reproduced from Roberts et al.[31] with permission of the American Chemical Society).*

The above case is an ideal example, as there is no spin–spin coupling to complicate the spectrum. In many cases extensive coupling makes the analysis of such rate processes extremely difficult, and this is particularly true for ring inversion processes in cyclic molecules, which have been studied by NMR. Consider a proton in cyclohexane (**5**):

Hax

Heq

5

The interconversion of the cyclohexane ring flips the Heq proton to Hax and as these protons have different chemical shifts (δeq 1.67, δax 1.19) this rate process is detectable by ¹H NMR. The ring inversion is rapid at room temperature and the ¹H spectrum of cyclohexane is a single sharp line. On cooling the peak broadens and at 193 K the spectrum consists of two broad humps (Figure 8.6). The spectrum is complex due to extensive coupling as now the spectrum is an A_6B_6 spin system. In order to analyse this exchange process in

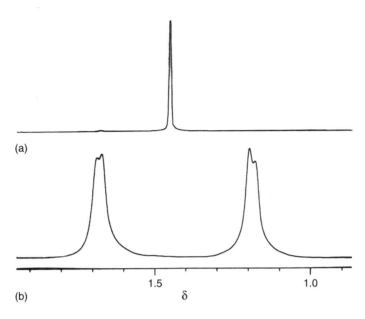

(a)

(b)

1.5   1.0

δ

Figure 8.6   *The 250 MHz ¹H spectrum of cyclohexane, (a) at room temperature and (b) at 193 K.*

detail the experiment was performed on the deuterated species C_6HD_{11} and the ¹H spectrum obtained whilst decoupling the deuterium.[32] Analysis of these spectra gave the value of ΔG^* as 10.2 kcal mol⁻¹ at 213 K, with ΔH^* 10.7 kcal mol⁻¹ and ΔS^* 2.2 cal mol⁻¹ deg⁻¹. This accurate value allowed the determination of the complete energy profile of the cyclohexane inversion, which goes via a half-chair transition state and a twist-boat intermediate.[33]

This simple case can be analysed completely, but for most cyclohexane derivatives the complexity of the ¹H spectrum precludes a complete rate analysis. Figure 8.7 shows the ¹H spectrum of chlorocyclohexane at 298 and 193 K. The molecule is interconverting rapidly between the equatorial and axial conformers at 298 K thus one averaged spectrum is observed but at 193 K the separate spectra of the two conformers is observed.

Note in particular H-1eq in the minor axial conformer at 4.6δ and H-1ax in the major equatorial conformer at 3.9δ. The intensities of the two signals gives directly the populations of the two isomers and thus the value of $\Delta G = 0.60$ kcal mol⁻¹, but an accurate value of ΔG^* is difficult to obtain from this spectrum. Most of the values of ΔG^* for such conformational inversions are obtained from ¹³C NMR where the absence of any couplings makes the computation much simpler.

In the low temperature spectrum the large geminal and axial–axial couplings are resolved but not the eq–eq couplings, thus the equatorial protons show only the geminal HH coupling (cf. the H-2,6eq doublets from the equatorial conformer at 2.22δ and the axial conformer at 2.0δ).

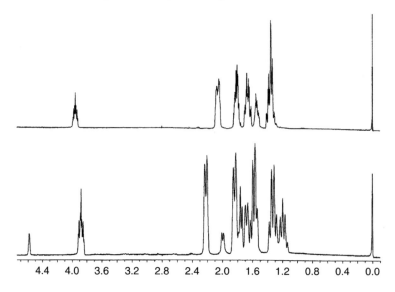

Figure 8.7   *The 400 MHz ¹H NMR spectra of chlorocyclohexane in 1:1 CDCl₃:CFCl₃ solution at 298 K (upper) and 193 K (lower) (from Warne [34]).*

8.6   Solvent Effects

8.6.1   Introduction

The solvent dependence of ¹H chemical shifts has been investigated since the beginning of high resolution ¹H NMR. In a seminal paper Buckingham *et al.*[35] defined four interactions responsible for solvent effects. These were hydrogen bonding, the anisotropy of the solvent molecules, polar and van der Waals effects (Equation (8.14)). This analysis has formed the basis for all subsequent investigations.

$$\Delta\delta = \delta_{HB} + \delta_A + \delta_E + \delta_W \tag{8.14}$$

The relative importance of these four contributions can vary considerably. When hydrogen bonding occurs with protic solutes, this is a major interaction with solvent effects of up to 5 ppm for the protic hydrogen.[35] Large anisotropy contributions of ca. 1 ppm have also been observed for non-polar anisotropic solvents such as benzene and CS_2.[35–37] The effect of the solvent on the solute chemical shifts due to the electric field of the polar solute molecule has been calculated using variations of the Onsager reaction field model.[35–39] despite its many limitations. In addition, van der Waals effects have been shown to be significant in gas to solvent shifts even for non-polar molecules in non-polar solvents.[40] This early work has been well summarized.[41–43]

Although the effect of solvent on chemical equilibria has been investigated in depth recently by both molecular modelling and quantum theory[44–47] there has been no quantitative treatment of differential solvent effects on the ¹H chemical shifts of organic solutes. The problems involved in the quantitative calculation of the four contributions of Equation (8.14) for polar, anisotropic and protic solvents are prohibitive. Barone *et al.*[44] have recently employed the polarizable continuum model (PCM) solvation routine.[48,49] to calculate ¹H and ¹³C chemical shifts in solution via the quantum mechanical GIAO approach in the Gaussian suite.[50] However this model is the quantum mechanical formulation of the Onsager reaction field model and does not include any solvent hydrogen bonding, van der Waals or anisotropy contributions.

The absence of any tested predictive package for ¹H chemical shifts in polar, anisotropic solvents severely limits the usefulness of such solvents for characterization purposes. The CHARGE program discussed so far has been developed to predict ¹H chemical shifts of organic compounds in $CDCl_3$ solution.[19,51–53] DMSO is the solvent of choice for pharmaceutical compounds due to its excellent solubility properties for many protic and charged molecules which are insoluble in $CDCl_3$. It is also non-toxic, water miscible, biodegradable and has a strong deuterium 'lock'. It is therefore important to include ¹H chemical shifts in DMSO as solvent in the CHARGE package. ¹H chemical shifts in DMSO can differ by up to 5 ppm from the corresponding shifts in $CDCl_3$ and therefore the calculations for $CDCl_3$ cannot be used with confidence to predict chemical shifts in DMSO.

Few detailed studies of the effects of DMSO on ¹H chemical shifts have been carried out. Gottlieb *et al.*[54] gave the ¹H (and ¹³C) chemical shifts of 36 common impurities in seven common solvents and more recently Jones *et al.*[55] recorded similar data for 60 common solvents in $CDCl_3$, DMSO, D_2O and CD_3OD. Hobley *et al.*[56] have given the ¹H (and ¹³C) shifts of eleven monosaccharides in both the α and β forms in D_2O and DMSO solution and an algorithm for predicting the shifts. Abraham *et al.*[19,53] observed for a selection of aromatic aldehydes and ketones small solvent effects of either sign for DMSO vs. $CDCl_3$ solvent and Perez[22] recorded the ¹H shifts of a number of aliphatic and aromatic amides in both solvents.

Abraham *et al.*[57] compiled an extensive data set of ¹H chemical shifts in DMSO vs. $CDCl_3$ solvent and showed how the CHARGE program can be simply extended to provide a satisfactory prediction of the shifts in DMSO. The analysis was particularly relevant for the protic hydrogen in alcohols, amides, etc. as the large concentration dependence of these chemical shifts in $CDCl_3$ due to intermolecular hydrogen bonding prohibited the use of these shifts for diagnostic purposes. In contrast the corresponding shifts in DMSO solvent show no concentration dependence and can be used for diagnostic purposes in a precisely similar manner to other ¹H shifts.

The CHARGE program was developed using shift data in $CDCl_3$ solvent. To obtain the corresponding shifts in DMSO the authors[57] corrected the shifts calculated in $CDCl_3$ by adding a contribution for DMSO solvent, i.e. $\Delta\delta = \delta(DMSO) - \delta(CDCl_3)$. The aim of their work was to measure and then simulate the values of $\Delta\delta$ for a variety of compounds and functional groups.

Thus for any solute molecule CHARGE calculated $\delta(CDCl_3)$ in the usual way. The value of $\Delta\delta$ was then calculated in the 'DMSO' subroutine. These calculations followed a similar procedure to the main program. For the fragment I–J–K–L there is an α effect on atom I from atom J, a β effect from atom K and a γ effect from atom L. All effects over more than three bonds are termed long-range effects. Each functional group can have in principle α, β, γ and long-range effects.

The only exceptions were the hydroxyl and amide protons. The large concentration dependence of their chemical shifts in $CDCl_3$ solution prevented any analysis of the effects of functional groups on these shifts. In contrast these ¹H chemical shifts in DMSO have no concentration dependence. Thus the effects of functional groups on the OH and amide NH chemical shifts in DMSO solution were treated in CHARGE in the same manner as any other proton. Also modern spectrometers can routinely detect the OH protons of alcohols and phenols at such low concentrations (ca. 1 mg ml^{-1}) that the OH chemical shift approximates to the limiting ∞ dilution value.[58,59] The alcohol and phenol OH shifts recorded are these limiting values. Due to exchange broadening acid and amide protic hydrogens are difficult to detect at low concentrations and this precludes a similar analysis for these protons.

For polyfunctional molecules, $\Delta\delta$ will be affected by all the functional groups. It was found that in practice short-range effects dominate, with any long-range effects reduced by the presence of the short-range functional group. Compounds with low molecular dipole moments either lack polar functional groups or are highly symmetrical. Experimental $\Delta\delta$ values of the first type of non-polar compound were low, whereas those having polar groups and high symmetry may have large $\Delta\delta$ values. The CHARGE program calculates the partial atomic charges and hence the dipole moment of the molecule considered. For molecules with dipole moments $< 0.5D$ the α,β and γ effects were included but long-range effects were omitted. The molecular geometries were obtained from the MMFF94 force field and the *ab initio* calculations at the B3LYP/6-31g(d,p) level. The data for the amides are from Perez[22] and for the diols and inositols from Abraham *et al.*[52]

The compounds were separated into four groups: non-polar, polar aprotic, protic and polyhydroxy compounds. The non-polar group includes alkanes, alkenes, alkynes and aromatics. The polar aprotic group includes aliphatic/aromatic tertiary amines, ethers, ketones, esters and halides. The protic group includes aliphatic and aromatic primary and secondary amines, alcohols, thiols, aldehydes, carboxylic acids and amides and the last group diols, triols and inositols. The compounds measured are shown in Figures 8.8–8.10 and 8.13 and the results for these compounds are given in Tables 8.4–8.6 and 8.9.

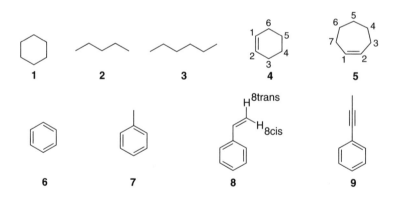

Figure 8.8 *Non-polar compounds.*

8.6.2   Non-polar Compounds

The $\Delta\delta$ values of the non-polar compounds (Table 8.4) were small, in many cases not much greater than the experimental error with the notable exception of the acetylenic

Table 8.4 ^1H chemical shifts of non-polar solutes in $CDCl_3$ and DMSO

Compound	Proton	δ(DMSO)		δ(CDCl₃)		Δδ(DMSO–CDCL₃)	
		Obs.	Calc.	Obs.	Calc.	Obs.	Calc.
Cyclohexane (**1**)ᵃ	CH₂	1.40	1.40	1.43	1.40	−0.03	0.00
n-Pentane (**2**)	Me	0.86	0.84	0.88	0.84	−0.02	0.00
	2,4	1.27	1.14	1.27	1.14	0.00	0.00
	3	1.27	0.94	1.27	0.94	0.00	0.00
n-Hexane (**3**)	Me	0.86	0.84	0.88	0.84	−0.02	0.00
	2,5	1.25	1.15	1.26	1.15	−0.01	0.00
	3,4	1.25	0.96	1.26	0.96	−0.01	0.00
Cyclohexene (**4**)	=CH	5.65	5.75	5.69	5.75	−0.04	0.00
	3,6	1.95	2.05	1.99	2.05	−0.04	0.00
	4,5	1.55	1.55	1.60	1.55	−0.05	0.00
Cycloheptene (**5**)	=CH	5.77	5.62	5.79	5.62	−0.02	0.00
	3,7	2.08	2.06	2.12	2.06	−0.04	0.00
	4,6	1.45	1.43	1.50	1.43	−0.05	0.00
	5	1.69	1.44	1.75	1.44	−0.06	0.00
Benzene (**6**)	CH	7.37	7.34	7.34	7.34	0.03	0.00
Toluene (**7**)	Me	2.30	2.33	2.36	2.33	−0.06	0.00
	o	7.18	7.13	7.17	7.13	0.01	0.00
	m	7.25	7.28	7.25	7.28	0.00	0.00
	p	7.18	7.17	7.17	7.17	0.01	0.00
Styrene (**8**)	o	7.47	7.77	7.42	7.77	0.05	0.00
	m	7.35	7.38	7.32	7.38	0.03	0.00
	p	7.28	7.34	7.25	7.34	0.03	0.00
	=CH	6.74	6.73	6.72	6.73	0.02	0.00
	8(*cis*)	5.26	5.26	5.24	5.26	0.02	0.00
	8(*trans*)	5.84	5.93	5.75	5.93	0.09	0.00
Phenylacetylene (**9**)	CH	4.20	4.28	3.07	3.15	1.13	1.13
	o	7.49	7.56	7.50	7.56	−0.01	0.00
	m	7.40	7.35	7.33	7.35	0.07	0.00
	p	7.40	7.36	7.33	7.36	0.07	0.00

ᵃ From Gottlieb *et al.*[54].

proton of phenylacetylene (**9**, Figure 8.8). The α, β and γ contributions for all protons of these compounds were therefore set equal to zero, except for the acetylenic proton. These compounds all have dipole moments < 0.5D and the long-range effects were also set to zero.

8.6.3   Polar Aprotic Compounds

These compounds have Δδ values ca. ± 0.1 ppm (Table 8.5), except for the *geminal* dihalo and trihalo compounds (e.g. CH_2Cl_2 and $CHCl_3$ for which Δδ is +0.49 and +1.06 res.) due to hydrogen bonding (see later). The Δδ values for the H.C.C=O protons were dependent on the H.C.C=O dihedral angle and the values for the four protons 2e, 2a and 3x, 3n in the rigid molecules **17** and **18** (Figure 8.9) were simulated by Equation (8.15).

$$\Delta\delta = -0.047 - 0.091\cos\theta - 0.085\cos2\theta \qquad (8.15)$$

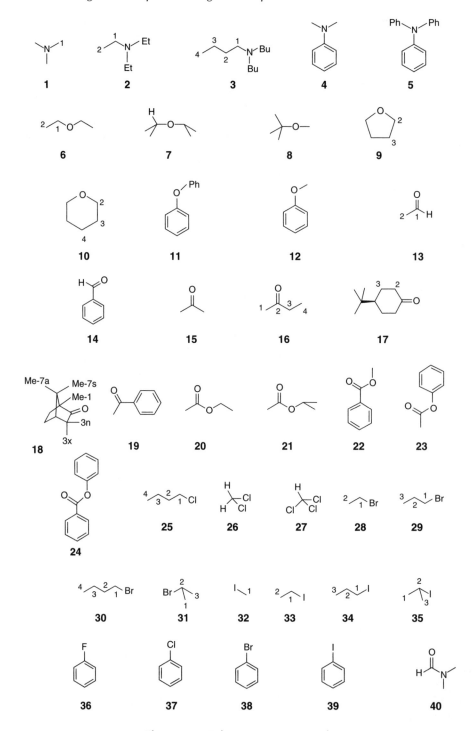

Figure 8.9   *Polar aprotic compounds.*

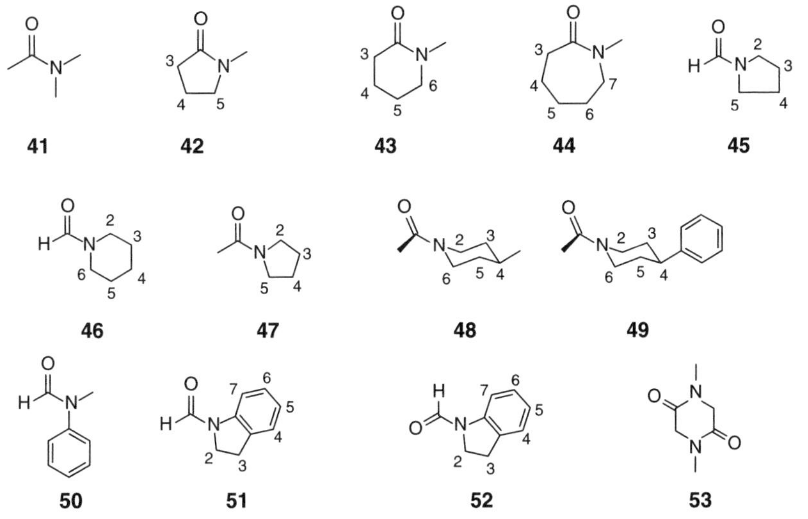

Figure 8.9 (Continued).

The rotationally averaged value was −0.09 ppm (cf. obs. values of −0.08 and −0.07 for the Me.CO groups in acetone and 2-butanone).

Table 8.5   ¹H chemical shifts of polar aprotic solutes in CDCl₃ and DMSO

Compound	Proton	δ(DMSO)		δ(CDCl₃)		Δδ(DMSO–CDCL₃)	
		Obs.	Calc.	Obs.	Calc.	Obs.	Calc.
Trimethylamine (**1**)	Me	2.09	2.19	2.22	2.28	−0.13	−0.09
Triethylamine (**2**)	Me	2.42	2.46	2.52	2.55	−0.10	−0.09
	CH₂	0.93	0.82	1.03	0.91	−0.09	−0.09
Tributylamine (**3**)	1	2.31	2.29	2.38	2.38	−0.07	−0.09
	2	1.34	1.20	1.41	1.28	−0.07	−0.08
	3	1.26	1.06	1.29	1.13	−0.03	−0.07
	Me	0.87	0.77	0.91	0.84	−0.04	−0.07
Dimethylaniline (**4**)	o	6.71	6.60	6.74	6.66	−0.03	−0.06
	m	7.17	7.10	7.24	7.10	−0.07	0.00
	p	6.63	6.66	6.72	6.66	−0.08	0.00
	Me	2.87	2.72	2.94	2.82	−0.06	−0.10
Triphenylamine (**5**)	o	6.99	7.07	7.08	7.13	−0.09	−0.06
	m	7.29	7.27	7.24	7.27	0.06	0.00
	p	7.03	7.05	7.00	7.05	0.03	0.00
Diethylether (**6**)	Me	1.09	1.08	1.21	1.17	−0.12	−0.09
	CH₂	3.38	3.45	3.48	3.55	−0.10	−0.10
Diisopropylether (**7**)	Me	1.04	1.08	1.13	1.17	−0.09	−0.09
	CH	3.60	3.63	3.65	3.74	−0.05	−0.11
t-Bu-methyl ether (**8**)	tBu-Me	1.11	1.16	1.19	1.25	−0.08	−0.09
	Me	3.08	3.17	3.21	3.28	−0.13	−0.11
THF (**9**)	2,5	3.60	3.60	3.74	3.71	−0.14	−0.11
	3,4	1.76	1.44	1.85	1.53	−0.09	−0.09

Table 8.5   (*Continued*)

Compound	Proton	δ(DMSO)		δ(CDCl₃)		Δδ(DMSO–CDCL₃)	
		Obs.	Calc.	Obs.	Calc.	Obs.	Calc.
THP (**10**)	2,6	3.53	3.55	3.65	3.65	−0.12	−0.10
	3,5	1.47	1.41	1.57	1.52	−0.10	−0.11
	4	1.58	1.48	1.64	1.55	−0.07	−0.07
Diphenylether (**11**)	o	7.01	7.17	7.01	7.17	0.00	0.00
	m	7.39	7.30	7.32	7.30	0.07	0.00
	p	7.14	7.18	7.09	7.18	0.05	0.00
Anisole (**12**)	o	6.93	6.80	6.90	6.80	0.02	0.00
	m	7.29	7.15	7.29	7.15	0.00	0.00
	p	6.93	6.87	6.94	6.87	−0.02	0.00
	Me	3.74	3.67	3.80	3.78	−0.06	−0.11
Acetaldehyde (**13**)	CHO	9.66	9.68	9.79	9.81	−0.14	−0.13
	Me	2.13	2.10	2.20	2.15	−0.08	−0.05
Benzaldehyde (**14**)	o	7.92	7.85	7.88	7.78	0.04	0.07
	m	7.62	7.55	7.53	7.48	0.09	0.07
	p	7.73	7.63	7.63	7.56	0.10	0.07
	CHO	10.03	9.82	10.03	9.82	0.00	0.00
Acetone (**15**)	Me	2.09	2.16	2.17	2.21	−0.08	−0.05
2-Butanone (**16**)	CO.Me	2.07	2.14	2.14	2.19	−0.07	−0.05
	CH₂	2.43	2.41	2.44	2.38	−0.01	0.03
	Me	0.91	0.78	1.06	0.85	−0.15	−0.07
4-t-Butylcyclohexanone (**17**)	2a/6a	2.36	2.18	2.31	2.20	0.05	−0.02
	2e/6e	2.19	2.15	2.40	2.35	−0.21	−0.20
	3a/5a	1.35	1.32	1.44	1.39	−0.10	−0.07
	3e/5e	1.99	1.99	2.08	2.06	−0.10	−0.07
	4a	1.50	1.22	1.46	1.29	0.04	−0.07
	tBu Me	0.89	0.85	0.92	0.92	−0.04	−0.07
Camphor (**18**)	3n	1.80	1.49	1.84	1.62	−0.04	−0.13
	3x	2.29	2.36	2.35	2.47	−0.06	−0.11
	4	2.05	2.04	2.09	2.11	−0.04	−0.07
	5n	1.30	1.05	1.34	1.12	−0.04	−0.07
	5x	1.87	2.05	1.95	2.12	−0.08	−0.07
	6n	1.26	1.34	1.40	1.41	−0.14	−0.07
	6x	1.65	1.82	1.68	1.89	−0.03	−0.07
	Me-1	0.80	0.96	0.92	1.03	−0.12	−0.07
	Me-7s	0.76	0.67	0.84	0.74	−0.08	−0.07
	Me-7a	0.91	0.81	0.96	0.88	−0.05	−0.07
Acetophenone (**19**)	Me	2.58	2.56	2.60	2.61	−0.02	−0.05
	o	7.96	7.86	7.96	7.79	0.00	0.07
	m	7.53	7.56	7.46	7.49	0.07	0.07
	p	7.64	7.63	7.56	7.56	0.08	0.07
Ethyl acetate (**20**)	CH₂	4.03	4.06	4.12	4.17	−0.09	−0.11
	CO.Me	1.99	2.01	2.04	2.07	−0.05	−0.06
	Me	1.18	1.00	1.26	1.16	−0.08	−0.16
Isopropyl acetate (**21**)	CO.Me	1.96	2.02	2.01	2.07	−0.05	−0.05
	CH	4.86	4.74	4.99	4.85	−0.13	−0.11
	Me	1.17	1.03	1.23	1.12	−0.06	−0.09
Methyl benzoate (**22**)	o	7.97	7.98	8.04	7.91	−0.07	0.07
	m	7.53	7.59	7.43	7.52	0.10	0.07
	p	7.66	7.50	7.55	7.50	0.00	0.11
	Me	3.86	3.78	3.92	3.89	−0.05	−0.11

Phenyl acetate (**23**)	Me	2.26	2.17	2.29	2.22	−0.03	−0.05
	o	7.12	7.09	7.08	7.02	0.04	0.07
	m	7.41	7.30	7.37	7.23	0.04	0.07
	p	7.25	7.20	7.22	7.13	0.03	0.07
Phenyl benzoate (**24**)	o (ben)	8.14	8.01	8.21	7.94	−0.07	0.07
	m(ben)	7.62	7.57	7.52	7.50	0.10	0.07
	p(ben)	7.76	7.54	7.64	7.47	0.12	0.07
	o(ph)	7.29	7.21	7.22	7.14	0.07	0.07
	m(ph)	7.49	7.35	7.44	7.28	0.05	0.07
	p(ph)	7.33	7.26	7.28	7.19	0.05	0.07
Chlorobutane (**25**)	1	3.62	3.62	3.54	3.54	0.08	0.08
	2	1.69	1.49	1.76	1.56	−0.06	−0.07
	3	1.40	1.09	1.47	1.16	−0.06	−0.07
	Me	0.89	0.80	0.94	0.87	−0.05	−0.07
Dichloromethane (**26**)	CH_2	5.79	5.73	5.30	5.27	0.49	0.46
Chloroform (**27**)	CH	8.32	8.32	7.26	7.28	1.06	1.06
Bromoethane (**28**)	CH_2	3.53	3.64	3.43	3.52	0.10	0.12
	Me	1.60	1.58	1.67	1.65	−0.08	−0.07
1-Bromopropane (**29**)	1	3.51	3.46	3.39	3.35	0.12	0.11
	2	1.81	1.78	1.88	1.84	−0.07	−0.06
	Me	0.96	0.80	1.03	0.87	−0.07	−0.07
1-Bromobutane (**30**)	1	3.53	3.50	3.42	3.38	0.11	0.12
	2	1.78	1.59	1.84	1.66	−0.06	−0.07
	3	1.40	1.08	1.47	1.15	−0.07	−0.07
	Me	0.89	0.78	0.94	0.85	−0.05	−0.07
2-Bromopropane (**31**)	Me	1.66	1.64	1.71	1.70	−0.05	−0.06
	CH	4.43	4.42	4.29	4.30	0.14	0.12
Iodomethane (**32**)	Me	2.18	2.28	2.16	2.21	0.02	0.07
Iodoethane (**33**)	CH_2	3.26	3.34	3.19	3.27	0.06	0.07
	Me	1.76	1.80	1.85	1.87	−0.08	−0.07
Iodopropane (**34**)	1	3.27	3.16	3.18	3.09	0.08	0.07
	2	1.76	1.92	1.83	1.99	−0.07	−0.07
	Me	0.93	0.77	0.99	0.84	−0.06	−0.07
2-Iodopropane (**35**)	Me	1.84	1.85	1.89	1.92	−0.05	−0.07
	CH	4.43	4.55	4.32	4.48	0.12	0.07
Fluorobenzene (**36**)	o	7.19	7.03	7.05	7.03	0.14	0.00
	m	7.42	7.29	7.33	7.29	0.09	0.00
	p	7.21	7.07	7.13	7.07	0.09	0.00
Chlorobenzene (**37**)	o	7.44	7.27	7.33	7.27	0.11	0.00
	m	7.40	7.21	7.28	7.21	0.13	0.00
	p	7.34	7.20	7.22	7.20	0.12	0.00
Bromobenzene (**38**)	o	7.57	7.44	7.49	7.44	0.08	0.00
	m	7.34	7.17	7.22	7.17	0.12	0.00
	p	7.39	7.23	7.28	7.23	0.11	0.00
Iodobenzene (**39**)	o	7.74	7.67	7.70	7.67	0.04	0.00
	m	7.19	7.02	7.10	7.02	0.09	0.00
	p	7.40	7.24	7.32	7.24	0.08	0.00
N,N-Dimethylformamide (**40**)	Me(*cis*)	2.89	2.89	2.88	2.95	0.01	−0.06
	Me(*trans*)	2.73	2.74	2.94	2.94	−0.21	−0.20
	C$\underline{H}$O	7.95	7.98	8.02	8.02	−0.07	−0.04
N,N-Dimethylacetamide (**41**)	Me(*cis*)	2.95	2.91	3.01	2.97	−0.06	−0.06
	Me(*trans*)	2.79	2.70	2.94	2.90	−0.15	−0.20
	MeCO	1.96	1.86	2.08	2.01	−0.12	−0.15

Table 8.5   (*Continued*)

Compound	Proton	δ(DMSO)		δ(CDCl₃)		Δδ(DMSO–CDCL₃)	
		Obs.	Calc.	Obs.	Calc.	Obs.	Calc.
N-Methyl pyrrolidinone (**42**)	N–Me	2.69	2.75	2.85	2.95	−0.17	−0.20
	3	2.16	2.15	2.37	2.37	−0.21	−0.22
	4	1.91	2.00	2.03	2.10	−0.13	−0.10
	5	3.29	3.28	3.39	3.34	−0.10	−0.06
N-Methyl valerolactam (**43**)	N–Me	2.80	2.65	2.94	2.92	−0.15	−0.27
	3	2.18	2.12	2.37	2.34	−0.19	−0.22
	4	1.72	1.82	1.82	1.89	−0.09	−0.07
	5	1.70	1.72	1.81	1.89	−0.11	−0.17
	6	3.23	3.02	3.29	3.15	−0.06	−0.13
N-Methyl caprolactam (**44**)	N–Me	2.83	2.74	2.98	2.94	−0.15	−0.20
	3	2.40	2.30	2.52	2.48	−0.12	−0.18
	4	1.52	1.57	1.66	1.64	−0.15	−0.07
	5	1.65	1.55	1.70	1.62	−0.05	−0.07
	6	1.55	1.58	1.65	1.68	−0.09	−0.10
	7	3.34	3.14	3.36	3.20	−0.02	−0.06
Formyl pyrrolidine (**45**)	2	3.22	3.07	3.43	3.34	−0.21	−0.27
	3	1.79	1.57	1.90	1.74	−0.11	−0.17
	4	1.81	1.59	1.92	1.76	−0.11	−0.17
	5	3.44	3.27	3.43	3.40	0.02	−0.13
	CHO	8.17	7.98	8.26	8.02	−0.09	−0.04
Formyl piperidine (**46**)	2	3.29	3.24	3.30	3.30	−0.01	−0.06
	3	1.61	1.57	1.58	1.67	0.03	−0.10
	4	1.42	1.51	1.54	1.58	−0.12	−0.07
	5	1.47	1.47	1.69	1.57	−0.22	−0.10
	6	3.33	3.36	3.48	3.56	−0.15	−0.20
	CHO	7.95	7.98	8.01	8.02	−0.05	−0.04
N-Acetyl pyrrolidine (**47**)	2	3.38	3.21	3.46	3.41	−0.08	−0.20
	3	1.78	1.62	1.86	1.72	−0.08	−0.10
	4	1.87	1.67	1.96	1.77	−0.09	−0.10
	5	3.25	3.37	3.42	3.43	−0.17	−0.06
	CO.Me	1.93	1.85	2.05	2.00	−0.12	−0.15
4-Methyl	2a	2.97	2.48	2.54	2.68	0.43	−0.20
N-acetylpiperidine (**48**)	2e	4.32	4.50	4.55	4.70	−0.23	−0.20
	3a	0.90	0.97	1.08	1.07	−0.17	−0.10
	3e	1.56	1.64	1.65	1.74	−0.09	−0.10
	4a	1.54	1.37	1.60	1.44	−0.06	−0.07
	5a	1.03	1.08	1.13	1.18	−0.10	−0.10
	5e	1.63	1.70	1.69	1.80	−0.06	−0.10
	6a	2.48	2.71	3.02	2.77	−0.54	−0.06
	6e	3.75	4.00	3.77	4.06	−0.02	−0.06
	Me(4e)	0.90	0.87	0.95	0.94	−0.06	−0.07
	CO.Me	1.96	1.80	2.08	1.96	−0.12	−0.16
4-Phenyl	2a	3.15	2.58	2.63	2.78	0.52	−0.20
N-acetylpiperidine (**49**)	2e	4.56	4.58	4.79	4.78	−0.23	−0.20
	3a	1.60	1.69	1.61	1.79	−0.01	−0.10
	3e	1.79	1.98	1.88	2.08	−0.09	−0.10
	4a	2.61	2.73	2.73	2.80	−0.13	−0.07
	5a	1.46	1.81	1.65	1.91	−0.19	−0.10
	5e	1.72	2.02	1.91	2.12	−0.20	−0.10
	6a	2.78	2.82	3.16	2.88	−0.38	−0.06

	6e	3.95	4.09	3.92	4.15	0.02	−0.06
	CO.Me	2.06	1.85	2.14	2.00	−0.08	−0.15
	o	7.27	7.19	7.21	7.19	0.06	0.00
	m	7.33	7.32	7.32	7.32	0.01	0.00
	p	7.23	7.20	7.22	7.20	0.00	0.00
N-Methylformanilide (**50**)	N.Me	3.29	3.15	3.32	3.35	−0.03	−0.20
	o	7.34	7.22	7.18	7.06	0.16	0.16
	m	7.44	7.59	7.41	7.59	0.03	0.00
	P	7.26	7.46	7.29	7.46	−0.03	0.00
	CHO	8.53	8.73	8.48	8.68	0.04	0.05
endo-Formylindoline (**51**)	2	4.10	3.90	4.12	3.96	−0.02	−0.06
	3	3.11	2.99	3.20	3.09	−0.09	−0.10
	4	7.01	7.19	7.24	7.19	−0.23	0.00
	5	7.15	7.20	7.07	7.20	0.08	0.00
	6	7.25	7.31	7.21	7.31	0.04	0.00
	7	8.05	8.11	8.08	7.95	−0.03	0.16
	CHO	8.81	8.75	8.53	8.50	0.28	0.25
exo-Formylindoline (**52**)	2	3.89	3.71	4.07	3.91	−0.18	−0.20
	3	3.08	2.94	3.16	3.04	−0.08	−0.10
	4	7.25	7.25	7.26	7.25	−0.01	0.00
	5	7.15	7.26	7.06	7.26	0.09	0.00
	6	7.40	7.38	7.21	7.38	0.19	0.00
	7	7.00	7.27	7.18	7.11	−0.18	0.16
	CHO	9.01	8.74	8.95	8.69	0.06	0.05
Sarcosine (**53**)	CH$_2$	3.98	3.95	3.98	3.95	0.00	0.00
	Me	2.96	3.00	2.96	3.00	0.00	0.00

8.6.4 Protic Compounds

Very large positive Δδ values (1–4 ppm) were found for the labile hydrogens of these compounds (Tables 8.6 and 8.9). In addition significant ($\sim$ 0.3 ppm) values were observed for some H.C.X protons. The NH protons of amides have an intriguing solvent dependence in that the *cis* and *trans* NH protons of formamide (**30**, Figure 8.10) change assignments in going from CDCl$_3$ to DMSO solvent (Table 8.6). These assignments were obtained from the H.C.NH couplings and NOE data.[22] These effects were simulated by the expression Δδ (H.N.CO) $= -0.20-0.14\cos \theta$ but in this case due to the approximate planarity of the amide group only dihedral angles of ca. 0 and 180° are found.

For both the aliphatic H.C.C.N protons of amines and the H.C.C.O protons of alcohols a dihedral angle dependence of Δδ was found. The relevant data and graphs are given in Table 8.7 and Figure 8.11 for the amines and Table 8.8 and Figure 8.12 for the alcohols.

The rotationally averaged value of the above function is −0.13 ppm (cf. obs. values of −0.12, −0.12, −0.11 and −0.11 ppm) for the methyl protons of diethylamine and the α CH$_2$ protons of propylamine, butylamine and amylamine.

The rotationally averaged correction from Figure 8.12 is −0.15 ppm (cf. methyl of ethanol and 2-propanol, obs. −0.18 and −0.17 res. and C$_{(2)}$H$_2$ of n-propanol and n-butanol, obs. −0.17 and −0.16, respectively).

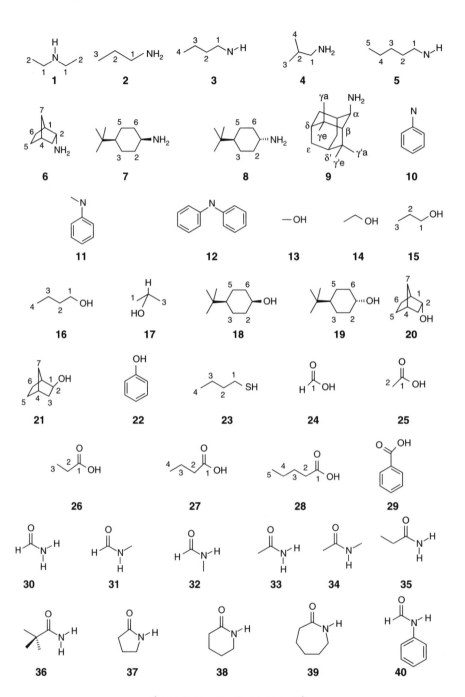

Figure 8.10   Protic compounds.

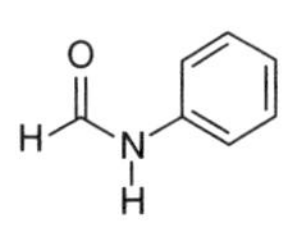

41 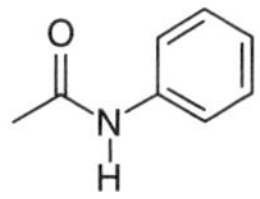   **42** 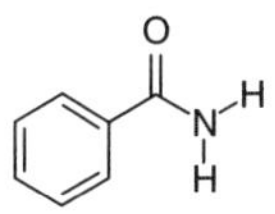   **43** 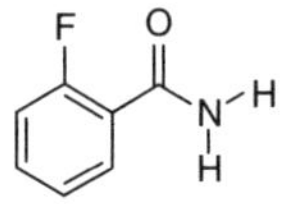   **44**

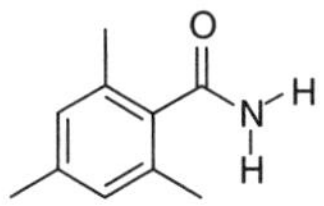

45 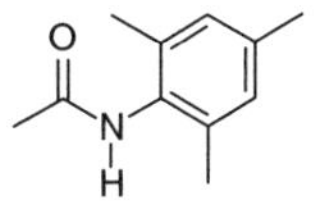   **46** 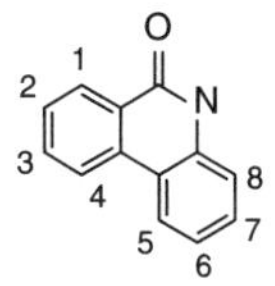   **47** 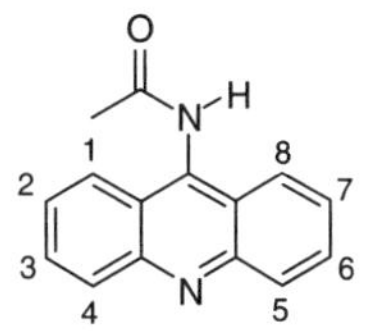   **48**

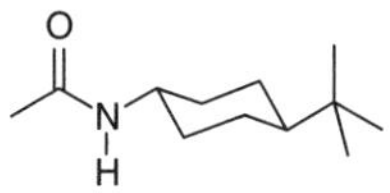

49 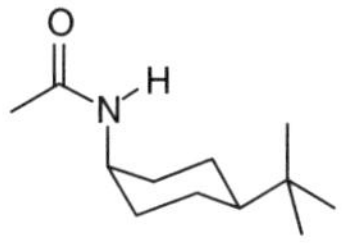   **50**

Figure 8.10   (Continued).

Table 8.6   ¹H chemical shifts of protic solutes in CDCl₃ and DMSO^a

Compound	Proton	δ(DMSO)		δ(CDCl₃)		Δδ(DMSO–CDCL₃)	
		Obs.	Calc.	Obs.	Calc.	Obs.	Calc.
Diethylamine (**1**)	NH	1.29	1.65	0.80	1.01	0.49	0.64
	CH₂	2.50	2.39	2.66	2.55	−0.15	−0.16
	Me	0.99	0.74	1.11	0.87	−0.12	−0.13
n-Propylamine (**2**)	NH₂	—	1.73	1.78^b	1.11	—	0.62
	1	2.48	2.43	2.65	2.67	−0.18	−0.16
	2	1.33	1.24	1.46	1.36	−0.12	−0.11
	Me	0.84	0.83	0.91	0.91	−0.08	−0.07
n-Butylamine (**3**)	NH₂	2.66	2.66	1.11^b	1.12	1.55	0.64
	1	2.52	2.50	2.69	2.68	−0.18	−0.16
	2	1.32	1.19	1.43	1.20	−0.11	−0.11
	3	1.28	1.17	1.35	1.24	−0.07	−0.07
	Me	0.86	0.83	0.92	0.89	−0.06	−0.07
Isobutylamine (**4**)	NH₂	2.93*	1.82	1.71*	1.18	1.22	0.64
	1	2.34	2.40	2.50	2.56	−0.16	−0.16
	2	1.48	1.66	1.58	1.77	−0.11	−0.11
	Me	0.83	0.85	0.90	1.02	−0.07	−0.07
Amylamine (**5**)	NH₂	1.22	1.74	1.00	1.10	0.22	0.64
	1	2.50	2.50	2.68	2.68	−0.18	−0.16
	2	1.33	1.10	1.44	1.22	−0.12	−0.11
	3	1.26	0.99	1.32	1.05	−0.06	−0.07
	4	1.26	1.19	1.32	1.25	−0.06	−0.07
	Me	0.86	0.80	0.90	0.84	−0.04	−0.07

Table 8.6   (*Continued*)

Compound	Proton	δ(DMSO)		δ(CDCl₃)		Δδ(DMSO–CDCL₃)	
		Obs.	Calc.	Obs.	Calc.	Obs.	Calc.
endo-2-Amino	NH₂	2.20*	1.79	—	1.15	—	0.64
norbornane (**6**)	1	1.95	2.14	2.07	2.26	−0.13	−0.12
	2x	3.12	2.99	3.26	3.15	−0.14	−0.16
	3n	0.50	0.53	0.59	0.62	−0.09	−0.09
	3x	1.80	1.77	1.96	1.91	−0.16	−0.14
	4	2.05	2.12	2.14	2.19	−0.09	−0.07
	5n	1.14	1.19	1.22	1.26	−1.22	−0.07
	5x	1.44	1.44	1.55	1.51	−1.55	−0.07
	6n	1.78	1.57	1.73	1.64	−1.73	−0.07
	6x	1.24	1.37	1.37	1.44	−1.37	−0.07
	7a	1.19	1.15	1.30	1.22	−1.30	−0.07
	7s	1.30	1.14	1.39	1.21	−1.39	−0.07
4-t-Butyclo	NH₂	1.24**	1.82	1.30**	1.18	—	0.64
hexylamine(*cis*) (**7**)	1e	3.03	3.05	3.15	3.21	−0.11	−0.16
	2a/6a	1.37	1.27	1.54	1.44	−0.17	−0.17
	2e/6e	1.56	1.51	1.65	1.61	−0.10	−0.10
	3a/5a	0.90	1.23	1.27	1.30	−0.37	−0.07
	3e/5e	1.37	1.62	1.53	1.69	−0.16	−0.07
	4a	0.94	1.00	0.96	1.07	−0.02	−0.07
	t-Bu Me	0.83	0.84	0.86	0.91	−0.03	−0.07
4-t-Butyclo	NH₂	1.24**	1.72	1.30**	1.08	—	0.64
hexylamine(*trans*) (**8**)	1a	2.40	2.60	2.55	2.76	−0.15	−0.16
	2a/6a	0.94	1.11	1.03	1.22	−0.09	−0.11
	2e/6e	1.78	1.57	1.89	1.68	−0.11	−0.11
	3a/5a	0.94	0.98	1.03	1.05	−0.09	−0.07
	3e/5e	1.66	1.69	1.76	1.76	−0.09	−0.07
	4a	0.94	1.03	0.96	1.10	−0.02	−0.07
	t-Bu Me	0.82	0.84	0.84	0.91	−0.03	−0.07
2-Adamantanamine (**9**)	NH₂	1.46	1.94	—	1.30	—	0.64
	α	2.85	2.92	2.98	3.08	−0.13	−0.16
	β	1.60	1.78	1.72	1.89	−0.12	−0.11
	γa	2.05	1.94	1.98	2.01	0.07	−0.07
	γe	1.37	1.53	1.53	1.60	−0.16	−0.07
	δ	1.70	1.85	1.79	1.92	−0.09	−0.07
	ε	1.65	1.61	1.72	1.68	−0.07	−0.07
	γ′a	1.66	1.59	1.73	1.66	−0.07	−0.07
	γ′e	1.76	1.63	1.85	1.70	−0.09	−0.07
	δ′	1.65	1.85	1.82	1.92	−0.17	−0.07
Aniline (**10**)	NH₂	4.94	4.87	3.61	3.54	1.34	1.33
	o	6.55	6.47	6.67	6.59	−0.12	−0.12
	m	6.98	7.03	7.14	7.11	−0.16	−0.08
	p	6.47	6.60	6.75	6.68	−0.13	−0.08
N-Methylaniline (**11**)	NH	5.52	5.29	3.66	3.64	1.86	1.65
	o	6.52	6.49	6.61	6.61	−0.09	−0.12
	m	7.07	7.03	7.18	7.11	−0.12	−0.08
	p	6.51	6.59	6.70	6.67	−0.20	−0.08
	Me	2.65	2.71	2.83	2.87	−0.18	−0.16

N-Phenylaniline (**12**)	NH	8.10	8.37	5.68	5.71	2.42	2.66
	o	7.06	6.91	7.07	7.03	−0.01	−0.12
	m	7.22	7.09	7.26	7.17	−0.04	−0.08
	p	6.81	6.69	6.92	6.77	−0.11	−0.08
Methanol (**13**)	OH	4.05	4.04	0.85	0.89	3.20	3.15
	Me	3.17	3.10	3.48	3.37	−0.31	−0.27
Ethanol (**14**)	OH	4.31	4.13	1.10	1.14	3.21	2.99
	CH_2	3.44	3.47	3.71	3.74	−0.27	−0.27
	Me	1.06	1.02	1.24	1.18	−0.18	−0.16
Propanol (**15**)	OH	4.31	4.13	1.22	1.14	3.09	2.99
	1	3.34	3.30	3.59	3.57	−0.25	−0.27
	2	1.42	1.29	1.59	1.46	−0.17	−0.17
	Me	0.84	0.78	0.94	0.85	−0.10	−0.07
n-Butanol (**16**)	OH	4.30	4.16	1.17	1.17	3.13	2.99
	1	3.38	3.32	3.64	3.59	−0.26	−0.27
	2	1.40	1.11	1.56	1.28	−0.16	−0.17
	3	1.30	1.08	1.39	1.15	−0.09	−0.07
	Me	0.87	0.77	0.94	0.84	−0.07	−0.07
2-Propanol (**17**)	OH	4.30	4.21	1.23	1.20	3.07	3.01
	Me	1.04	1.09	1.21	1.24	−0.17	−0.15
	CH	3.77	3.71	4.02	3.98	−0.25	−0.27
(*cis*) 4-t-Butylcyclohexanol (**18**)	OH	4.11	4.29	1.40***	1.27	2.77	3.02
	1e	3.80	3.75	4.03	4.02	−0.23	−0.27
	2a/6a	1.34	1.38	1.49	1.52	−0.15	−0.14
	2e/6e	1.67	1.61	1.83	1.78	−0.16	−0.17
	3a/5a	1.34	1.28	1.35	1.35	−0.01	−0.07
	3e/5e	1.46	1.60	1.54	1.67	−0.08	−0.07
	4a	0.93	0.96	0.99	1.03	−0.06	−0.07
	tBu Me	0.82	0.83	0.86	0.90	−0.04	−0.07
(*trans*) 4-t-Butylcyclohexanol (**19**)	OH	4.39	4.26	1.40***	1.25	2.99	3.01
	1a	3.25	3.34	3.52	3.61	−0.27	−0.27
	2a/6a	1.07	1.18	1.22	1.35	−0.15	−0.17
	2e/6e	1.84	1.70	2.01	1.86	−0.17	−0.16
	3a/5a	0.96	0.99	1.05	1.06	−0.09	−0.07
	3e/5e	1.68	1.69	1.78	1.76	−0.11	−0.07
	4a	0.92	1.04	0.97	1.11	−0.05	−0.07
	tBu Me	0.82	0.84	0.85	0.91	−0.03	−0.07
endo- 2-Norborneol (**20**)	OH	4.43	4.19	—	1.18	—	3.01
	1	2.06	2.12	2.25	2.30	−0.19	−0.18
	2x	4.00	3.77	4.23	4.04	−0.23	−0.27
	3n	0.72	0.82	0.84	0.93	−0.13	−0.11
	3x	1.76	1.77	1.95	1.95	−0.19	−0.18
	4	2.06	2.12	2.17	2.19	−0.11	−0.07
	5n	1.22	1.21	1.35	1.28	−0.13	−0.07
	5x	1.46	1.43	1.57	1.50	−0.12	−0.07
	6n	1.86	1.68	1.87	1.75	−0.01	−0.07
	6x	1.22	1.34	1.35	1.41	−0.13	−0.07
	7a	1.22	1.16	1.35	1.23	−0.13	−0.07
	7s	1.22	1.16	1.35	1.23	−0.13	−0.07
exo-2-Norborneol (**21**)	OH	4.38	4.26	—	1.26	—	3.00
	1	1.99	1.99	2.14	1.96	−0.14	0.03
	2n	3.52	3.51	3.76	3.78	−0.23	−0.27
	3n	1.49	1.46	1.66	1.64	−0.18	−0.18
	3x	1.16	1.13	1.29	1.24	−0.12	−0.11

Table 8.6   (*Continued*)

Compound	Proton	δ(DMSO)		δ(CDCl₃)		Δδ(DMSO–CDCL₃)	
		Obs.	Calc.	Obs.	Calc.	Obs.	Calc.
	4	2.14	2.13	2.25	2.20	−0.12	−0.07
	5n	0.94	1.13	1.02	1.20	−0.08	−0.07
	5x	1.35	1.45	1.44	1.52	−0.09	−0.07
	6n	0.94	1.18	1.02	1.25	−0.08	−0.07
	6x	1.35	1.47	1.44	1.54	−0.09	−0.07
	7a	0.98	1.08	1.12	1.15	−0.14	−0.07
	7s	1.49	1.49	1.57	1.56	−0.09	−0.07
Phenol (**22**)	OH	9.29ᶜ	9.28	4.69ᶜ	4.71	4.60	4.57
	o	6.75	6.78	6.83	6.78	−0.08	0.00
	m	7.15	7.05	7.24	7.13	−0.09	−0.08
	p	6.76	6.78	6.93	6.86	−0.17	−0.08
n-Butylthiol (**23**)	SH	2.17	2.18	1.31	1.32	0.86	0.86
	1	2.47	2.32	2.53	2.37	−0.05	−0.05
	2	1.52	1.30	1.60	1.38	−0.08	−0.08
	3	1.36	1.09	1.41	1.16	−0.05	−0.07
	Me	0.87	0.77	0.91	0.84	−0.04	−0.07
Formic acid (**24**)	COOH	12.50	12.54	10.85	10.89	1.65	1.65
	CH	8.13	8.00	8.05	8.07	0.08	−0.08
Acetic acid (**25**)	COOH	11.91	11.92	11.51	11.07	0.40	0.85
	Me	1.91	1.98	2.10	2.08	−0.19	−0.10
Propionic acid (**26**)	COOH	11.90	11.95	10.35	11.10	1.55	0.85
	CH₂	2.21	2.28	2.39	2.30	−0.17	−0.02
	Me	1.00	1.13	1.16	1.20	−0.16	−0.07
Butyric acid (**27**)	COOH	11.91	11.96	11.10	11.10	0.81	0.86
	2	2.17	2.11	2.34	2.13	−0.16	−0.02
	3	1.51	1.76	1.67	1.83	−0.16	−0.07
	Me	0.88	0.90	0.98	0.97	−0.10	−0.07
Valeric acid (**28**)	COOH	—	11.96	—	11.10	—	0.86
	2	2.19	2.13	2.36	2.15	−0.16	−0.02
	3	1.48	1.58	1.63	1.65	−0.15	−0.07
	4	1.29	1.20	1.38	1.27	−0.09	−0.07
	Me	0.87	0.83	0.93	0.90	−0.06	−0.07
Benzoic acid (**29**)	COOH	—	13.72	—	12.07	—	1.65
	o	8.02	8.00	8.13	7.93	−0.10	0.07
	m	7.57	7.56	7.48	7.49	0.09	0.07
	p	7.69	7.53	7.62	7.46	0.07	0.07
Formamide (**30**)	NH (*cis*)	7.14	7.11	5.80	5.77	1.34	1.34
	NH (*trans*)	7.41	7.41	5.48	5.45	1.93	1.96
	CHO	7.98	8.00	8.23	8.25	−0.25	−0.25
N-Methylformamide	NH (*trans*)	7.90	7.92	5.55	5.51	2.35	2.41
(*trans*) (**31**)	NMe	2.59	2.47	2.86	2.74	−0.27	−0.27
	CHO	8.01	7.97	8.19	8.13	−0.18	−0.16
N-Methylformamide	NH (*cis*)	7.90	7.62	5.86	5.83	2.04	1.79
(*cis*) (**32**)	NMe	2.72	2.68	2.94	2.95	−0.22	−0.27
	CHO	7.81	7.97	8.06	8.13	−0.25	−0.16
Acetamide (**33**)	NH (cis)	6.70	6.66	5.42	5.81	1.29	0.85
	NH (*trans*)	7.30	7.11	5.42	5.22	1.88	1.89
	CO.Me	1.76	1.88	2.03	2.03	−0.27	−0.15

N-Methyl acetamide (**34**)	NH (*trans*)	7.70	7.58	5.53	5.25	2.17	2.33
	NMe	2.50	2.51	2.80	2.78	−0.30	−0.27
	CO.Me	1.78	1.87	1.98	2.02	−0.20	−0.15
Propionamide (**35**)	NH (*cis*)	6.62	6.71	6.14	5.87	0.48	0.84
	NH (*trans*)	7.16	7.14	5.38	5.25	1.78	1.89
	2	2.04	2.22	2.26	2.29	−0.22	−0.07
	3	0.97	1.00	1.17	1.07	−0.20	−0.07
tri-Methyl acetamide (**36**)	NH (*cis*)	6.64	6.75	5.56	5.91	1.08	0.84
	NH (*trans*)	6.97	7.15	5.22	5.25	1.74	1.90
	CO.Me	1.07	1.06	1.23	1.13	−0.16	−0.07
2-Pyrrolidinone (**37**)	NH	7.46	7.36	6.06	5.96	1.39	1.40
	3	2.07	2.15	2.30	2.38	−0.24	−0.23
	4	1.96	1.92	2.14	2.11	−0.18	−0.19
	5	3.20	3.28	3.40	3.42	−0.20	−0.14
Valerolactam (**38**)	NH	7.34	7.39	6.33	5.99	1.00	1.40
	3	2.11	2.13	2.36	2.36	−0.25	−0.23
	4	1.66	1.84	1.81	1.91	−0.15	−0.07
	5	1.65	1.64	1.79	1.90	−0.13	−0.26
	6	3.11	3.10	3.31	3.31	−0.21	−0.21
Caprolactam (**39**)	NH	7.34	7.33	6.35	5.98	0.99	1.35
	3	2.28	2.30	2.46	2.49	−0.17	−0.19
	4	1.52	1.60	1.69	1.66	−0.17	−0.06
	5	1.66	1.56	1.75	1.63	−0.10	−0.07
	6	1.49	1.49	1.65	1.68	−0.15	−0.19
	7	3.04	3.13	3.21	3.29	−0.16	−0.16
Formyl anilide (*cis*) (**40**)	NH	10.14	9.58	7.50	7.32	2.64	2.26
	o	7.15	6.98	7.09	6.92	0.06	0.06
	m	7.32	7.44	7.37	7.52	−0.05	−0.08
	p	7.07	7.28	7.21	7.36	−0.14	−0.08
	CHO	8.79	8.77	8.69	8.67	0.10	0.10
Formyl anilide (*trans*) (**41**)	NH	10.19	10.28	8.40	7.47	1.79	1.81
	o	7.59	7.32	7.55	7.26	0.05	0.06
	m	7.32	7.36	7.33	7.44	−0.01	−0.08
	p	7.09	7.16	7.15	7.24	−0.06	−0.08
	CHO	8.28	8.51	8.40	8.63	−0.13	−0.12
Z-Acetanilide (**42**)	NH	9.88	10.15	—	7.39	—	2.76
	o	7.56	7.66	7.49	7.60	0.07	0.06
	m	7.28	7.35	7.32	7.43	−0.04	−0.08
	p	7.02	7.13	7.11	7.21	−0.09	−0.08
	CO.Me	2.03	1.96	2.18	2.12	−0.14	−0.16
Benzamide (**43**)	NH (*cis*)	7.34	7.30	6.08	5.86	1.26	1.44
	NH (*trans*)	7.93	7.59	6.08	5.90	1.85	1.69
	o	7.87	7.63	7.82	7.71	0.05	−0.08
	m	7.44	7.45	7.45	7.53	−0.01	−0.08
	p	7.51	7.49	7.53	7.57	−0.02	−0.08
2-Fluorobenzamide (**44**)	NH (*cis* to O)	7.59	7.42	6.09	5.99	1.50	1.43
	NH (*trans* to O)	7.66	8.01	6.72	6.31	0.94	1.70
	3	7.29	7.23	7.15	7.31	0.14	−0.08
	4	7.55	7.51	7.51	7.59	0.04	−0.08
	5	7.32	7.24	7.29	7.32	0.03	−0.08
	6	7.69	8.14	8.14	8.22	−0.45	−0.08
2,4,6-Trimethyl benzamide (**45**)	NH (*cis*)	7.34	7.66	6.22	6.23	1.12	1.43
	NH (*trans*)	7.59	7.57	5.74	5.88	1.85	1.69
	Me (o)	2.22	2.21	2.32	2.28	−0.10	−0.07

Table 8.6   (*Continued*)

Compound	Proton	δ(DMSO)		δ(CDCl₃)		Δδ(DMSO–CDCL₃)	
		Obs.	Calc.	Obs.	Calc.	Obs.	Calc.
	m	6.81	6.81	6.84	6.89	−0.03	−0.08
	Me (p)	2.22	2.32	2.27	2.39	−0.05	−0.07
Trimethylacetanilide (**46**)	NH (*trans*)	9.20	8.85	6.66	6.10	2.54	2.75
	Me (o)	2.08	2.11	2.20	2.18	−0.12	−0.07
	m	6.85	6.78	6.90	6.86	−0.05	−0.08
	Me (p)	2.21	2.32	2.26	2.39	−0.05	−0.07
	CO.Me	2.01	1.97	2.20	2.21	−0.19	−0.24
Phenanthridone (**47**)	NH	11.65	11.55	9.17	9.25	2.48	2.30
	1	8.33	8.66	8.55	8.74	−0.22	−0.08
	2	7.65	7.56	7.62	7.64	0.02	−0.08
	3	7.86	7.74	7.82	7.82	0.04	−0.08
	4	8.51	8.47	8.31	8.55	0.20	−0.08
	5	8.39	8.46	8.24	8.54	0.15	−0.08
	6	7.27	7.46	7.32	7.54	−0.05	−0.08
	7	7.49	7.59	7.52	7.67	−0.03	−0.08
	8	7.38	7.38	7.22	7.24	0.16	0.14
Acridine amide (**48**)	NH	—	9.02	—	7.32	—	1.70
	1	8.08	7.82	7.86	7.82	0.22	0.00
	2	7.74	7.67	7.63	7.67	0.11	0.00
	3	7.93	7.83	7.81	7.83	0.12	0.00
	4	8.28	8.47	8.31	8.47	−0.03	0.00
	5	8.28	8.47	8.31	8.47	−0.03	0.00
	6	7.93	7.83	7.81	7.83	0.12	0.00
	7	7.74	7.67	7.63	7.67	0.11	0.00
	8	8.08	7.82	7.86	7.82	0.22	0.00
	CO.Me	2.25	1.95	2.29	2.17	−0.04	−0.22
cis-4 -t-Bu	NH	7.66	7.20	5.56	5.90	2.11	1.30
N−acetyl -1-	1e	3.85	3.75	4.11	4.07	−0.26	−0.32
aminocyclohexane (**50**)	2a/6a	1.47	1.37	1.51	1.56	−0.03	−0.19
	2e/6e	1.70	1.64	1.85	1.83	−0.15	−0.19
	3a/5a	0.94	1.12	1.03	1.19	−0.09	−0.07
	3e/5e	1.51	1.74	1.66	1.81	−0.15	−0.07
	4a	1.22	1.05	1.04	1.12	0.18	−0.07
	tBu Me	0.84	0.87	0.87	0.94	−0.03	−0.07
	CO.Me	1.83	1.84	1.99	1.99	−0.16	−0.15
trans-4-t-Bu N−acetyl	NH	7.61	7.61	5.42	5.27	2.19	2.34
1-aminocyclohexane (**49**)	1a	3.40	3.39	3.67	3.70	−0.28	−0.31
	2a/6a	0.98	1.09	1.07	1.28	−0.09	−0.19
	2e/6e	1.81	1.68	2.01	1.87	−0.21	−0.19
	3a/5a	1.06	1.14	1.11	1.21	−0.05	−0.07
	3e/5e	1.71	1.73	1.78	1.80	−0.06	−0.07
	4a	0.93	1.04	1.01	1.11	−0.08	−0.07
	tBu Me	0.83	0.86	0.84	0.93	−0.02	−0.07

* Indicates a broad peak.

** Compounds **7** and **8** were run as a mixture of isomers and gave one NH₂ peak.

*** Compounds **18** and **19** were run as a mixture of isomers and gave one OH peak.

ᵃ '—' indicates peak not observed.

ᵇ From Pouchert and Behnke.[60].

ᶜ From Mobli.[61].

Table 8.7   *Solvent contributions (Δδ) versus dihedral angles for the H.C.C.NH group*

θ (degrees)	δ (DMSO)	δ (CDCl₃)	Δδγ (exp.)	Δδγ (calc.)	Exp.–Calc.	Source (¹H/compound)
7	0.50	0.59	−0.09	−0.09	0.00	3n, (**5**), Figure 8.11
54	1.56	1.65	−0.10	−0.11	0.01	2e, (**6**), Figure 8.11
56	0.94	1.03	−0.09	−0.11	0.02	2a, (**7**), Figure 8.11
57	1.60	1.72	−0.12	−0.11	−0.01	βH, (**8**), Figure 8.11
62	1.78	1.89	−0.11	−0.11	0.00	2e, (**7**), Figure 8.11
74	1.95	2.07	−0.13	−0.12	−0.01	1H, (**5**), Figure 8.11
114	1.80	1.96	−0.16	−0.15	−0.01	3x, (**5**), Figure 8.11
170	1.37	1.54	−0.17	−0.17	0.00	2a, (**6**), Figure 8.11

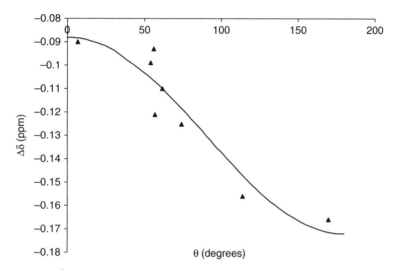

Figure 8.11   *Δδ versus θ (H̲.C.C.NH) for protons in conformationally fixed amines. Observed points versus the function* $\Delta\delta_\gamma = -0.13 + 0.042\cos\theta$.

Table 8.8   *Solvent contributions (Δδ) versus dihedral angles for the H.C.C.OH group*

θ (degrees)	Δδ (exp.)	Δδ (calc.)	Source (compound/proton)
3	−0.12	−0.11	3x, (**20**), Figure 8.12
3	−0.13	−0.11	3n, (**19**), Figure 8.12
39	−0.14	−0.14	1H, (**20**), Figure 8.12
53	−0.16	−0.15	2e, (**17**), Figure 8.12
53	−0.15	−0.15	2a, (**18**), Figure 8.12
62	−0.17	−0.17	2e, (**18**), Figure 8.12
75	−0.19	−0.18	1H, (**19**), Figure 8.12
118	−0.19	−0.18	3x, (**19**), Figure 8.12
119	−0.18	−0.18	3n, (**20**), Figure 8.12
171	−0.15	−0.14	2a, (**17**), Figure 8.12

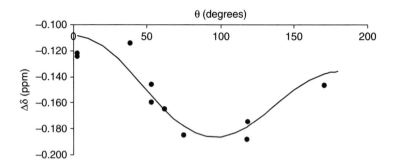

Figure 8.12 $\Delta\delta_y$ versus θ (*H*-C-C-OH) for protons in conformationally fixed alcohols. Observed points versus the function $\Delta\delta_y = -0.186 + 0.014\cos\theta + 0.064\cos^2\theta$.

8.6.5   Diols and Polyhydroxy Compounds

These compounds include diols, triols and inositols. The polyhydroxy compounds are insoluble in CDCl$_3$ but soluble in D$_2$O. It was shown (Chapter 6, Section 6.3 plus Abraham *et al.*[52,53] and Gottlieb *et al.*[54]) that ^{1}H chemical shifts in D$_2$O for a wide variety of alcohols are identical to the corresponding shifts in CDCl$_3$. Thus the solvent shifts $\Delta\delta = \delta$(DMSO)–δ(D$_2$O) were treated in precisely the same manner as the solvent effects of DMSO vs. CDCl$_3$ above.

These compounds may contain protons affected by β, γ and long-range effects from one or more hydroxyl and ether groups simultaneously. The observed experimental $\Delta\delta$ values were much less than the cumulative additions of all these factors. For example the observed $\Delta\delta$ value for the 3a/5a protons in *myo*-inositol (**9**, Figure 8.13) is −0.27 ppm. If this correction was simply an additive function of all the effects present (i.e.one β OH, two γ OHs at ca. 60° plus one long-range effect from the C$_1$OH) $\Delta\delta$ is calculated as −0.66 ppm. Thus a priority system was included in the program. For these protons only the β-OH correction was made, the remaining effects were ignored. This gave $\Delta\delta$ equal to −0.27 ppm, in good agreement with the observed value. For protons with no β OH but γ OH interactions such as 2a/4a/6a of *cis,cis,cis* 1,3,5 cyclohexane triol (**8**, Figure 8.13), both the γ-OH corrections were made but long-range effects ignored. This gave $\Delta\delta$ equal to −0.29 ppm (cf. the observed value of −0.27 ppm). This approach was applied to all the multifunctional compounds, with short-range effects taking priority over long-range effects. Where there was more than one of the same short-range effects, i.e. α, β or γ, all contributions were included. Also for those compounds insoluble in CDCl$_3$ the only solvent in which the OH proton can be observed is DMSO. Thus Table 8.9 for these compounds gives the observed vs. calculated shifts for the OH proton in DMSO solvent. The data in Table 8.9 show that this approach gives generally good agreement.

8.6.6   Chemical Shift Contributions

The discrete contributions (a, β, γ and long-range) to $\Delta\delta$ discussed above are collected in Table 8.10 for the individual functional groups examined.

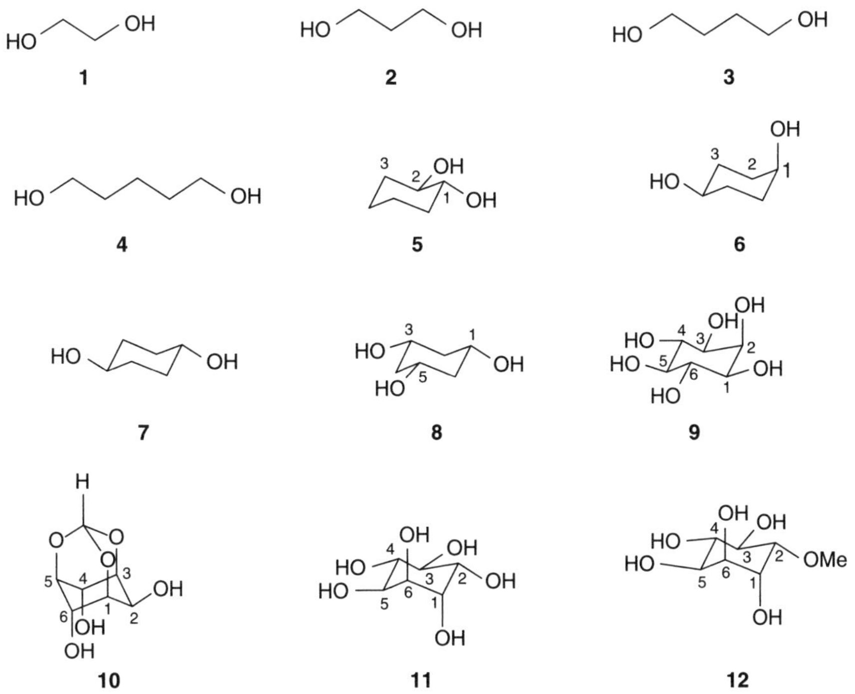

Figure 8.13   *Diols and polyhydroxy compounds.*

The prediction of the ¹H shifts in DMSO solution is dependent on both the accuracy of the original prediction of the shifts in CDCl₃ and on the DMSO correction. The accuracy of the CHARGE prediction has been dealt with in previous chapters. The observed vs. calculated correction Δδ(DMSO–CDCl₃) is shown in Figure 8.14 for all the shifts in this study. The linear correlation through the origin shown has slope equal to 0.991 with an rms error of 0.05 ppm for the 1138 protons. In Figure 8.14 the largest errors are for the labile protons of the acids, amides and diols as expected. This may be due to the large concentration dependence of these protons due to intermolecular hydrogen bonding.[62] Large errors also occur in some diols where intramolecular hydrogen bonding occurs in CDCl₃ solvent which is not observed in DMSO. Thus in propane-1,3-diol and butane-1,4-diol the OH chemical shift in CDCl₃ solution is less shielded than in the simple alcohols due to the presence of an intramolecular hydrogen bond. In DMSO solution the OH shift is normal, as expected. In contrast diols with no intramolecular hydrogen bond such as *cis* cyclohexane-1,4-diol, (**6**, Figure 8.13) give large Δδ values which are well reproduced. Intramolecular hydrogen bonding in these diols is considered in more depth elsewhere.[63]

The dilution shifts of other labile protons (e.g amides, acids, etc.) are also large but these protons often gave broad signals which were difficult to observe. For these reasons only the shifts in DMSO solvent (in which there is no appreciable dilution shift) were predicted and the observed vs. calculated shifts in DMSO were in good agreement.

The ¹H shifts of the NH protons of the aliphatic amines showed little consistency in either solvent and often the signals are too broad to observe. The NH₂⁺ protons of dimethylamine

Table 8.9   *¹H chemical shifts of diols and polyhydroxy solutes in CDCl₃ (D₂O) and DMSOᵃ*

Compound	Proton	δ(DMSO)		δ(CDCl₃)		Δδ(DMSO–CDCl₃)	
		Obs.	Calc.	Obs.	Calc.	Obs.	Calc.
Ethylene glycol (**1**)	OH	4.47	4.17	1.78	1.82*	2.69	2.35
	CH₂	3.38	3.40	3.65	3.67	−0.27	−0.27
1,3-Propane diol (**2**)	OH	4.32	4.16	1.89	1.17	2.43	2.99
	1,3	3.45	3.32	3.69	3.59	−0.24	−0.27
	2	1.56	1.27	1.8	1.60	−0.24	−0.33
1,4-Butanediol (**3**)	OH	4.38	4.16	1.83	1.17	2.55	2.99
	1,4	3.39	3.33	3.63	3.60	−0.24	−0.27
	2,3	1.43	1.14	1.6	1.31	−0.17	−0.17
1,5-Pentanediol (**4**)	OH	4.34	4.16	1.24	1.17	3.10	2.99
	1,5	3.38	3.33	3.62	3.60	−0.24	−0.27
	2,4	1.41	1.08	1.58	1.31	−0.17	−0.23
	3	1.31	0.93	1.41	1.00	−0.10	−0.07
Cyclohexane	1a/2a	3.11	3.20	3.37	3.47	−0.26	−0.27
1,2-diol (*trans*) (**5**)	3a/6a	1.13	1.14	1.25	1.33	−0.12	−0.19
	3e/6e	1.74	1.65	1.92	1.85	−0.18	−0.20
	4a/5a	1.13	1.16	1.25	1.22	−0.12	−0.06
	4e/5e	1.56	1.60	1.67	1.67	−0.11	−0.07
	OH	4.42	4.25	2.07	1.89	2.35	2.36
cis Cyclohexane	1e/4a	3.51	3.52	3.81	3.79	−0.29	−0.27
1,4-diol (**6**)	2a/6a/3e/5e	1.56	1.47	1.66	1.66	−0.11	−0.19
	2e/6e/3a/5a	1.40	1.48	1.66	1.67	−0.26	−0.19
	OH	4.25	4.27	1.55**	1.09	2.70	3.18
trans Cyclohexane	1a/4a	3.36	3.33	3.66	3.60	−0.29	−0.27
1,4-diol (**7**)	2a/6a/3a/5a	1.16	1.20	1.34	1.36	−0.18	−0.16
	2e/6e/3e/5e	1.74	1.68	1.93	1.85	−0.20	−0.17
	OH	4..38	4.26	1.55**	1.08	2.83	3.30

Compound	Proton	δ(DMSO)		δ(D₂O)		Δδ(DMSO–D₂O)	
		Obs.	Calc.	Obs.	Calc.	Obs.	Calc.
cis,cis,cis 1,3,5-	1/3/5	3.34	3.36	3.70	3.63	−0.37	−0.27
Cyclohexane triol	2a/4a/6a	0.97	1.13	1.24	1.48	−0.27	−0.35
(eq) (**8**)	2e/4e/6e	1.96	1.66	2.22	2.02	−0.26	−0.36
	OH	4.49	4.26	—	—	—	—
myo Inositol	1e	3.72	3.67	4.07	3.94	−0.34	−0.27
(1ax-5eq) (**9**)	2a/6a	3.14	3.36	3.54	3.63	−0.40	−0.27
	3a/5a	3.37	3.40	3.64	3.68	−0.27	−0.28
	4a	2.93	3.22	3.27	3.49	−0.34	−0.27
	OH (1,3)	4.51	4.59	—	—	—	—
	OH(2)	4.55	4.83	—	—	—	—
	OH(4,6)	4.46	4.76	—	—	—	—
	OH(5)	4.31	4.77	—	—	—	—
1,3,5-o-Methylidine	1e/3e	3.95	3.91	4.25	4.21	−0.29	−0.30
myo inositol	2	4.01	3.94	4.28	4.21	−0.27	−0.27
(1eq-5ax) (**10**)	4e/6e	4.28	4.13	4.59	4.40	−0.32	−0.27
	5	4.07	3.92	4.34	4.21	−0.27	−0.29
	CH	5.46	5.32	5.61	5.47	0.00	−0.15
	OH (2)	5.32	5.51	—	—	—	—
	OH(4,6)	5.48	5.44	—	—	—	—

(+) *chiro*-Inositol	1,6	3.65	3.67	4.03	3.94	−0.37	−0.27
(4eq-2ax) (**11**)	2,5	3.42	3.51	3.77	3.79	−0.34	−0.28
	3,4	3.27	3.37	3.60	3.64	−0.33	−0.27
	OH (1,6)	4.36	4.99	—	—	—	—
	OH(2,5)	4.03	4.51	—	—	—	—
	OH(3,4)	4.15	4.75	—	—	—	—
Quebrachitol	1	3.87	3.88	4.27	4.15	−0.41	−0.27
(4eq-2ax) (**12**)	2	3.10	2.87	3.40	3.17	−0.30	−0.30
	3	3.38	3.61	3.62	3.88	−0.24	−0.27
	4	3.29	3.34	3.60	3.61	−0.31	−0.27
	5	3.43	3.54	3.75	3.81	−0.32	−0.27
	6	3.68	3.71	4.06	3.98	−0.38	−0.27
	OMe	3.41	3.32	3.47	3.43	−0.06	−0.11
	OH(1)	4.60	5.39	—	—	—	—
	OH(3)	4.39	5.13	—	—	—	—
	OH(4)	4.41	4.80	—	—	—	—
	OH(5)	4.26	4.78	—	—	—	—
	OH(6)	4.62	5.10	—	—	—	—

[a] '—' indicates peak not observed.
* O.C.C.O gauche.
** Compounds **6** and **7** were run as a mixture to give one broad peak.

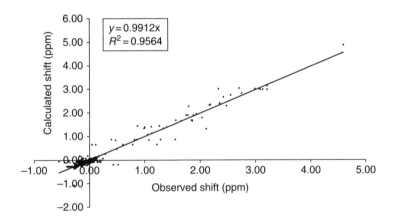

Figure 8.14 *Calculated versus observed solvent shifts,* $\Delta\delta(DMSO–CDCl_3)$.

hydrochloride occur at 9.2 ppm in $CDCl_3$ compared to the free base at ca. 1 ppm,[60] hence traces of acid will produce large shifts of the NH protons and also broaden the signal due to exchange. This is the most probable reason for the lack of consistency of these shifts in either solvent. These protons are thus not useful for diagnostic purposes.

The aliphatic halo compounds are aprotic but large $\Delta\delta$ values were observed for *geminal*, di and tri-halo compounds. All the haloalkanes examined exhibit similar behaviour, with the β proton(s) deshielded and the γ and long-range protons shielded (Table 8.10). For

Table 8.10   Contributions to the DMSO solvent effect (Δδ) in ppm for functional groups

Aliphatic group	α(H.X)	β(H.C.X)	γ(H.C.C.X)	LR(H.C. . . .X)
NH₂, NHR	0.64	−0.16	Figure 8.11	−0.05
OH	2.88	−0.27	Figure 8.12	−0.08
SH	0.86	−0.05	−0.08	−0.08
CO₂H	1.13	0.08	−0.17	−0.08
NH.CO	–	1.63	See text	−0.08
NR₂	–	−0.09	−0.09	−0.05
OR	–	−0.11	−0.09	−0.06
CHO	–	−0.14	−0.08	−0.06
CO.R	–	–	Equation (8.15)	−0.06
CO₂R	–	−0.09	−0.05	−0.06
Halogenᵃ	–	0.10	−0.07	−0.06
Aromatic group				
NH₂	1.34	–	−0.12	−0.20
NHR	1.65	–	−0.07	−0.16
H.N.Ph₂	2.66	–	−0.07	−0.07
OH	4.60	–	−0.08	−0.13
CO₂H	b	–	−0.11	0.08
≡CH	1.13	–	–	0.00
NR₂	–	–	−0.07	−0.07,0.05 (NPh₂)
OR	–	–	−0.01	0.03
CHO	–	–	0.00	0.09
CO.R	–	–	0.00	0.07
CO₂R	–	–	0.06	0.11
F	–	–	0.14	0.09
Cl	–	–	0.11	0.12
Br	–	–	0.08	0.12
I	–	–	0.04	0.09

ᵃ *CHCl₂* and HCCl₃ have additional contributions of 0.49 and 1.08 ppm, respectively.
ᵇ The acid proton was not observed in either solvent.

monohalo substitution the β HCX contribution is small (ca. 0.10 ppm, Table 8.10). This increases to 0.49 ppm in dichloromethane and 1.08 ppm in chloroform. The solvation site of these solute molecules appears to be the β-protons suggesting that these protons are hydrogen bonding to the DMSO (see below).

The long-range effects for protons in polar compounds were all small. There was a constant value of ca. −0.06 ppm for aliphatic protons. For aromatic protons the values are ca. ±0.1 ppm, but there is no obvious trend and each aromatic substituent was considered separately.

The above model of the solvent effect of DMSO on the solute ¹H chemical shifts does not give any chemical explanation of these effects. It is thus pertinent to consider the possible causes of these shifts with reference to Equation (8.14). For protic solutes hydrogen bonding is the generally accepted mechanism for the solvent effects in DMSO solution. All the labile protons are deshielded from CDCl₃ to DMSO. In a separate investigation it was shown that the solvent effect of DMSO vs. CDCl₃ on the protic hydrogen chemical shifts (at limiting dilution in CDCl₃) is closely correlated with the hydrogen bond acidity (*A* value) of the solute.[64] This correlation is valid not only for the simple alcohols and amides but also for the carbon hydrogen bond acceptors of $CHCl_3$ and CH_2Cl_2. Thus the large positive shifts of these protons in DMSO vs. $CDCl_3$ can be attributed to the direct effect of the hydrogen bond. This large positive effect on the α proton is accompanied in most cases by a negative effect on the β (H.C.XH) proton (Table 8.10). This could be due to the effect of the hydrogen bond being transmitted to the β proton or to the direct effect of the attached DMSO molecule. The ^1H shifts of all the alcohols in Tables 8.6 and 8.9 which are soluble in D_2O and $CDCl_3$ are identical in the two solvents.[52] These alcohols will be strongly hydrogen bonded to the D_2O solvent and this suggests that the primary cause of the β, γ and long-range proton shifts in DMSO is due to the direct effect of the attached DMSO molecule and not to any transmission via the hydrogen bond.

The accepted effects of the solvent from Equation (8.14) are the anisotropy, the electric field and the van der Waals shielding due to the solvent molecule. Of these three effects the van der Waals shielding can be eliminated as again if this is present it should also occur in D_2O solvent. There was also no noticeable electric field effect in D_2O (due to the identity of the shifts in D_2O and $CDCl_3$). However the dipole moment of DMSO (3.9D) is much greater than that of D_2O (1.85D). These considerations suggest that the major cause of the β, γ and long-range proton shifts in DMSO is the electric field and magnetic anisotropy of the DMSO solvent.

8.7   Summary

In this chapter we have considered the various factors which can affect both the calculated and the observed ^1H chemical shifts of molecules in solution. The calculated shifts are defined by the molecular geometry and in the first section we detail the various approaches to determining the geometries of molecules in solution. For the *ab initio* calculations both the level of theory and basis set variations should be considered before obtaining an acceptable geometry. In the MM calculations large errors may occur when the force field has not been parameterized for this type of molecule. Thus it is important to check the force field used against well known examples. The problems caused by incorrect geometries in the calculation of the ^1H chemical shifts are illustrated both statistically and with three selected examples.

An introduction to rate processes in NMR is presented with examples illustrating the effect of such rate processes on the ^1H NMR spectra, with particular emphasis on conformational and proton exchange processes.

An extensive data set of ^1H chemical shifts in $CDCl_3$ and DMSO solution is presented together with the model used to predict the differential solvent effect $\delta(DMSO)–\delta(CDCl_3)$ on the solute ^1H shifts. This included the OH proton shifts of alcohols and phenols in DMSO vs. dilute $CDCl_3$ solution, but not the protic hydrogen of acids and amides. An alternative method of approach was used to directly calculate protic ^1H shifts in DMSO solution and this gave reasonably accurate shifts for the OH protons of alcohols and NH protons of amides in DMSO solution. Neither approach was tenable for the NH protons of amines as these shifts were scattered in both solvents probably due to traces of acid in the medium.

References

1. Simons, J., *J. Phys. Chem.* 1991, **95**, 1017.
2. Foresman, J. B.; Frisch, Æ., *Exploring Chemistry with Electronic Structure Methods.* 2nd ed.; Gaussian, Inc.: Pittsburgh, PA, 1996.
3. Atkins, P. W., *Physical Chemistry.* 6th ed.; Oxford University Press: Oxford, 1998.
4. Atkins, P. W.; Friedman, R. S., *Molecular Quantum Mechanics.* 3rd ed.; Oxford University Press: New York, 1997.
5. Zielinski, T. J.; Swift, M. L., *Using Computers in Chemistry and Chemical Education.* American Chemical Society: Washington, DC, 1997.
6. Cross, G., Basic ab initio quantum chemistry, 2004, http://www.chem.swin.edu.au/.
7. Clark, T., *A Handbook of Computational Chemistry.* John Wiley & Sons, Ltd: Chichester, 1985.
8. Cramer, C. J., *Essentials of Computational Chemistry: Theories and Models.* John Wiley & Sons, Ltd: Chichester, 2002.
9. Schlecht, M. F., *Molecular Modeling on the PC.* John Wiley & Sons, Inc.: New York, NY, 1998.
10. Atkins, P. W.; de Paula, J., *Elements of Physical Chemistry.* Oxford University Press: Oxford, 2005.
11. Weiner, S. J.; Kollman, P. A.; Nguyen, D. T.; Case, D. A., *J. Comput. Chem.* 1986, **7**, 230.
12. Blondel, A.; Karplus, M., *J. Comput. Chem.* 1996, **17**, 1132.
13. Cross, G., Molecular Mechanics and dynamics, 2004, http://www.chem.swin.edu.au/.
14. Serena Software: Bloomington, IN, 1998.
15. Howard, A. E.; Kollman, P. A., *J. Med. Chem.* 1988, **31**, 1669.
16. Stote, R., http://www.ch.embnet.org, 2004.
17. Saunders, M.; Houk, K. N.; Wu, Y.; Still, W. C.; Lipton, M.; Chang, G.; Guida, W. C., *J. Am. Chem. Soc.* 1990, **112**, 1419.
18. Modgraph Consultants Ltd. 2.14.0 ed.; http://www.modgraph.co.uk/, 2005.
19. Abraham, R. J.; Mobli, M.; Smith, R. J., *Magn. Reson. Chem.* 2004, **42**, 436.
20. Heller, E.; Schmidt, G. M. J., *Isr. J. Chem.* 1971, **9**, 449.
21. Bourn, A. J. R.; Gillies, D. G.; Randall, E. W., *Tetrahedron* 1964, **20**, 1811.
22. Perez, M., *Ph.D. Thesis.* University of Liverpool, 2004.
23. Epsztajn, J.; Bieniek, A.; Brzezinski, J. Z.; Kalinowski, H., *Tetrahedron* 1986, **42**, 3559.
24. Binsch, G.; Kessler, H., *Angew. Chem. Int. Ed. Engl.* 1980, **19**, 411.
25. Günther, H., *NMR Spectroscopy.* 2nd ed.; John Wiley & Sons, Ltd: Chichester, 1995.
26. Jackman, L. M., *Dynamic Nuclear Mangetic Resonance Spectroscopy.* Academic Press: New York, NY, 1975.
27. Kessler, H., *Angew. Chem. Int. Ed. Engl.* 1970, **9**, 219.
28. Sandstrom, J., *Dynamic NMR Spectroscopy.* Academic Press: New York, NY, 1982.
29. Abraham, R. J.; Fisher, J.; Loftus, P., *Introduction to NMR Spectroscopy.* 2nd ed.; John Wiley & Sons, Ltd: Chichester,1988.
30. McFarlane, W.; White, R. F. M., *Techniques of High Resolution NMR Spectroscopy.* Butterworths: London, 1972.
31. Roberts, J. D.; Hawkins, B. L.; Bremser, W.; Borcic, S., *J. Am. Chem. Soc.* 1971, **93**, 4472.
32. Bovey, F. A., NMR Spectroscopy. Academic Press: New York, NY, 1969.
33. Eliel, E. L.; Wilen, S. H., *Stereochemistry of Carbon Compounds.* J. Wiley & Sons, Inc.: New York, NY, 1994.
34. Warne, M.A., Ph.D thesis, University of Liverpool, 1996.
35. Buckingham, A. D.; Schaefer, T.; Schneider, W. G., *J. Chem. Phys.* 1960, **32**, 1227.
36. Abraham, R. J., *Mol. Phys.* 1961, **4**, 369.
37. Abraham, R. J., *J. Chem. Phys.* 1961, **34**, 1062.
38. Buckingham, A. D., *Can. J. Chem.* 1960, **38**, 300.

39. Onsager, L., *J. Am. Chem. Soc.* 1936, **58**, 1486.
40. Raynes, W. T.; Buckingham, A. D.; Bernstein, H. J., *J. Chem. Phys.* 1962, **36**, 3481.
41. Homer, J., *The Chemical Society Specialist Periodical* 1978, **7**, 318.
42. Jameson, C., *The Chemical Society Specialist Periodical* 1980, **9**, 61.
43. Lazlo, P., *Prog. Nucl. Magn. Reson. Spectrosc.* 1968, **3**, 203.
44. Benzi, C.; Grexenzi, O.; Pavone, M.; Barone, V., *Magn. Reson. Chem.*, 2004, **S57**, 42.
45. Cossi, M.; Barone, V.; Tomasi, J., *J. Chem. Phys.* 1996, **255**, 327.
46. Cramer, C. J.; Truhlar, D. G., *Chem. Rev.* 1999, **99**, 2161.
47. Helgaker, T.; Jaszunski, M.; Ruud, K., *Chem. Rev.* 1999, **99**, 293.
48. Tomasi, J.; Mennucci, B.; Cammi, R., *Chem. Rev.* 2005, **105**, 1999.
49. Tomasi, J.; Persico, M., *Chem. Rev.* 1994, **94**, 2027.
50. Frisch, M. J.; Trucks, G. W.; Schlegel, H. B.; Scuseria, G. E.; Robb, M. A.; Cheeseman, J. R.; Montgomery, J. A.; Jr., T. V.; Kudin, K. N.; Burant, J. C.; Millam, J. M.; Iyengar, S. S.; Tomasi, J.; Barone, V.; Mennucci, B.; Cossi, M.; Scalmani, G.; Rega, N.; Petersson, G. A.; Nakatsuji, H.; Hada, M.; Ehara, M.; Toyota, K.; Fukuda, R.; Hasegawa, J.; Ishida, M.; Nakajima, T.; Honda, Y.; Kitao, O.; Nakai, H.; Klene, M.; Li, X.; Knox, J. E.; Hratchian, H. P.; Cross, J. B.; Bakken, V.; Adamo, C.; Jaramillo, J.; Gomperts, R.; Stratmann, R. E.; Yazyev, O.; Austin, A. J.; Cammi, R.; Pomelli, C.; Ochtersk, J. W.; Ayala, P. Y.; Morokuma, K.; Voth, G. A.; Salvador, P.; Dannenberg, J. J.; Zakrzewski, V. G.; Dapprich, S.; Daniels, A. D.; Strain, M. C.; Farkas, O.; Malick, D. K.; Rabuck, A. D.; Raghavachari, K.; Foresman, J. B.; Ortiz, J. V.; Cui, Q.; Baboul, A. G.; Clifford, S.; Cioslowski, J.; Stefanov, B. B.; Liu, G.; Liashenko, A.; Piskorz, P.; Komaromi, I.; Martin, R. L.; Fox, D. J.; Keith, T.; Al-Laham, M. A.; Peng, C. Y.; Nanayakkara, A.; Challacombe, M.; Gill, P. M. W.; Johnson, B.; Chen, W.; Wong, M. W.; Gonzalez, C.; Pople, J. A., *Gaussian 03* D. Gaussian Inc.: Wallingford, CT, 2006.
51. Abraham, R. J., *Prog. Nucl. Magn. Reson. Spectrosc.* 1999, **35**, 85.
52. Abraham, R. J.; Byrne, J. J.; Griffiths, L.; Koniotou, R., *Magn. Reson. Chem.* 2005, **43**, 611.
53. Abraham, R. J.; Mobli, M.; Smith, R. J., *Magn. Reson. Chem.* 2003, **41**, 26.
54. Gottlieb, H. E.; Kotlyar, V.; Nudelman, A., *J. Org. Chem.* 1997, **62**, 7512.
55. Jones, I. C.; Sharman, G. J.; Pidgeon, J., *Magn. Reson. Chem.* 2005, **43**, 497.
56. Hobley, P.; Howarth, O.; Ibbett, R. N., *Magn. Reson. Chem.* 1996, **34**, 755.
57. Abraham, R. J.; Byrne, J. J.; Griffiths, L.; Perez, M., *Magn. Reson. Chem.* 2006, **44**, 491.
58. Pearce, C. M.; Sanders, K. M., *J. Chem. Soc. Perkin Trans. 1* 1994, 1119.
59. Abraham, R. J.; Mobli, M., *Magn. Reson. Chem.* 2007, **45**, 865.
60. Pouchert, C. J.; Behnke, J., *Aldrich Library of 13C and 1H FT NMR Spectra.* Aldrich Chemical Company: Milwaukee, WI, 1993.
61. Mobli, M., *Ph.D. Thesis.* University of Liverpool, 2004.
62. Farrar, T. C.; Ferris, T. D.; Zeidler, M. D., *Mol. Phys.* 2000, **98**, 737.
63. Byrne, J. J., *Ph.D.Thesis.* University of Liverpool, 2007.
64. Abraham, M. H.; Abraham, R. J.; Byrne, J. J.; Griffiths, L., *J. Org. Chem.* 2006, **71**, 3389.
65. Abraham, R. J.; Bardesley, B.;Mobli, M.; Smith, R. J., *Magn. Reson. Chem.* 2005, **43**, 3.

9

A Practical Approach to ^{1}H NMR Calculation and Prediction

9.1 Introduction

In Chapters 1 and 2 the two fundamental quantities in ^{1}H NMR spectra, namely the chemical shifts and coupling constants, are defined and described. Chapter 3 shows how chemical shifts may be obtained by various methods including the use of a number of semi-empirical equations and Chapters 4–7 describe in detail how these equations can be applied to any organic molecule from simple alkanes to multi-substituted aromatics. In Chapter 8 effects which indirectly contribute to the chemical shift, such as the averaging of multiple conformations and solvent effects, are considered. Thus all the tools necessary for calculating the complete ^{1}H NMR spectrum of any given compound containing the molecular fragments that have been characterized here have been given.

In this chapter we present some computational tools which apply the theory discussed earlier, show how this software may be used practically and highlight some common pitfalls.

The process for generating a ^{1}H NMR spectrum can be described by the flowchart in Figure 9.1. The molecule for which the ^{1}H NMR spectrum is to be calculated must be drawn, the 3D molecular geometry/geometries must be determined and both the ^{1}H chemical shifts and J couplings calculated. At this point if multiple geometries are generated the corresponding calculated chemical shifts and coupling constants must be averaged according to their energies (Equation (8.4)) and finally this information can be used to generate a spectrum.

The first two steps of Figure 9.1 (from left) are usually achieved by using a molecular modelling program. Here we will use the PCModel (Section 8.3) program (a restricted version 'pcm92_small' is included in the software CD). The third box in Figure 9.1 (from left) represents the chemical shift and coupling calculations based on the 3D structures. This is achieved here using the HNMRSPEC program (a restricted version 'HNMRSPEC_s'

Modelling ^{1}H NMR Spectra of Organic Compounds: Theory, Applications and NMR Prediction Software
Raymond Abraham and Mehdi Mobli   © 2008 John Wiley & Sons, Ltd

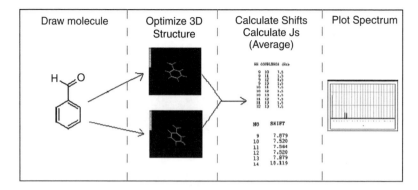

Figure 9.1   *Figure illustrating the flow in generating a spectrum from a chemical structure.*

is included in the software CD). Finally we will use the 1HPlot program to plot the ¹H spectrum. The entire process is achieved seamlessly from the leftmost section of Figure 9.1 to the rightmost section using the NMRPredict software (Section 8.4) (a molecule limited version of this software is provided in the software CD).

The following section of this chapter (Section 9.2) covers the use of PCModel for drawing and minimizing chemical structures using molecular mechanics force fields (as discussed in Chapter 8). Next the three-dimensional (3D) structures generated by PCModel are input to the HNMRSPEC software which calculates the ¹H chemical shifts and J coupling constants (using the CHARGE routine, cf. Chapters 2 and 3). We also discuss how HNMRSPEC can be used to calculate the ¹H NMR spectrum from user specified chemical shifts and coupling constants. Finally the 1HPLOT software is used to generate a complete spectrum.

The next section of this chapter will introduce the NMRPredict software; this software automates all of the steps discussed in Section 9.2. This automation allows for minimal user intervention resulting in fast ¹H NMR spectrum prediction. We hope that the knowledge gained in Section 9.2 will allow for critical assessment of these fast predictions and help the user make the best use of the available software.

It should also be noted that the two approaches, i.e. PCModel-HNMRSPEC-1HPLOT and NMRPredict, are not mutually exclusive and that both offer unique features which will also be discussed in the respective sections of this chapter.

9.2   A Step-by-Step Description of Calculating ¹H NMR Spectra

Most of the software included in this section requires that files be written to disk during execution, therefore it is important to copy the files from the CD onto a media where the user has write access to the disk.

9.2.1   Molecular Modelling and Conformational Searching – PCModel

The molecular mechanics software PCModel comes with an extensive manual (pcmodel.pdf), therefore the account given here will be brief and the reader is encouraged to consult the manual for additional information.

The program does not require installation under windows and is simply initialized by double clicking on the pcm92.exe icon –Pcm92.exe. Once the program has opened the user will be presented with a blank screen. This is the canvas on which the molecules of interest will be drawn. This can be achieved by either free-hand drawing by clicking on the draw button on the left hand side toolbar or by clicking on one of the templates on the same toolbar. These are Rings – which contain both aromatic and aliphatic rings, AA – which are amino acids, Su – which are sugars and Nu – which are nucleotides. Other options are available but these templates provide an excellent starting point for drawing most organic compounds.

We start by drawing two molecules which will be used to demonstrate various aspects of the chemical shift calculations. These are o-chlorobenzaldehyde and α-tetralone (if you are familiar with molecular modelling and the PCModel program you may wish to skim through this section).

o-Chlorobenzaldehyde. To draw this molecule we start by choosing the most appropriate template substructure, which in this case is a phenyl ring found by clicking on the rings templates on the left hand toolbar. Once the rings button is clicked a number of rings will be present, the phenyl ring here is abbreviated as Ph. If we now click on Ph the appropriate aromatic benzene ring will appear on the drawing canvas. We now need to substitute this ring with a chlorine group and an aldehyde group. We start by substituting on the ring protons with an aldehyde group. This can be done in a number of ways but most simply this is done by clicking on the Sel-Atm icon on the left hand toolbar and clicking on the hydrogen atom we wish to substitute for an aldehyde. We will now find the aldehyde group under the menu heading 'Templates' and 'Functional Groups' (see Figure 9.2).

We identify the aldehyde group in the resulting window as –CHO; once the substitution has been done we can simply close the 'Function Groups' window. Next we will in a similar procedure substitute one of the ortho protons with a chlorine atom. In this case we wish to substitute the proton with a chemical element rather than a functional group, therefore we will click on the PT icon on the left hand toolbar (note we don't need to select the atom before hand). The PT icon will open up a reduced Periodic Table window with common chemical elements. If we now click on Cl, to choose the chlorine atom, we may substitute any hydrogen atom on the structure with this atom by simply clicking on the atom to be substituted. In this case the proton substituted is the one ortho to the aldehyde group (it doesn't matter which side is chosen as will become apparent). If the above instructions are followed we will at this point have an o-chlorobenzaldehyde structure on the drawing canvas, albeit with incorrect 3D structure (bond lengths, etc. will be incorrect at this point). To rotate the structure to better appreciate the 3D structure elements we can right click anywhere on the drawing canvas and simply move the mouse around to find the appropriate viewing angle.

Geometry optimization. Next the drawn structure has to be optimized to find a reasonable structure. This is done first by choosing the appropriate force field (the MMX and MMFF94 force fields are included as they have been used throughout this book). This is done by

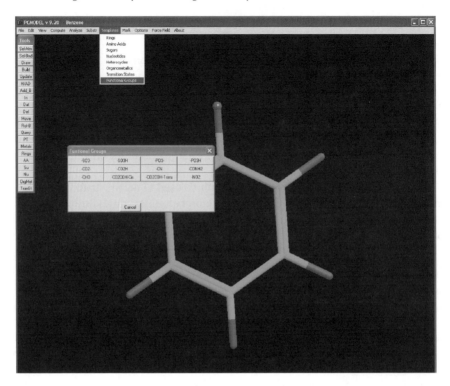

Figure 9.2   *Substituting a selected atom with a functional group in PCModel (see color Plate 5).*

clicking on the menu options 'Force Field' and choosing one of the given force fields. The default force field is MMX. We will optimize the structure using this force field by clicking on the 'Compute' menu button and then choosing 'Minimize' from the resulting scroll down menu. The calculation is now performed and a minimum energy structure is found. Depending on which side of the aldehyde group we chose to put the chlorine atom we will get two different structures (see Figure 9.3).

These are the *cis/trans* conformers of o-chlorobenzaldehyde. These are both minimum energy structures but only one of these is *global* energy minimum. In this case we can find which of these represents the global energy minimum by noting the energy of the two conformers from the right hand 'Output' window. We find that the *trans* (right hand conformer in Figure 9.3) conformer has significantly lower energy and this is the global energy minimum for this compound.

In this example there are only two stable conformations as there is only one rotatable bond (the bond connecting the benzene ring to the aldehyde group). Most compounds have several freely rotating bonds and hence multiple conformations. In such cases it is worth applying a search algorithm to find all minimum energy conformations for a specific compound. In PCModel this is done by the GMMX search engine.

GMMX. We will here demonstrate its use, using the above compound. If we return to the menu heading 'Compute' and instead of choosing 'Minimize' choose GMMX a dialog

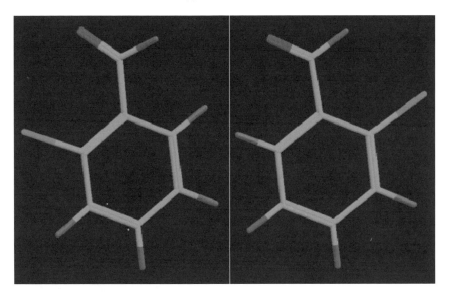

Figure 9.3 *Optimized cis and trans o-chlorobenzaldehyde resulting from optimization starting from different starting geometries (see color Plate 6).*

window will appear asking us to save the file, this will be where the result of our search will be stored. We can in this call the file '2CLBENZAL' by typing it into the 'File name' field (note this may be stored in any folder of our chose but the default folder is the PCModel folder). Once we have written in the file name and saved it we will be presented with the 'Interface to GMMX' window. Here we will only consider three of the options. These are 'Setup Rings', 'Setup Bonds' and 'Options' buttons. For the above molecule we choose first 'Setup Bonds' and in the resulting window choose the appropriate bond. In this case there is only one rotatable bond and we can simply highlight this bond by clicking on it and then clicking on the 'Add' button (see Figure 9.4).

Note: At any time we may alter the view by clicking on the 'View' menu options and altering the settings therein. For example we may add the atomic numbers which will help us identify rotatable bonds etc. by clicking 'View' > 'Lables' and choosing 'Atom Numbers'.

We will now accept the addition of this bond by clicking on the 'OK' button and will then return to the GMMX options window. We will next click on 'Options'. We have a number of options to choose from (we will leave it to the reader to experiment with various settings) and we can choose which force field we wish to use and various other parameters, most importantly we can define the number of cycles for the search. This is done by manipulating the Stop Criteria, most dramatically this is done by simply limiting the number of total cycles allowed (Maximum Conf minimized); this parameter should be set to a high enough number to allow the program to find all conformers (for 1 rotatable bond 50 would be more than sufficient), alternatively we may change the other stop criteria (see PCModel manual). We can now make sure that the MMX force field is chosen and that the stop criteria (Maximum Conf minimized) is set to 50 and accepting the changes by clicking on the 'OK' button. All that remains now is to click on the 'Run Gmmx' button. Once the calculations are finished the minimum energy conformation will be displayed, which in

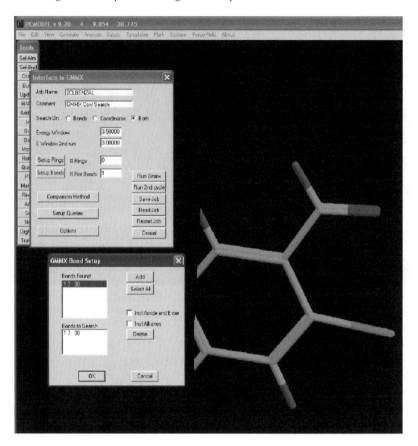

Figure 9.4   *Setting up rotatable bonds in GMMX (see color Plate 7).*

our case is the *trans* conformer. However if we click on the 'File' menu heading and click on 'Open', we should be presented with two files having the names '2CLBENZAL1' and '2CLBENZAL2', these are the results from the first and second round of optimization in the calculations (see PCModel manual for further details). We will open the '2CLBENAL2' file and depending on the options we chose in the GMMX calculations we may be presented with up to four entries in the resulting 'Structure List' window. We note that the minimum energy is ca. 2 kcal lower than the other conformers. This difference in energy difference means that this conformer will be the most abundant conformer contributing to the NMR spectrum (see Equation (8.4)). We will for the subsequent NMR calculation thus be interested in the lowest energy conformation and will thus choose this conformer form the structure list (topmost entry).

The other higher energy conformations result from the fact that three different conformers have been found in the *cis* state, two of these have the aldehyde group bent out of the plane of the benzene ring (on either side due to symmetry) and one where the *trans* conformer is flat.

For the subsequent NMR calculations we will need to save the lowest energy conformer in the sdf format. This is done by clicking on the 'File' menu item and clicking on 'Save' from the scroll down menu; in the resulting 'Save as' window we change the 'Save as type' option to *.sdf and give the file an appropriate name (see Figure 9.5). Ideally we would save this file in the HNMRSPEC folder (or HNMRSPEC_S since we are using the restricted version of the program). This is done by first navigating to this folder prior to clicking on the 'Save' button. The molecule is now saved in the appropriate place for NMR calculation using the HNMRSPEC_S program.

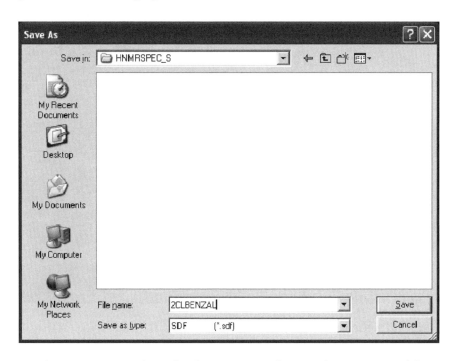

Figure 9.5   *Saving the molecular geometry in the 'SDF' format in PCModel.*

α-Tetralone. Before showing how the NMR calculations may be executed we will give another example of a molecule, namely α-tetralone. This molecule consists of two, six-membered, rings: one aromatic and one aliphatic. To draw this molecule we will start by choosing the appropriate ring template (as in the previous example), which in this case is the naphthalene ring. We will now need to remove the two double bonds in the aliphatic ring (to be). This is done by clicking on the 'Del' button on the tool bar and then clicking on the double bond which we wish to reduce to a single bond. Next we need to convert one of the hydrogen atoms to a carbonyl group (see Figure 9.6). We will first click on 'PT' in the toolbar and convert the appropriate hydrogen atom to an oxygen atom; next we will click on 'Draw' and click first on that oxygen atom and then on the attached carbon atom. Next we will click on the 'H/AD' button on the left hand toolbar (resulting in the drawing in Figure 9.6). If we now click on the 'H/AD' button again the appropriate number of hydrogen atoms will be added to the to each heteroatom. This procedure is a very useful method to draw molecules very quickly and once mastered the user should be able to draw very complex structures very rapidly.

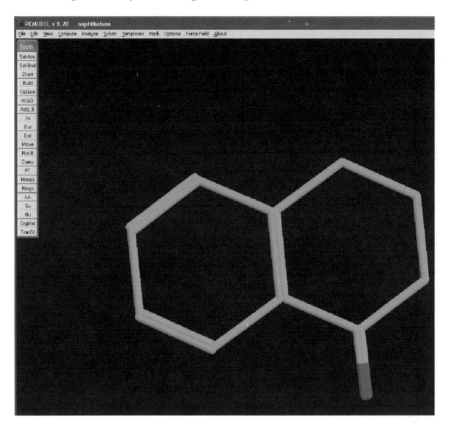

Figure 9.6   *Heteroatom representation of α-tetralone in PCModel (see color Plate 8).*

We should now have the correct atomic representation of α-tetralone, and what remains is to find the minimum energy conformation. We may now either simply minimize the structure or use the GMMX procedure to find the optimal conformation of the labile aliphatic ring. Note that in this case we need to choose the aliphatic ring from the GMMX interface by clicking on 'Setup Rings'. The result from the GMMX calculation should be two minimum energy conformations; these will have the exact same energy and will be the mirror image of each other differing only in the puckering of the aliphatic ring (being in opposite directions with respect to the plane of the aromatic ring). From this result we can assume that these two species will be present with equal populations in solution and hence will contribute equally to the NMR spectrum. However this is only true if the energy barrier between the two conformers is low, with respect to the available energy determined by the temperature of the sample. In fact if we lower the temperature we will at some point reach a point where we can stop this rapid interconversion of conformers. The result on the NMR spectrum will be that the two protons in each of the CH_2 groups will have unique chemical shifts, as a single chemical environment is sampled on the NMR time scale (see also Chapter 8, Section 8.5). However, at room temperature there is enough energy available (through kinetic energy) that this conformational averaging will result in one averaged chemical shift for each CH_2 group. It is difficult to know *a priori* which cases will have a high enough energy barrier for conformational exchange to either give averaged spectra or not. This is indeed one of the most difficult things to predict in the approach we have taken here. Often it is obvious from the NMR spectrum whether conformational averaging is occurring and the simplest way to find out may be to compare the experimental spectrum with the calculated averaged spectrum and the calculated non-averaged spectrum. This is exactly what we shall discuss further in the next section. What remains here is to save the resulting minimum energy conformation file in the HNMRSPEC_S folder as before in the SDF format.

9.2.2   Calculating ¹H Chemical Shifts and J-Coupling Constants – HNMRSPEC

We should at this point have two 'sdf' files in our HNMRSPEC folder, one of o-chloro-benzaldehyde and one of α-tetralone. The use of the HNMRSPEC program is straight forward and we refer the reader to the HNMRSPEC.DOC and INPUT_PARAMETERS.txt files for details not covered here.

o-Chlorobenzaldehyde. To calculate the chemical shifts and J coupling constants relevant to this molecule we simply click and drag the 2CLBENZAL.sdf file on top of the HNMRPSEC_S.exe file (see Figure 9.7).

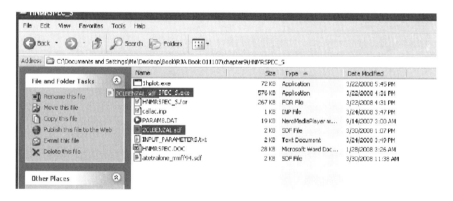

Figure 9.7   *Drag and drop execution of HNMRSPEC_S.exe in Windows XP.*

This action will produce two results files '2CLBENZAL.RES' and '2CLBENZAL.spct' (note: alternatively the program may be executed from the command line in 'Commander'– see HNMRSPEC.DOC for more details). We will first inspect the .RES file (Figure 9.8) which is the main output from the program containing the calculated chemical shifts and coupling constants. This is a text file which may be viewed using any text processing program (e.g. notepad, wordpad, etc.).

This file contains information regarding the executed calculation. To best interpret the calculations we may wish to open the .sdf file in PCModel to re-inspect the atomic numbering and we find that for the above case there are five atoms and the atom numbers consistent with the 2-chlorobenzaldehyde are as follows: '9' corresponds to 6, '10' to 5, '11' to 4, '12' to 3 and '14' to the aldehyde proton. These atoms are listed at the top of the file as atoms 9–14 and detailed information is given as to all the through space effects

```
Title: 1      7.079    -20.694
Number of atoms:  14
AROMATIC SYSTEM No   1      9 ATOMS AND 10 ELECTRONS   ATOM No   1  2  6   7  3 13   5  8  4

   No    HSTER    CSTER    XOSTER    XSTER    CC+CTST    STERSH   CANIS     XANIS     XOANIS    CHEF      CXEF      RINGCT    PSHIFT

    9    0.017    0.000    0.153     0.006    0.000      0.159    0.000     -0.049    0.123     -0.123    0.403     1.683     0.095
   10    0.007    0.000    0.003     0.003    0.000      0.006    0.000     -0.043    0.019     -0.064    0.094     1.666     -0.044
   11    0.006    0.000    0.001     0.007    0.000      0.007    0.000     -0.046    0.021     -0.045    0.094     1.663     0.086
   12    0.011    0.000    0.001     0.000    0.000      0.001    0.000     0.000     0.047     -0.096    0.064     1.683     -0.107
   14    0.011    0.000    0.000     0.219    0.000      0.219    0.000     0.162     0.000     -0.045    0.090     0.556     1.261

  HH COUPLINGS (Hz)

      9   10    7.5
      9   11    1.5
      9   12    0.5
     10   11    7.5
     10   12    1.5
     11   12    7.5

          GEOMETRY (Z-MATRIX)

  NO AN           BL           ALPHA          BETA           CHARGE      PIEXS      SHIFT

   1 55    7    0.0000    0    0.0000    0    0.0000    0    0.0031    -0.0137    147.790
   2 55    1    0.0000    0    0.0000    0    0.0000    0   -0.0454     0.0142    140.029
   3 55    2    0.0000    0    0.0000    0    0.0000    0   -0.0807    -0.0096    134.390
   4 55    3    0.0000    0    0.0000    0    0.0000    0   -0.0562     0.0120    138.316
   5 55    4    0.0000    0    0.0000    0    0.0000    0   -0.0430    -0.0075    140.416
   6 55    5    0.0000    0    0.0000    0    0.0000    0    0.0296    -0.0284    152.037
   7 55    1    0.0000    0    0.0000    0    0.0000    0    0.1756     0.1289    175.400
   8 60    7    0.0000    0    0.0000    0    0.0000    0   -0.3207    -0.1693      0.000
   9  1    2    0.0000    0    0.0000    0    0.0000    0    0.0767     0.0000      7.964
  10  1    3    0.0000    0    0.0000    0    0.0000    0    0.0772     0.0000      7.386
  11  1    4    0.0000    0    0.0000    0    0.0000    0    0.0767     0.0000      7.449
  12  1    5    0.0000    0    0.0000    0    0.0000    0    0.0783     0.0000      7.515
  13 17    6    0.0000    0    0.0000    0    0.0000    0   -0.0639     0.0734      0.000
  14  1    7    0.0000    0    0.0000    0    0.0000    0    0.0926     0.0000     10.467

  DIPOLE MOMENT =    2.705 DEBYES       COMPONENTS    X  -1.468    Y  -2.159    Z   0.709

  *****  END OF JOB   *****
```

Figure 9.8 *The main output from the HNMRSPEC_S program showing the details of the chemical shift calculations.*

which contribute to the chemical shifts of these protons. For example, we see that the ring protons experience a much higher ring current than the aldehyde proton ('14'). Also it can be seen that atom 9 suffers a large steric and XO (in this case C=O) anisotropy due to its proximity to the carbonyl group. After this output section, the different calculated couplings are given. These will be used in calculating the ^{1}H spectrum. Below the couplings we find the z-matrix (note: when using sdf files geometric information is not given in this matrix); in this matrix we find the atomic and π charges that are used to calculate the chemical shifts which are given in the rightmost column in this table (note that the carbon shifts are not well parameterized and should therefore not be used).

The different effects discussed in the previous chapters can now be investigated in detail and we can use these effects to rationalize the observed shifts. We encourage the reader to try the *cis* conformer of this molecule and see how all of the above effects change.

a-Tetralone. As mentioned in the previous section this molecule can potentially produce two very different ^{1}H spectra, depending on whether the chemical shifts of the labile aliphatic ring are being averaged or not. If we follow the same procedure as above we similarly should find a resulting .RES file where a similar output is given (Figure 9.9).

In this output file it can be seen that the CH_2 groups are not averaged; from experimental data we know that this is not the case and that they indeed should be averaged (and indeed if you compare the .spct file with the experimental spectrum you will find that these are very different). There are two ways of averaging shifts; the more elegant way is by including

```
Title: SDF   1      39.794    -18.698
Number of atoms:  21
AROMATIC SYSTEM No  1      8 ATOMS AND  8 ELECTRONS  ATOM No  1  2  6  3  5  4  7 11

 No    HSTER    CSTER    XOSTER    XSTER    CC+CTST   STERSH    CANIS    XANIS    XOANIS    CHEF     CXEF    RINGCT   PSHIFT

 12    0.038    0.000    0.001    0.000    0.000    0.001    0.000    0.000    0.047   -0.174    0.061    1.667   -0.052
 13    0.007    0.000    0.001    0.000    0.000    0.001    0.000    0.000    0.021   -0.060    0.044    1.642    0.039
 14    0.006    0.000    0.003    0.000    0.000    0.003    0.000    0.000    0.019   -0.057    0.054    1.631   -0.072
 15    0.010    0.000    0.135    0.000    0.000    0.135    0.000    0.000    0.114   -0.115    0.330    1.656    0.055
 16   -0.119    0.000    0.002    0.000    0.000    0.000    0.000    0.000    0.087   -0.157    0.140    0.602    0.120
 17   -0.141    0.000    0.005    0.000    0.000    0.004    0.000    0.000    0.108   -0.132    0.139    0.505    0.120
 18   -0.002    0.000    0.005    0.000    0.000   -0.000    0.000    0.000    0.096   -0.012    0.181    0.223    0.000
 19   -0.006    0.000    0.012    0.000    0.075    0.010   -0.000    0.000    0.086   -0.013    0.191    0.232    0.000
 20   -0.019    0.000    0.000    0.000    0.000    0.000    0.000    0.000    0.000   -0.062    0.000    0.230    0.143
 21   -0.118    0.000    0.000    0.000    0.003    0.000    0.000    0.000    0.000   -0.090    0.000    0.282    0.143

HH COUPLINGS (Hz)

 12  13    7.5
 12  14    1.5
 12  15    0.5
 13  14    7.5
 13  15    1.5
 14  15    7.5
 16  17  -12.5
 16  18    2.7
 16  19    3.4
 16  20    1.3
 17  18    3.2
 17  19   13.0
 18  19  -12.5
 18  20    2.3
 18  21    3.6
 19  20    3.9
 19  21   12.9
 20  21  -12.5

       GEOMETRY (Z-MATRIX)

NO AN        BL         ALPHA          BETA         CHARGE    PIEXS    SHIFT

 1 55  12   0.0000   0   0.0000   0   0.0000   0   -0.0861   -0.0180   133.527
 2 55   1   0.0000   0   0.0000   0   0.0000   0   -0.0631    0.0097   137.197
 3 55   2   0.0000   0   0.0000   0   0.0000   0   -0.0830   -0.0109   134.022
 4 55   3   0.0000   0   0.0000   0   0.0000   0   -0.0506    0.0089   139.199
 5 55   4   0.0000   0   0.0000   0   0.0000   0   -0.0195   -0.0062   144.187
 6 55   5   0.0000   0   0.0000   0   0.0000   0   -0.0139    0.0545   145.080
 7 55   5   0.0000   0   0.0000   0   0.0000   0    0.1541    0.1297   171.952
 8  6   6   0.0000   0   0.0000   0   0.0000   0   -0.0579    0.0000    47.975
 9  6   8   0.0000   0   0.0000   0   0.0000   0   -0.0811    0.0000    37.836
10  6   9   0.0000   0   0.0000   0   0.0000   0   -0.0264    0.0000    61.770
11 60   7   0.0000   0   0.0000   0   0.0000   0   -0.3125   -0.1677     0.000
12  1   1   0.0000   0   0.0000   0   0.0000   0    0.0780    0.0000     7.460
13  1   2   0.0000   0   0.0000   0   0.0000   0    0.0781    0.0000     7.574
14  1   3   0.0000   0   0.0000   0   0.0000   0    0.0781    0.0000     7.464
15  1   4   0.0000   0   0.0000   0   0.0000   0    0.0775    0.0000     7.974
16  1   8   0.0000   0   0.0000   0   0.0000   0    0.0569    0.0000     3.142
17  1   8   0.0000   0   0.0000   0   0.0000   0    0.0559    0.0000     2.908
18  1   9   0.0000   0   0.0000   0   0.0000   0    0.0528    0.0000     2.299
19  1   9   0.0000   0   0.0000   0   0.0000   0    0.0501    0.0000     1.952
20  1  10   0.0000   0   0.0000   0   0.0000   0    0.0573    0.0000     2.825
21  1  10   0.0000   0   0.0000   0   0.0000   0    0.0553    0.0000     2.431

DIPOLE MOMENT =   3.258 DEBYES      COMPONENTS   X   0.284   Y   0.342   Z  -3.228

*****  END OF JOB  *****
```

Figure 9.9   *Output from HNMRSPEC_S for α-tetralone.*

a few modifications to the sdf file. A more laborious method is by manually averaging the shifts and then calculating the spectrum using a different mode of HNMRSPEC_S as discussed in Section 9.2.4.

In Figure 9.10 we show the necessary modifications to the sdf file and as can be seen, at the top of the file the number of equivalent groups must be given, and at the bottom of the file two lines are included for each equivalent group; the first of these lines will specify the number of atoms in the equivalent group and the second line will contain the corresponding atom numbers. This is repeated three times in the current case.

The information regarding equivalence will further be used in calculating the spectrum and therefore is the recommended way of averaging shifts when using HNMRSPEC_S.

```
SDF   1      39.794    -18.698
 PCMODEL   v9.0   1.00000      0.00000

 21 22003 0  0  0  0              1 V2000
    -1.3025     0.2922   -1.4786 C   0  0  0  0  0  0
    -2.4627     0.3937   -0.8151 C   0  0  0  0  0  0
    -2.4745     0.3149    0.6648 C   0  0  0  0  0  0
    -1.3219     0.1419    1.3261 C   0  0  0  0  0  0
    -0.0384     0.0295    0.5871 C   0  0  0  0  0  0
    -0.0189     0.1007   -0.7552 C   0  0  0  0  0  0
     1.2129    -0.1637    1.3664 C   0  0  0  0  0  0
     1.2631    -0.0040   -1.5405 C   0  0  0  0  0  0
     2.4467     0.5300   -0.7299 C   0  0  0  0  0  0
     2.5271    -0.2083    0.6081 C   0  0  0  0  0  0
     1.2166    -0.3024    2.5700 O   0  0  0  0  0  0
    -1.2914     0.3504   -2.5806 H   0  0  0  0  0  0
    -3.4130     0.5359   -1.3606 H   0  0  0  0  0  0
    -3.4307     0.3985    1.2123 H   0  0  0  0  0  0
    -1.3166     0.0806    2.4290 H   0  0  0  0  0  0
     1.1802     0.5696   -2.4937 H   0  0  0  0  0  0
     1.4259    -1.0753   -1.8046 H   0  0  0  0  0  0
     3.3981     0.4027   -1.3006 H   0  0  0  0  0  0
     2.3129     1.6236   -0.5469 H   0  0  0  0  0  0
     3.3295     0.2433    1.2383 H   0  0  0  0  0  0
     2.7884    -1.2803    0.4447 H   0  0  0  0  0  0
  1  2  2  0
  1  6  1  0
  1 12  1  0
  2  3  1  0
  2 13  1  0
  3  4  2  0
  3 14  1  0
  4  5  1  0
  4 15  1  0
  5  6  2  0
  5  7  1  0
  6  8  1  0
  7 10  1  0
  7 11  2  0
  8  9  1  0
  8 16  1  0
  8 17  1  0
  9 10  1  0
  9 18  1  0
  9 19  1  0
 10 20  1  0
 10 21  1  0
M END
1     39.794    -18.698

002
020021
002
019018
002
017016
$$$$
```

Figure 9.10   *SDF file containing modification for averaging of chemical shifts (the modifications are given in bold and underlined for clarity and should not actually be formatted).*

9.2.3 Displaying the Calculated ^{1}H Spectrum – 1HPLOT

In the above section we mentioned that a file ending with '.spct' will also be outputted from HNMRSPEC_S when chemical shifts are calculated (this can be turned off by modifying the INPUT_PARAMTERS.txt file). In this section we will show how to display this spectrum. The program 1HPLOT works much in the same way as HNMRSPEC_S in that one can simply click and drag the '.spct' file onto the 1HPLOT program (more information on 1HPLOT can be found in the HNMRSPEC.DOC document). If we do this for the non-averaged and the averaged α-tetralone spectrum we can see the dramatic effect of averaging (see Figure 9.11).

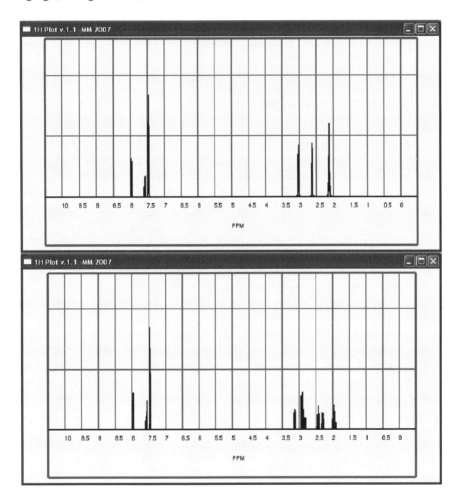

Figure 9.11 *The effect of averaging of the CH$_2$ groups of a-Tetralone on the calculated ^{1}H NMR spectrum.*

Using the keyboard we may navigate around the spectrum as well as zooming in and out and changing the amplitude of the peaks; details of the specific commands are given by right clicking on the spectrum and clicking on the help button. The instructions will appear on the accompanying 'Commander' window.

9.2.4   Advanced Use of HNMRSPEC_S

In this section we shall describe some of the advanced options in the HNMRSPEC_S program. These are primarily accessed through the INPUT_PARAMETERS.txt file:

9	*Input file type.*
0	*Output file type.*
0	*Solvent.*
0	*Calculate spectrum (i.e. .spct file)? (1 = no, 0 = yes).*
1	*Calculate coupling constants? (0 = no, 1 = yes).*
400.0	*Spectrometer frequency.*
0.2	*Line width.*
0	*User specified file (given in line below) for spectrum calcu-lation (1 = YES, 0 = NO).*

callac.inp.

Information on file types can also be found in the INPUT_PARAMETERS.txt file and will not be discussed any further here. The solvent option allows the user to specify in which solvent they wish to calculate their spectrum. The two solvents considered are $CDCl_3$ (the default option) and DMSO. The chemical shifts for DMSO are based on the results in Chapter 8. There is also a switch for calculating the .spct file or not, but perhaps of greater interest is the option which defines whether the coupling constants should be calculated or not. If this option is set to 0 no coupling constants will be calculated and 'non' will be used in the calculation of the spectrum and the result is a ¹H decoupled ¹H spectrum, a feat which is much harder to achieve experimentally! The spectrometer frequency and line width parameters are used when calculating the spectrum and may serve as an ideal tool to visualize the gain in resolution and the loss of 2nd order effects at higher frequencies.

Finally we have the 'User specified file option'; note that if this option is set to 1 the program will only work if the file given in the line below this line (in this case callac.inp) is present in the same folder. If the file is present we may simply double click on the HNMRSPEC_S icon and the specified operation will be carried out. Note that whilst this option is set to 1, dragging and dropping sdf files onto HNMRSPEC_S will not work! These complications aside, this option will allow the user to calculate a '.spct' file from a set of shifts and coupling constants. Below we find an example of an input file for this type of calculation (this is the callac.inp file included in the HNMRPSEC_S folder).

```
TITLE
5                       (NUMBER OF ATOMS)
4                       (NUMBER OF COUPLINGS)
Shifts (Do not remove this line)
1       100.0           (ATOM LABEL, CHEMICAL SHIFT)
2       1.2
3       2.4
4       5.2
5       5.2
```

```
Couplings (Do not remove this line)
1     2        50.0           (ATOM LABLE 1, ATOM LABEL 2, COUPLING)
1     3        50.0
2     3        10.0
4     5        -6.0
Equivalent nuclei (Do not remove this line)
0001                          (NUMBER OF EQUIVALENT SETS OF ATOMS)
0002                          (NUMBER OF ATOMS INCLUDED IN 1ST SET)
0004                          (ATOM 1 OF FIRST SET)
0005                          (ATOM 2 OF FIRST SET)
0000                          (NUMBER OF ATOMS INCLUDED IN 2ND SET)
0000                          (ATOM 1 OF 2ND SET)
0000                          (ATOM 2 OF 2ND SET, ETC)
```

It should be noted that the above example is not a real case and is just given to highlight the functionality of the calculations.

Firstly we note the Title line, where the user may specify the desired filename, and it should be borne in mind that this is the name that will be used to name the .spct file (in this case the output file will be TITLE.SPCT). The next two lines contain the number of atoms and couplings and in the following lines these are given. It is important to make sure that the number of atoms and couplings specified agree with the number of lines under the Shifts and Couplings headings. In the above example we note that the chemical shift of one of the atoms is 100 ppm (all shifts are given in ppm and couplings in Hz) and this simply indicates that there is a coupling to a nucleus other than a proton. In most cases this may be useful in simulating the spectrum of fluorinated compounds, etc. It should also be noted that the *averaged* shifts should be given for equivalent protons and that the program does not average shifts prior to calculating the spectrum.

Finally we note that by opening the '.spct' file we can see the various frequency lines and their amplitudes which result from the 2nd order spectral calculation.

9.2.5 Iteration of 2nd Order ^{1}H Spectra from Pre-specified δs and Js – LAOCOON

LAOCOON. The LAOCOON program has been described and illustrated with examples in Chapter 2. It therefore provides a good starting place for the calculation of 2nd order spectra. As discussed in Chapter 2, the primary inputs required are the nuclear chemical shifts and couplings However there are in practice other input parameters and also as the program is written in FORTRAN there is the additional constraint of the use of fields and the differentiation of real numbers (F), integers (I) and characters (A) in the data file. In practice this can be overcome most easily by following the examples given. We will first outline the data file format and then discuss it with some examples. The data file consists of the following lines:

(1) The case number (N), the number of spin $\frac{1}{2}$ nuclei in the system (NN) and the title. Format (2I3,10A6).
(2) The range of frequencies required and the minimum intensity. Format (3F10.3).
(3) The isotope identification for each nucleus. Format (7I1).
(4) The chemical shifts $W(I)$ of the nuclei in the same order as the isotope numbers in (3). Format (7F10.3).

(5) The coupling constants $A(I, J)$ starting with $A(1, J)$, $J = 1, NN$, next line $A(2, J)$, etc. Format (7F10.3).

(6) The control variable NI: $NI = 0$, no iteration; $NI = 9$ iterates on observed frequencies. Format (I1).

(7) The assigned lines. For each peak in the spectrum which can be matched with a calculated transition this line contains the calculated transition identifier obtained from a previous non-iterative run and the observed peak frequency. Continue for all matching lines. Format (I4,F20.3).

(8) A blank line signals the end of the assignments.

(9) Parameter Sets. Each line contains the parameter to be varied to fit the spectrum, These can be single parameters or they may be varied together due to symmetry. Format (6(I1,I1,2x)).

(10) A blank line signals the end of the parameter sets.

(11) Another blank line ends the run.

An illustrated data set is shown below and the file is also given in the CD as cysteine.dat. This was used in the analysis of the cysteine spectrum (Figure 2.8).

```
   1  3  cysteine HCl monohydrate in D2O (Aldrich data)
     0.000   1500.000   0.0100
  1110000
    935.0       958.0      1305.5
   -15.30       4.4
      5.5
   9
  0011        925.88
  0001        930.29
  0015        940.95
  0007        945.37
  0012        947.57
  0002        953.09
  0014        963.02
  0004        968.53
  0013       1300.57
  0009       1304.98
  0005       1306.08
  0003       1310.50
  0000
   1
   2
   3
  12
  13
  23
  0000
  0000
```

The interpretation is straightforward. Reading from the first line the case number can be any integer, there are three nuclei in the system, the range of frequencies required is 0.0 to

1500 Hz and the minimum intensity is 0.01. The iso numbers are all the same as the nuclei are all ¹H. The chemical shifts and couplings are in Hz and the couplings are in the order 12,13 and then 23.

NI = 9 and thus this is an iteration and the assigned transitions numbers are given with the observed frequencies. 0000 ends the assignments and all the chemical shifts and couplings are being iterated. The results of the iteration are shown in Chapter 2, Section 2.2.5. Also note that the program requires that the data file is called lac.dat and the result file is always lac.res. It is always preferable to copy the data file into lac.dat to keep a record.

The data file for another example considered in Chapter 2, the spectrum of 1.1-difluoroethylene (Chapter 2, Figure 2.4) is also given in the CD (CH2CF2.dat) and this illustrates the use of the *iso* numbers as the spin system contains ¹H and ¹⁹F nuclei and also the use of symmetry as the spectrum is a symmetric AA'XX' system.

9.3 Automated Spectral Prediction Using the NMRPredict Software

In this section we shall very briefly show how the above manual calculations have been automated in the commercial NMRPredict software. In the version included in the attached CD spectra both ¹H and heteronuclear chemical shifts can be predicted. The prediction of ¹H spectra in this software follows two different methods defined as conformer and increment. The conformer method uses the chemical shift calculations described in this book. The increments algorithm is an SCS/data-based approach developed by Professor Ernö Pretsch and coworkers. It is a very useful complementary approach to the calculations presented here but we shall not describe this method any further.

After installing the software we will be presented with a window where we may choose to predict other nuclear chemical shifts in addition to proton chemical shifts. By default this is set to C-13 and it will be set to this in the examples below. The program once launched only allows prediction of molecules included in the list given by pressing the

'open molecule' icon 📁 MOL. The compounds listed are those tabulated in Chapter 8, Section 8.6 which includes the ¹H chemical shifts in both CDCl₃ and DMSO solvents. We choose cyclohexane as an example. We see first that the molecule is represented by a two dimensional drawing, and this is an integral part of the philosophy of the NMR-Predict software. That is for the user to be able to draw the molecules of interest in a familiar 2D drawing program (e.g. ISIS draw or ChemDraw). At this point we may simply click on the predict icon and the program will produce the calculated spectrum but before doing so we shall have a brief look at the options available to us from the ¹H setup icon (Figure 9.12). First of all we find the 'Solvent' options and as before these will allow the spectrum to be calculated in either CDCl₃ or DMSO. In addition the force field will include the appropriate dielectric constant in its calculation of the molecular energies (see also Chapter 8).

As the program uses PCModel for its 3D structure generation and conformational searching we will recognize many of the options from the previous sections. We can choose the spectrometer frequency and line width (as was specified in the INPUT_PARAMETERS.txt file previously). In addition there is an option to set a cut-off value for the lowest coupling

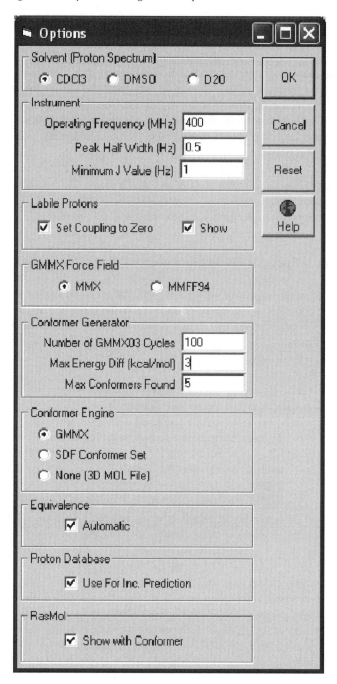

Figure 9.12   *¹H setup options in NMRPredict.*

included in the spectrum calculation, the default value being 1 Hz. There is an option for labile protons. This is mainly for OH and NH protons which may be broadened due to exchange and thus not detected in the experimental spectrum. Similarly the coupling from labile protons can also be manipulated. Below this field there are options for the force field used and below that we find the stop criteria described previously (Section 9.2) in the GMMX routine in the PCModel program. We can control the number of the cycles in the conformational search which is being conducted. Below this the program offers the user the alternatives of using conformers from an SDF file or to simply calculate the spectrum from a predefined mol file (perhaps generated using *ab initio* calculations). The next set of options allows the user to turn the automatic equivalence on or off. As described with α-tetralone this routine will try to set the correct equivalence but if the user finds that the nuclear parameters are being averaged contrary to experiment this routine can be turned off. The other options offered are not relevant to the discussion here and shall not be discussed further.

We are now ready to predict our spectrum by clicking on the Predict icon; this operation will predict both the proton and C-13 spectra for the cyclohexane. The carbon predictions provided by Professor Wolfgang Robien use a neural network trained on an extensive database. Once the calculations are finished, a C-13 and a proton spectrum are shown and a table containing the calculated chemical shifts using the various techniques. In addition a 'Best' prediction will be given; this value is based on extensive testing of the capabilities of the two ¹H prediction methods and utilizes this information to weigh the two predictions in order to find the best calculated value.

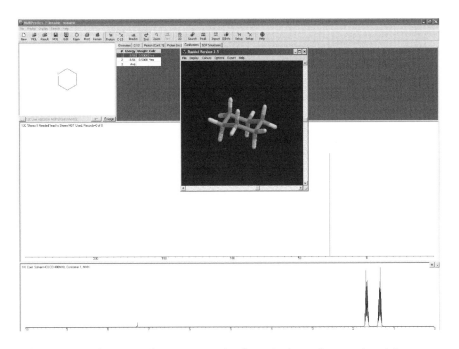

Figure 9.13 *The output from NMRPredict for a single conformer of cyclohexanes.*

The calculations from the CHARGE routine described in this book can be found under the tab 'Proton (Average of 2)'. We can see what the two conformers are by clicking on the 'Conformers' tab, where we find two entries with identical energies. These are the two chair conformations of cyclohexanes with the only difference between them being a ring inversion. At room temperature this inversion is rapid on the NMR time scale and the protons on the ring are constantly interconverting between an axial and equatorial conformation. The result is a single chemical shift for this compound (Chapter 8, Section 8.5). However if we double click on one of the entries in the 'Conformers', the 3D structure for that specific species will appear and the corresponding 'frozen' spectrum will be shown.

This spectrum will contain distinct chemical shifts for the axial and equatorial protons, indeed if the spectrum of cyclohexane is measured at a temperature of $-80\,^{\circ}$C (in CFCl$_3$) the ring inversion will be effectively stopped and the single conformation can be observed (compare Figure 9.13 and Figure 8.6 in Chapter 8). It is apparent from the above that this process allows even the novice user to produce a predicted spectrum for the compound being studied in a few clicks.

9.4   Concluding Remarks

The theory of ^{1}H chemical shifts has been under constant development and refinement ever since the beginning of NMR. The wealth of information that can potentially be extracted from chemical shifts and coupling constants in a ^{1}H NMR spectrum has always been a great driving force for this type of research.

It is however only recently that we have to our disposal the tools necessary to calculate ^{1}H NMR spectra which are of practical use to the experimentalist. Also there has recently been renewed interest in this field as the relatively high sensitivity afforded by this nucleus makes it invaluable for detecting anything from metabolite ^{1}H signatures for diagnosing disease to identifying low level impurities in drug preparations. Another area of recent attention is the use of ^{1}H chemical shift calculations in automated structure verification procedures. Since the sensitivity of spectrometers is ever increasing, 2D ^{1}H–^{13}C correlation spectra can now be collected in a matter of minutes (if not seconds!) in favourable cases and this coupled with accurate chemical shift prediction may allow for fully automated structure verification of vast libraries of chemicals.

We have attempted in this book firstly to give an introduction to the analysis of ^{1}H NMR spectra and subsequently a compilation of observed and assigned ^{1}H chemical shifts and coupling constants of molecules in known conformations. The theoretical base of a semi-empirical ^{1}H chemical shift calculation is presented and the ^{1}H chemical shifts are considered in the light of these semi-empirical shift calculations. Finally we show how the ^{1}H NMR spectrum of any given compound can be simply predicted and presented. We hope that this presentation will give the reader an appreciation and understanding of the phenomena contributing to the observed ^{1}H NMR spectra. This provides both a critical insight into the prediction of ^{1}H NMR spectra and also the facility for the reader to successfully undertake such analysis themselves.

Index

References to page numbers for illustrations are in italics, to tables in bold

---

Modelling ¹H NMR Spectra of Organic Compounds: Theory, Applications and NMR Prediction Software
Raymond Abraham and Mehdi Mobli © 2008 John Wiley & Sons, Ltd